ClearRevise™

AQA GCSE
Combined Science
8464

Illustrated revision and practice

Trilogy Course
Foundation and Higher

Published by
PG Online Limited
The Old Coach House
35 Main Road
Tolpuddle
Dorset
DT2 7EW
United Kingdom

sales@pgonline.co.uk
www.clearrevise.com
www.pgonline.co.uk
2021

PG ONLINE

PREFACE

Absolute clarity! That's the aim.

This is everything you need to ace the examined component in this course and beam with pride. Each topic is laid out in a beautifully illustrated format that is clear, approachable and as concise and simple as possible.

Content and questions for the Higher tier are clearly indicated. The checklist on the contents pages will help you keep track of what you have already worked through and what's left before the big day.

We have included worked exam-style questions with answers for almost every topic. This helps you understand where marks are coming from and to see the theory at work for yourself in an exam situation. There is also a set of exam-style questions at the end of each section for you to practise writing answers for. You can check your answers against those given at the end of the book.

LEVELS OF LEARNING

Based on the degree to which you are able to truly understand a new topic, we recommend that you work in stages. Start by reading a short explanation of something, then try and recall what you've just read. This has limited effect if you stop there but it aids the next stage. Question everything. Write down your own summary and then complete and mark a related exam-style question. Cover up the answers if necessary but learn from them once you've seen them. Lastly, teach someone else. Explain the topic in a way that they can understand. Have a go at the different practice questions – they offer an insight into how and where marks are awarded.

ACKNOWLEDGEMENTS

The questions in the ClearRevise textbook are the sole responsibility of the authors and have neither been provided nor approved by the examination board.

Every effort has been made to trace and acknowledge ownership of copyright. The publisher would like to thank the following companies and individuals who granted permission for the use of their images in this textbook.

Design and artwork: Jessica Webb & Mike Bloys / PG Online Ltd
Potato solute experiment P18: © Simon Underhill
Asthma statistics P43: Cumella, A. 2020. The Great Asthma Divide: The Annual Asthma Survey 2019. Asthma UK
Chromatography P225 © sciencephotos/Alamy Stock Photo. Calorimeter heat test P195 © Thermtest inc., thermtest.com
Resistance experiment image P264 © iStock. Frictionless ramp image P316 © Adobe Stock. Icons: Icons8.com
All other images: © Shutterstock

First edition 2021   10 9 8 7 6 5 4 3 2 1
A catalogue entry for this book is available from the British Library
ISBN: 978-1-910523-34-6
Contributors: Dr J Gledhill, N Saunders, H Sayers, M Smith. Editing: Jim Newall
Copyright © PG Online 2021

Printed on FSC certified paper by Bell and Bain Ltd, Glasgow, UK.

THE SCIENCE OF REVISION

Illustrations and words

Research has shown that revising with words and pictures doubles the quality of responses by students.[1] This is known as 'dual-coding' because it provides two ways of fetching the information from our brain. The improvement in responses is particularly apparent in students when asked to apply their knowledge to different problems. Recall, application and judgement are all specifically and carefully assessed in public examination questions.

Retrieval of information

Retrieval practice encourages students to come up with answers to questions.[2] The closer the question is to one you might see in a real examination, the better. Also, the closer the environment in which a student revises is to the 'examination environment', the better. Students who had a test 2–7 days away did 30% better using retrieval practice than students who simply read, or repeatedly reread material. Students who were expected to teach the content to someone else after their revision period did better still.[3] What was found to be most interesting in other studies is that students using retrieval methods and testing for revision were also more resilient to the introduction of stress.[4]

Ebbinghaus' forgetting curve and spaced learning

Ebbinghaus' 140-year-old study examined the rate in which we forget things over time. The findings still hold true. However, the act of forgetting things and relearning them is what cements things into the brain.[5] Spacing out revision is more effective than cramming – we know that, but students should also know that the space between revisiting material should vary depending on how far away the examination is. A cyclical approach is required. An examination 12 months away necessitates revisiting covered material about once a month. A test in 30 days should have topics revisited every 3 days – intervals of roughly a tenth of the time available.[6]

Summary

Students: the more tests and past questions you do, in an environment as close to examination conditions as possible, the better you are likely to perform on the day. If you prefer to listen to music while you revise, tunes without lyrics will be far less detrimental to your memory and retention. Silence is most effective.[5] If you choose to study with friends, choose carefully – effort is contagious.[7]

1.   Mayer, R. E., & Anderson, R. B. (1991). Animations need narrations: An experimental test of dual-coding hypothesis. *Journal of Education Psychology*, (83)4, 484–490.

2.   Roediger III, H. L., & Karpicke, J.D. (2006). Test-enhanced learning: Taking memory tests improves long-term retention. *Psychological Science*, 17(3), 249–255.

3.   Nestojko, J., Bui, D., Kornell, N. & Bjork, E. (2014). Expecting to teach enhances learning and organisation of knowledge in free recall of text passages. *Memory and Cognition*, 42(7), 1038–1048.

4.   Smith, A. M., Floerke, V. A., & Thomas, A. K. (2016) Retrieval practice protects memory against acute stress. *Science*, 354(6315), 1046–1048.

5.   Perham, N., & Currie, H. (2014). Does listening to preferred music improve comprehension performance? *Applied Cognitive Psychology*, 28(2), 279–284.

6.   Cepeda, N. J., Vul, E., Rohrer, D., Wixted, J. T. & Pashler, H. (2008). Spacing effects in learning a temporal ridgeline of optimal retention. *Psychological Science*, 19(11), 1095–1102.

7.   Busch, B. & Watson, E. (2019), *The Science of Learning*, 1st ed. Routledge.

CONTENTS

Biology Paper 1

Topic 1   Cell biology

Topic 2   Organisation

Topic 3 — Infection and response

Topic 4 — Bioenergetics

Biology Paper 2

Topic 5 — Homeostasis and response

Topic 6 Inheritance, variation and evolution

Topic 7 Ecology

Chemistry Paper 1

Topic 1   Atomic structure and the periodic table

Topic 2   Bonding, structure, and the properties of matter

Chemistry Paper 2

Topic 6 — The rate and extent of chemical change

☑

Topic 7 — Organic chemistry

☑

Topic 8 — Chemical analysis

☑

Physics Paper 1

Topic 1   Energy

Topic 2   Electricity

Topic 3   Particle model of matter

Topic 6 Waves

☑

Topic 7 Magnetism and electromagnetism

☑

MARK ALLOCATIONS

Green mark allocations[1] on answers to in-text questions throughout this guide help to indicate where marks are gained within the answers. A bracketed '1' e.g.[1] = one valid point worthy of a mark. In longer answer questions, a mark is given based on the whole response. In these answers, a tick mark[✓] indicates that a valid point has been made. There are often many more points to make than there are marks available so you have more opportunity to max out your answers than you may think.

BIOLOGY
TOPICS FOR PAPER 1

Information about Paper 1:

Trilogy 8464:

Written exam: 1 hour 15 minutes
Foundation and Higher Tier
70 marks
16.7% of the qualification grade
All questions are mandatory

Specification coverage

The content for this assessment will be drawn from topics on: Cell biology; Organisation; Infection and response; and Bioenergetics.

Questions

A mix of calculations, multiple-choice, closed short answer and open response questions assessing knowledge, understanding and skills.

Questions assess skills, knowledge and understanding of Biology.

CELL SIZE

All living things are made up of at least one cell. Cells are very small.

The size and scale of cells

Look at the mm scale on a ruler. Most animal and plant cells are between 0.1 and 0.01 mm in diameter and are too small to be seen with the human eye. We need a microscope to see them. Bacterial cells are even smaller. Cells are usually measured in micrometres, represented by the Greek letter mu (μ). One μm is 1000th of a mm or 0.001 mm.

Units and order of magnitude

Standard form is used to make very big or very small numbers more manageable. Standard form is expressed as a value between 0 and 10 × 10 to a power. For example, the thickest human hairs are roughly 180 micrometres (μm) in diameter. This is written in standard form as 1.8×10^2 μm.

You will need to convert between different units. The following are compared to one metre.

SI prefixes	Symbol	Power	Scale	Standard form example
Centi	C	10^{-2}	Hundredth	$0.0135 = 1.35 \times 10^{-2}$
Milli	m	10^{-3}	Thousandth	$0.005\ 34 = 5.34 \times 10^{-3}$
Micro	μ	10^{-6}	Millionth	$0.000\ 006\ 75 = 6.75 \times 10^{-6}$
Nano	n	10^{-9}	Billionth	$0.000\ 000\ 005\ 07 = 5.07 \times 10^{-9}$

Atom	Glucose molecule	DNA	Virus	Bacterium	Pollen	Human hair	Grain of sand	Ant	Grapefruit
10^{-1}	1	10	10^2	10^3	10^4	10^5	10^6	10^7	10^8

Nanometers

1. Write the following measurements in standard form.
   (a) The length of blood vessels in the human body: 96 500 000 metres. [1]
   (b) The length of a sperm cell: 0.000 055 metres. [1]
2. A red blood cell is 5 μm wide. Write 5 μm in nanometres (nm). [1]
3. A frog egg is 1 mm wide. A bacterium is 1 μm wide. State how many times smaller the bacterium is compared with the egg. [1]

1. (a) 9.6×10^7 m. [1]   (b) 5.5×10^{-5} m. [1]
2. 5000 nm. [1]
3. 1000 times smaller. [1]

EUKARYOTES AND PROKARYOTES

Plant and animals are known as **eukaryotes**. Bacteria are **prokaryotes** and have a different cell structure.

Eukaryotic cells

Eukaryotic cells have a **cell membrane** containing **cytoplasm**. Within the cytoplasm is a **nucleus**. The genetic material, made of **DNA** organised into chromosomes, is enclosed within the membrane of the nucleus.

Animal cell

Cell membrane
Nucleus
Mitochondrion
Ribosome
Cytoplasm

Plant cell

Chloroplast
Vacuole
Cell wall

1. The genetic material of prokaryotic and eukaryotic cells is organised in different ways. State **two** differences between the genetic material of prokaryotic and eukaryotic cells. [2]

2. Give **two** similarities between the structure of prokaryotic and eukaryotic cells. [2]

3. State the function of the bacterial plasmid. [1]

   1. *Genetic material is found within a nucleus only in eukaryotes.*[1] *DNA is organised in a single loop and smaller plasmids in bacteria, but in eukaryotes it is found in chromosomes.*[1]

   2. *Both have cytoplasm*[1]*, a cell membrane*[1] *and ribosomes*[1]*.*

   3. *Carries additional genes.*[1] *Allows genes to be passed from one bacterial cell to another.*[1]

Prokaryotic cells

Prokaryotic cells are much smaller than **eukaryotic** cells. They have a cell wall, which surrounds a cell membrane. The membrane contains cytoplasm but there is no nucleus. Instead, the genetic material is a single **DNA loop** floating freely within the cytoplasm. There may also be one or more small rings of DNA called **plasmids**.

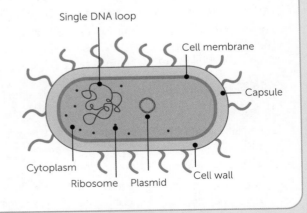

Single DNA loop
Cell membrane
Capsule
Cytoplasm
Ribosome   Plasmid   Cell wall

ANIMAL AND PLANT CELLS

Animal and plant cells contain many structures that are the same. Plants cells have some additional structures that animal cells do not have.

Structure and function of cell parts

All cells have **structures** within them. The structure of each of these **sub-cellular parts** is related to the job that they do or their function.

Cell part	Plant	Animal	Structure	Function
Nucleus	✓	✓	Contains chromosomes made of DNA	DNA controls cell processes including growth and development
Cytoplasm	✓	✓	Gel-like substance composed of mainly water, proteins, lipids and salts	Suspends cell structures; site of most cell reactions; place where most cell expansion and growth occurs
Cell membrane	✓	✓	Partially permeable	Controls what goes into and out of the cell
Mitochondria	✓	✓	Large internal surface area for reactions	Where aerobic respiration takes place
Ribosomes	✓	✓	Very small structures	Where proteins are made
Chloroplasts	✓		Contains chlorophyll pigment	Traps light energy to carry out photosynthesis
Cell wall	✓		Made of cellulose	Strengthens the cell
Permanent vacuole	✓		Made of a membrane and filled with cell sap	Stores nutrients and supports the cell

1. Algae are eukaryotes and have a cell wall. Name the substance that algal cell walls are made of.   [1]
2. Cell membranes are partially permeable. Explain why this is important for the cell.   [2]
3. The diagram shows an animal cell with a nucleus. The animal cell measures 20 μm in length, shown by the blue dotted line. Estimate the diameter of the nucleus.   [2]

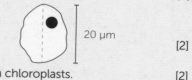
20 μm

4. Root cells are found in plants. Explain why root cells do not contain chloroplasts.   [2]

   1. Cellulose.[1]

   2. Partially permeable means that some substances can pass through the membrane and others cannot.[1] This enables control of the substances that enter and leave the cell.[1]

   3. 4 μm (One fifth of the length of the cell).[1]

   4. Root cells are below ground where there is no light.[1] Photosynthesis cannot take place without light, so chloroplasts are not needed in these cells.[1]

REQUIRED PRACTICAL 1

Using a light microscope

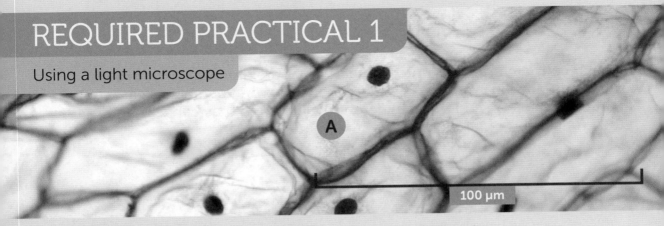

A

100 μm

This activity helps you measure, observe, draw and label cells.

Using the microscope

1. Place tissue on a glass slide with a coverslip on top.
2. Put the slide on the microscope stage and look through the **eyepiece lens**.
3. First use the lowest power **objective lens** to find cells.
4. Turn the coarse adjustment knob to focus on the cells.
5. Use the fine adjustment knob to focus more clearly.
6. Switch to a higher power lens to see finer detail, using only the fine adjustment knob to focus.

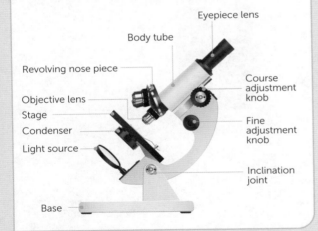

Eyepiece lens

Body tube

Revolving nose piece

Objective lens

Stage

Condenser

Light source

Course adjustment knob

Fine adjustment knob

Inclination joint

Base

1. What is the purpose of the different objective lenses on a light microscope? [2]

2. The image above shows stained onion cells, viewed under a light microscope.

   (a) Using the scale bar, estimate the length of cell A, shown above. [1]

   (b) Draw plant cell A. Label three cell structures that can be seen. [3]

   (c) The cells were viewed with a ×10 eyepiece and a ×40 objective lens. Write the magnification used to view the cells onto your drawing. [1]

1. To magnify[1] the image. Different objectives vary magnification levels.[1]

2. (a) 50 μm.[1] ~Half the length of the scale bar

   (b) Large drawing, with clear unbroken lines and no shading.[1] Shape and proportions match cell A.[1] Nucleus, cell wall and cytoplasm labelled using clear label lines / If cell membrane is labelled, the end of the label line must be clearly just inside the cell wall.[1]

   (c) 40 × 10 = 400 = ×400.[1]

Making a scientific drawing of cells

An examination question may ask you to draw cells from a microscope photograph. Draw exactly what you see, don't draw 'textbook' cells. The drawing should be as large as space allows. Use a sharp pencil to draw lines that are clear and not sketchy. Don't use shading. Label only those parts that you can see. Use a ruler to draw label lines with no arrowheads. Write the label at the end of the label line. Write the magnification used to view the cells onto the drawing (multiply the objective **magnification** by the eyepiece magnification).

CELL SPECIALISATION

Similar **cells** are organised into **tissues**, tissues are organised into **organs** and organs into **organ systems**. These components work together to carry out a particular function. Cells have different structures so they can carry out their function.

Specialised plant cells

Phloem

Phloem cells transport sugars made in leaves to the rest of the plant.

End walls have pores to allow sugar solution to pass from cell to cell easily

Phloem has elongated cells to transport sugars over long distances

Companion cells have many mitochondria for active transport

Xylem

Xylem cells transport water and mineral ions through the plant.

Hollow tubes (lumen) with no end walls to allow water and minerals to flow easily

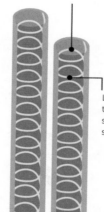

Lignin in walls to provide strength and support

Root hair cell

Root hair cells absorb water and mineral ions from soil.

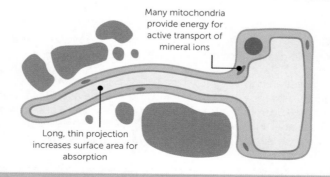

Many mitochondria provide energy for active transport of mineral ions

Long, thin projection increases surface area for absorption

Specialised animal cells

Sperm cell

The function of sperm cells is to carry genetic material to the egg cell.

Nucleus

Many mitochondria to provide energy for swimming

Tail for swimming

Enzymes in head to digest coating of egg

Muscle cell

The function of muscle cells is to produce movement.

Protein fibres shorten to make the muscle cells contract, shortening the whole muscle

Mitochondria provide energy for contraction

Nerve cell (neurone)

The role of nerve cells is to carry electrical impulses around the body. The pattern of these impulses represents a 'message'.

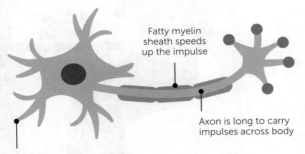

Fatty myelin sheath speeds up the impulse

Axon is long to carry impulses across body

Dendrites make connections with other cells

1. Phloem cells are elongated. How does this help them to carry out their function? [2]

2. Explain how one structure of a sperm cell helps it to reach the egg cell. [2]

   1. *It helps transport sugars[1] over a distance[1].*

   2. *Tail for swimming.[1] Lots of mitochondria to release energy for swimming.[1] OR Enzymes in head[1] help to digest coating of egg[1].*

CELL DIFFERENTIATION

As an organism develops, cells **differentiate** to form different types of **specialised cells**.

Differentiation in animal cells

Most types of animal cell differentiate at an early stage. The cells in this early embryo are unspecialised, but as cells continue to divide, they soon begin to change. Different cells develop different sub-cellular structures to become specialised and carry out different functions. This is called differentiation.

In mature animals, cell division is mainly restricted to **repair** of damaged tissues and **replacement** of worn out cells.

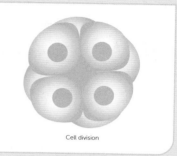

Cell division

Differentiation in plant cells

In plants, many types of cells continue to **divide** and differentiate throughout the life of the plant. Plants have regions called **meristems** that are responsible for cell division.

1. Red blood cells form from cells in the bone marrow by differentiation.

   Choose the sentence that describes differentiation.
   Tick **one** box. [1]

   ☐ Different types of red blood cells form.

   ☐ Cells divide rapidly to make exact copies.

   ☐ Cells become specialised to carry out a particular function.

2. When trees are pruned, small branches are cut off. Within a few months new shoots and leaves will begin to grow. Explain how the new leaf cells develop. [3]

3. Sperm cells form by differentiation from simple round cells. Describe **two** changes that must take place as these cells differentiate. [2]

   1. Cells become specialised to carry out a particular function.[1]

   2. Cells divide[1] and then differentiate[1] to make specialised leaf cells[1].

   3. Any two from: A tail develops.[1] The cells develop more mitochondria.[1] (for energy for swimming). The cells develop digestive enzymes in the head.[1]

MICROSCOPY

The power of a microscope depends on its magnification and resolution.

Resolution and magnification

Magnification means how many times bigger an image is than the object that is viewed. The magnifying power of a light microscope is worked out by multiplying the values of the eyepiece and objective lens together.

The **resolution** of a microscope is a measure of the smallest distance between two points that can still be distinguished. An image that has poor resolution will be blurred if it is magnified further.

Light microscopes and electron microscopes

An electron microscope

The first **light microscopes** were developed around 400 years ago. Over time these were improved, but it was still not possible to view smaller cell structures. In the 1930s, a microscope was developed that uses a beam of electrons rather than light. The **electron microscope** has much higher magnification and **resolving power** than a light microscope. Its invention led to the discovery of many new sub-cellular structures that could not be seen with the light microscope. See how to use a light microscope in Required Practical 1 on page 5.

Calculating the magnification of images

Microscopes are used to produce pictures or images of an object such as a cell. The magnification of an image can be calculated from measurements of the size of the image and of the object.

$$\text{magnification} = \frac{\text{size of image}}{\text{size of real object}}$$

1. A student viewed a cell under a microscope using a ×10 objective and a ×10 eyepiece lens. State the magnification that was used to view the cell [1]

2. A leaf cell measures 20 mm on a photograph. The original cell is 50 μm long. Calculate the magnification used to produce the image. [2]

3. An image of a bacterium measures 55 mm at a magnification of ×50,000. Calculate the size of the bacterium in mm. Give your answer in standard form. [3]

   1. ×100.[1]

   2. Convert to the same units: 20 mm = 20,000 μm.[1] Magnification = 20000 ÷ 50 = ×400.[1]

   3. Rearrange the equation to make size of real object the subject.[1]
      Size of bacterium = 55 ÷ 50,000 = 0.0011 mm.[1] Standard form 1.1 ×10^{-3} mm.[1]

CHROMOSOMES

The genetic material of eukaryotes is organised into chromosomes.

Chromosomes and genes are made of DNA

- The nucleus of a cell contains chromosomes.

- The chromosomes are made of DNA molecules.

- Each chromosome carries many genes.

- A gene is a small section of DNA on a chromosome.

- Genes carry instructions for making proteins. They determine the characteristics of every living thing.

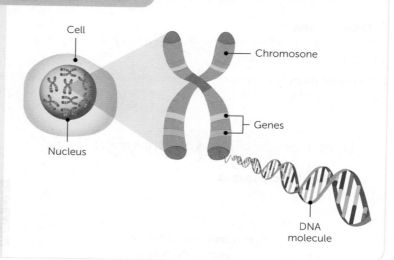

Cell

Chromosone

Genes

Nucleus

DNA molecule

You can revise more about genes and the structure of DNA (4.6.1.3 and 4.6.1.4) on pages 85 to 87.

Pairs of chromosomes

In body cells, chromosomes are normally found in pairs. Each chromosome in a pair is usually of a similar size and shape. The genes at the same locations on each of these matching chromosomes code for similar things. Humans have 23 pairs of chromosomes. Different species often have different numbers of chromosomes.

The diagram above shows chromosomes in a cell that is about to divide. The DNA has coiled up so that chromosomes become visible. The DNA molecule has been copied so that each chromosome has two DNA molecules that are joined together. In a cell that is not dividing, the chromosomes would each be made of only one DNA molecule and would not be clearly visible.

1. Write these structures in order of size, from smallest to largest:
   Chromosome   Nucleus   Gene   Cell   [1]
2. How many chromosomes are usually found in a human body cell?   [1]
3. Explain why chromosomes sometimes appear as two DNA strands.   [1]

   1. *Gene, chromosome, nucleus, cell.*[1]

   2. *46 (23 Pairs).*[1]

   3. *The DNA is replicated (copied) as a cell prepares to divide.*[1]

THE CELL CYCLE

Cells divide in a series of stages called the cell cycle.

Before a cell divides it must grow. The genetic material must double. It can then divide into two identical new cells called daughter cells. The cell cycle can be divided into three main stages. The diagram shows the cycle in a cell with just one pair of chromosomes:

Cell cycle

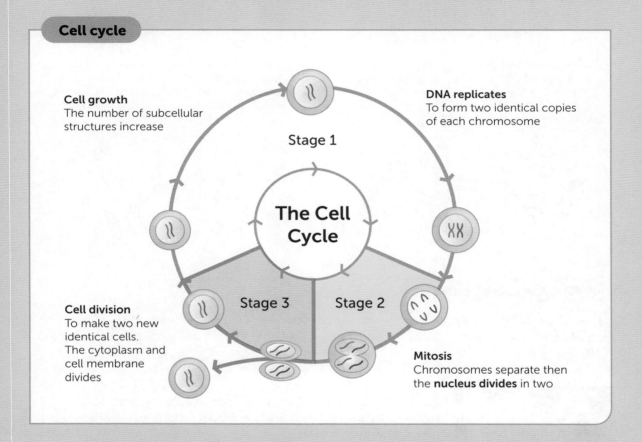

Cell growth
The number of subcellular structures increase

Stage 1

DNA replicates
To form two identical copies of each chromosome

The Cell Cycle

Cell division
To make two new identical cells.
The cytoplasm and cell membrane divides

Stage 3

Stage 2

Mitosis
Chromosomes separate then the **nucleus divides** in two

1. Describe what happens in the first stage of the cell cycle. [4]
2. At the start of the first stage of the cell cycle, the mass of DNA in a human cell is 4×10^{-12} g.
   State the mass of DNA that would be present:
   (a) Just before mitosis begins. [1]
   (b) In a new cell formed following cell division. [1]
3. Compare stage two and stage three of the cell cycle as shown on the diagram above. [2]

   1   Cell grows in size.[1] The numbers of sub-cellular structures increase [1], for example, more mitochondria and ribosomes are made[1]. DNA is replicated.[1]

   2. (a) 8×10^{-12} g.[1]   (b) 4×10^{-12} g.[1]

   3. In stage two only the nucleus divides.[1] In stage three the cytoplasm and cell membrane divide so the whole cell splits in two.[1]

MITOSIS

Mitosis is the process of cell division. The replicated chromosomes separate and then two new nuclei form.

Stages of mitosis

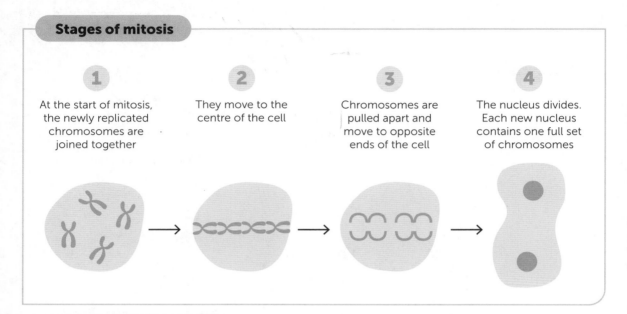

1	2	3	4
At the start of mitosis, the newly replicated chromosomes are joined together	They move to the centre of the cell	Chromosomes are pulled apart and move to opposite ends of the cell	The nucleus divides. Each new nucleus contains one full set of chromosomes

Cell division by mitosis

Cells divide by mitosis to make new, genetically identical cells. It is how multicellular organisms grow. Mitosis must happen many millions of times for a fertilised egg to grow into an adult human. Mitosis also makes new cells for repair and replacement of cells.

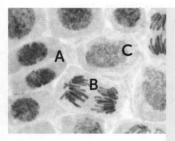

1. Look at the photograph of cells taken down a microscope. Cells A, B and C are all at different stages in the cell cycle. Describe what is happening in each cell. [3]

2. A scientist counted 100 cells on the slide, 11 of these were in stage B. The cell cycle takes 20 hours in this tissue.
   Calculate how much time in minutes is spent in stage B. [1]

3. Skin cells are continuously lost. Skin has several layers of differentiated cells and one layer of undifferentiated cells. Explain how new skin cells form. [3]

1. *The cytoplasm and cell membrane of cell A are dividing; it is undergoing cell division.[1] Mitosis is taking place in cell B[1], chromosomes are moving to opposite ends of the cell[1]. Cell C is preparing to divide, it will grow in size and replicate its DNA.[1]*

2. $\frac{11}{100} \times 20 \text{ hours} \times 60 \text{ minutes}$[1] *= 132 minutes.[1]*

3. *Division by mitosis makes new cells[1] in the undifferentiated layer.[1] These cells then differentiate to form other layers.[1]*

STEM CELLS

Stem cells are undifferentiated cells that divide to make new cells.

Cloning means making genetically identical copies of cells or of whole organisms.

Stem cells in human embryos

Cells in the early embryo are unspecialised. They will eventually differentiate to produce all the different types of cell needed to make the adult organism. Scientists can **clone** cells from early human embryos in a laboratory. Cloned cells can be made to differentiate into most different types of human cells.

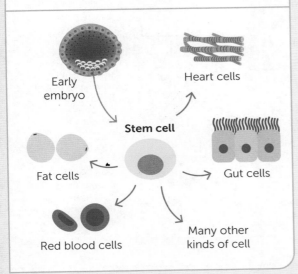

Early embryo

Stem cell

Heart cells

Gut cells

Fat cells

Red blood cells

Many other kinds of cell

Stem cells in adult animals

Some stem cells remain in adult animals. Stem cells in fully developed animals can only produce a limited range of cell types. Their function is to replace cells that become worn out or damaged. They differentiate to produce just a few cell types, usually within one tissue. Stem cells in **bone marrow** are unusual because they can produce many types of cells, including blood cells.

Stem cells in plants

Plants have stem cells in specialised tissues called **meristems**. The cells produced by plant meristems can differentiate into any type of plant cell throughout the life of the plant.

See **page 38** for more detail on meristems.

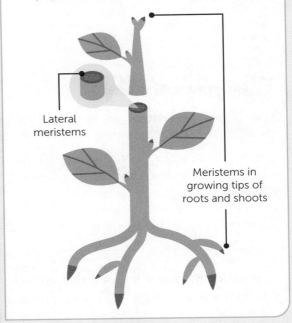

Lateral meristems

Meristems in growing tips of roots and shoots

1. Describe the function of stem cells in animal embryos. [2]
2. Give **one** difference between the function of stem cells in mature plants and those in animals. [1]

1. *They divide and differentiate[1] to give the many different cell types in the mature animal[1].*

2. *Mature plant stem cells can differentiate into all cell types, allowing the plant to grow, form new organs and repair damage. Mature animal stem cells make only a few types of cell (within a tissue) for repair.[1]*

USING STEM CELLS

Stems cells can be useful in agriculture, conservation and medicine.

Stem cells for medical treatment

Treatments are being developed that use stem cells to replace faulty or damaged cells. Conditions such as diabetes and paralysis may be treated this way in the near future. Therapeutic cloning is one way of producing stem cells from cloned embryos, as shown in the diagram opposite. These stem cells are not rejected by the patient's body because they have the same genes as the patient.

Advantages of therapeutic cloning	Disadvantages of therapeutic cloning
May treat diseases and replace faulty organs	Stem cells may transfer viral infection
Produces a wider range of cells than is possible from collecting adult stem cells	Some people believe using an embryo kills a potential life, which creates ethical or religious objections
Stem cells produced this way are not rejected by the patient's immune system	Human egg cells are in short supply. Collecting them has risks
Many cells are produced. Some could be used in research	There may be side effects of stem cells which are yet unknown

Cultivation process

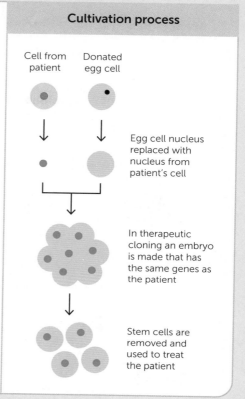

Cell from patient   Donated egg cell

Egg cell nucleus replaced with nucleus from patient's cell

In therapeutic cloning an embryo is made that has the same genes as the patient

Stem cells are removed and used to treat the patient

Using plant stem cells

Small groups of **meristem** cells can be cut from a plant and cultured in the laboratory to produce identical new plants. The new plants are clones of the parent plant.

The benefits of growing plants from stem cells are:
- Large numbers of new plants can be produced quickly and cheaply.
- Rare species can be cloned to protect them from extinction.
- Crop plants with useful features such as disease resistance can be cloned from meristem cells. This makes large numbers of identical useful plants available for farmers.

1. State **one** advantage of using stem cells to cultivate rare plants. [1]
2. Suggest **one** disadvantage of producing agricultural plants using stem cells. [1]

   1. Only one plant is needed to make many new plants; two parents are not needed.[1]

   2. All the new plants will be genetically identical, meaning they will be susceptible to the same diseases.[1]

DIFFUSION

The kinetic energy of the particles in a solution or a gas causes them to move about randomly. This movement causes **diffusion**, which is the spreading out of moving particles. There is a **net** (overall) movement from an area of higher concentration to an area of lower concentration. This is down a concentration gradient.

Arrows show the direction of the **concentration gradient**

Cells need to exchange substances with their surroundings to survive. Some substances can move into and out of cells across the cell membranes by diffusion. Movement by diffusion is always down the concentration gradient.

Example

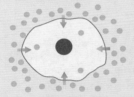

Oxygen and digested food molecules can move into cells by diffusion

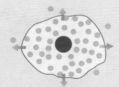

Waste products like **carbon dioxide** and **urea** move out of the cells by diffusion

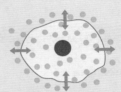

When the concentration inside and outside the cell is the same, particles still move across the cell membrane, but there is no **net** movement in or out

Factors increasing the rate of diffusion

- A steeper **concentration gradient** increases the chance of particles moving from the area of high concentration to the area of low concentration, than in the opposite direction.
- A higher **temperature** increases the kinetic energy of the particles, so they move faster.
- A larger **surface area** provides more membrane for the particles to move across.

Smaller organisms have a larger surface area to volume ratio. This increases exchange by diffusion. The large surface area to volume ratio of a single-celled organism allows enough exchange of molecules to meet its needs.

Model organism A: Sides 1 × 1 cm. Surface area 6 cm². Volume 1 cm³. Surface area to volume ratio 6:1.

Model organism B: Sides 2 × 2 cm. Surface area 24 cm². Volume 8 cm³.

1. Calculate the surface area to volume ratio of model organism B. [2]
2. Describe how urea produced inside cells travels to the kidneys for excretion. [2]

1. $\frac{SA}{Vol}$ = $\frac{28}{8}$ [1] *giving a surface area to volume ratio of 3:1[1].*

2. *Urea diffuses out of cells[1] into the blood plasma[1] to be carried to the kidneys for excretion.*

DIFFUSION AND EXCHANGE SURFACES

Bigger organisms have a smaller surface area to volume ratio. **Multicellular organisms** must have specially adapted **exchange surfaces**, connected to a **transport system**, to supply all the cells. Without these specialised structures the movement of substances into and out of internal cells would be too low for survival.

How diffusion is increased across exchange surfaces

- They have a large **surface area** – this increases the area of membrane available for molecules to diffuse across.
- They have **thin membranes** – so that the diffusion distance for particles is short.
- In animals, exchange surfaces have an **efficient blood supply**, and gas exchange surfaces are **ventilated**. This increases diffusion by keeping the concentration gradient high.

Examples of exchange organs

Small intestine

In the **small intestine** - projections of the wall (villi) and epithelial cells (microvilli) increase the surface area. The wall of the villi is thin and they are well supplied with blood from capillaries.

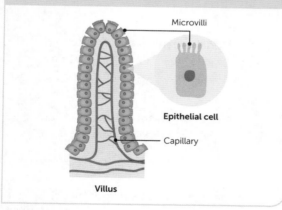

Fish gills

The **gills** of fish are adapted for gas exchange in water. Many gill filaments increase the surface area. Filaments are well supplied with blood. They are thin-walled to reduce the length of the diffusion pathway. The fish moves the floor of its mouth down to draw water in and moves it up to push water out through the gill slits – this ventilates the gills.

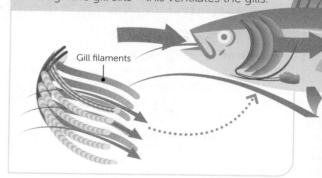

Lungs

In the **lungs** of mammals, large numbers of alveoli increase the surface area. Each alveolus is thin-walled and surrounded by capillaries. Muscles between the ribs and in the diaphragm move air in and out, ventilating the exchange surface. See more on **page 31**.

Roots and leaves of plants

In plants, many root hair cells increase the surface area of **roots** for absorption of mineral ions and water. **Leaves** are adapted for gas exchange. They have a large internal surface area for diffusion of gases and are thin to reduce the diffusion distance to cells. See more on **pages 39-41**.

OSMOSIS

Water moves across cell membranes by osmosis.

Definition

Osmosis is the diffusion of water from a dilute solution to a more concentrated solution through a partially permeable membrane. The partially permeable membrane allows water to diffuse through it, but the solute does not. It is a passive process. No energy input is needed by the cell.

Note: Concentrations of solutions are described in terms of solute molecules. So a dilute solution contains many water molecules and not many solute particles (high concentration of water) and a concentrated solution has a high concentration of solute molecules (lower concentration of water molecules). So water is in effect moving down its concentration gradient.

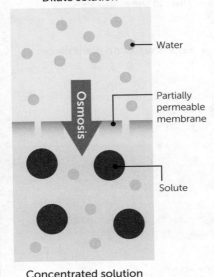

Dilute solution

Water

Osmosis

Partially permeable membrane

Solute

Concentrated solution

Measuring osmosis

Cells can change in size and mass due to osmosis. We can measure these changes in pieces of plant tissue to investigate osmosis.

Cell in dilute solution
Net movement of water into the cell. Cell volume and mass increase.

Cell in solution of same concentration
No net movement of water. Cell volume and mass stays the same.

Cell in concentrated solution
Net movement of water out of the cell. Cell volume and mass decreases.

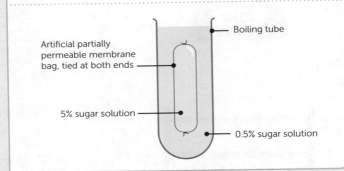

Artificial partially permeable membrane bag, tied at both ends

Boiling tube

5% sugar solution

0.5% sugar solution

1. Explain the term 'partially permeable membrane'. [1]
2. Animal cells placed in distilled water swell then burst. Plant cells placed in distilled water swell but do not burst. Suggest a reason for this difference. [2]
3. Explain why the bag of sugar solution shown in the diagram would increase in mass. [3]

   1. *A membrane which some substances can pass through and others cannot.[1]*
   2. *Water enters the cells by osmosis causing the cells to swell.[1] The plant cell wall prevents the cell from bursting. The animal cell does not have a cell wall.[1]*
   3. *Water moves into the bag by osmosis[1] from the dilute to the concentrated solution[1]. The extra water increases the mass of the bag.[1]*

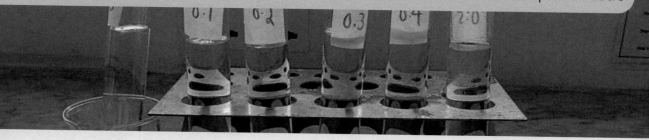

REQUIRED PRACTICAL 2

Investigating the effect of solute concentration on the mass of plant tissue

Pieces of plant tissue, such as potato or apple, are weighed using an electronic balance. They are then placed in a range of concentrations of salt or sugar solutions. They are reweighed after allowing time for osmosis to occur.

Equipment

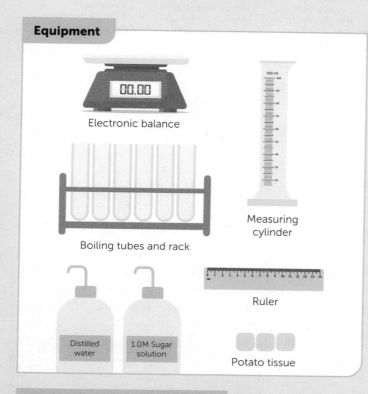

Electronic balance

Measuring cylinder

Boiling tubes and rack

Ruler

Distilled water

1.0M Sugar solution

Potato tissue

Variables

Other variables that affect osmosis or the mass of the plant material must be controlled:

- **Surface area of the plant tissue** – Use a cork borer or knife and ruler to cut plant tissue to the same-sized pieces.
- **Volume of solution** – Use a measuring cylinder to measure a set volume.
- **Time in the solution** – Use a stopwatch and leave for the same time e.g. 24 hours.
- **Excess water on plant pieces** – Pat dry using a paper towel before weighing.
- **Temperature** – Use a water bath to keep the temperature the same for all solutions.

Measuring the rate of osmosis

Plant tissue will lose mass in the experiment if it is placed in a solution that is more concentrated than the cells. It will gain mass if placed in a more dilute solution. The change in mass during the experiment is calculated. This is then converted into a percentage change in mass.

$$\text{percentage change in mass} = \frac{final\ mass - initial\ mass}{initial\ mass} \times 100$$

Recording results

Salt concentration (M)	Initial mass (g)	Final mass (g)	Change in mass (g)	% change in mass
0.00	2.63	2.88		
0.20	2.57	2.72	0.15	5.8
0.40	2.75	2.78	0.03	1.1
0.60	2.47	2.38	−0.09	−3.6
0.80	2.60	2.42	−0.18	−6.9
1.00	2.59	2.40	−0.19	−7.3

Change in mass of potato pieces at different salt concentrations

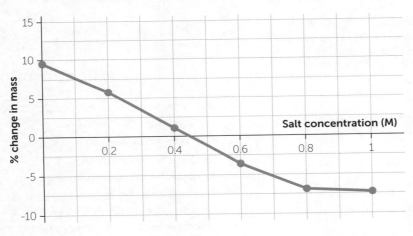

1. Explain why percentage change in mass is calculated. [2]
2. Calculate the **two** missing values in first line of the results table. [2]
3. (a) Use the graph to estimate the salt concentration that is equal to the concentration in the potato cells. [1]

   (b) Explain how you arrived at this answer. [2]
4. Explain the trends shown by the table. [4]

   1. So that results can be compared[1] even if initial mass is not identical[1].
   2. Change in mass = 2.88 − 2.63 = 0.25[1]. $\frac{\text{Change in mass } 0.25}{\text{Initial mass } 2.63} \times 100 = 9.5\ \%$.[1]

   3. (a) 0.45 M (mol).[1] (b) This is the point where change in mass was zero.[1] There is no net movement of water[1] so the concentration inside and outside the cell must be equal[1].
   4. In salt solutions from 0.0 to 0.4 M the potato pieces increased in mass[1] because water moved in by osmosis.[1] In solutions from 0.6 to 1.0 M the potato lost mass[1] because water moved out of cells by osmosis[1].

ACTIVE TRANSPORT

Active transport moves solute substances against a concentration gradient

Active transport

Cells may need to move substances from a more dilute solution to a more concentrated solution. They can do this using active transport. This active process requires energy from respiration. Substances are transported using carrier proteins in the cell membrane. Active transport can move substances into or out of a cell.

Active transport moves solute substances, whereas osmosis causes water molecules to move freely through a partially permeable membrane.

Outside cell
Low concentration

Cell membrane

Energy from respiration

Inside cell
High concentration

In plants

Active transport allows mineral ions to be absorbed into plant root hairs from the very dilute solutions in the soil. Plants need mineral ions for healthy growth.

In the gut

Active transport is also important for absorbing sugar from the gut (small intestine). It is needed because sugar is at a lower concentration in the gut than in the blood stream.

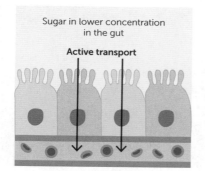

Sugar in lower concentration in the gut

Active transport

Sugar in higher concentration in the blood

1. State why all cells need the sugar molecules that are absorbed by the gut. [1]
2. Give **two** differences between:
   (a) Osmosis and diffusion. [2]
   (b) Active transport and diffusion. [2]
3. Root hair cells absorb water by osmosis and mineral ions by active transport. Explain why different processes are used to transport water and ions. [4]

1. *Sugar molecules are used for cell respiration.*[1]
2. (a) *Osmosis transports only water and is through a partially permeable membrane.*[1] *Other substances can move by diffusion which does not always involve a membrane.*[1]

   (b) *Active transport requires energy from respiration, diffusion does not.*[1] *Diffusion is always from a concentrated to a dilute solution, but active transport can move substances against the concentration gradient.*[1]
3. *The soil solution is more dilute than that in root cells*[1] *and water moves from a dilute to a concentrated region (by osmosis).*[1] *The concentration of ions is lower in the soil than in root hair cells*[1] *so energy is needed to transport them against the concentration gradient of ions*[1].

EXAMINATION PRACTICE

01 The table lists some sub-cellular structures that may be found in cells.

01.1 For each type of cell, fill in the table boxes with a tick if a sub-structure is present and a cross if it is not. One row has been completed for you. [2]

01.2 Describe where the genetic material is found in prokaryotic cells. [2]

Sub-cellular structure	Eukaryotic cell	Prokaryotic cell
Cytoplasm	✓	✓
Nucleus		
Cell membrane		

02 Name **three** cell parts belonging to plant cells that animal cells do not have. [3]

03 The diagram below shows a drawing of a red blood cell. The actual red blood cell was 8 μm in length.

03.1 The function of these cells is to carry oxygen. They are packed with a substance called haemoglobin that binds to oxygen. Red blood cells have no nucleus. Suggest why red blood cells have no nucleus. [2]

03.2 Calculate what the length of the red blood cell would be in metres. Give your answer in standard form. [2]

03.3 A student drew the red blood cell. The red blood cell on the drawing was 6 cm in length. Calculate the magnification of the drawing. [2]

03.4 Muscle cells bring about movement of the body. Explain why muscle cells contain many mitochondria. [2]

03.5 Describe **one** other feature of muscle cells that allows them to carry out their function. [1]

04 Ribosomes are small structures found in cells. They were discovered after the invention of the electron microscope.

04.1 Explain why an electron microscope was needed to observe ribosomes. [3]

04.2 Suggest why cells are often stained before observing with a light microscope. [1]

04.3 Light microscopes have a coarse focus and a fine focus knob.
Explain how these features are used when observing cells on a slide. [2]

05 Micro-organisms can be grown in the laboratory on agar plates. The agar contains added substances.

05.1 Explain the purpose of the agar. [2]

05.2 Aseptic technique is important to ensure that cultures of microorganisms are uncontaminated. Contamination can happen in different ways. Explain how contamination of cultures with unwanted or harmful microorganisms can be prevented. Your answer should mention the sources of contamination and the methods used to prevent it.

[6]

06 Plants continue to grow throughout their life. New cells form by division and then specialise into each cell type.

06.1 Name the process by which new cells become specialised. [1]

06.2 Describe the differences between the production of new specialised cells in mature plants and animals.

[3]

07 A scientist investigated the effectiveness of two disinfectants used in hospitals. The disinfectants were diluted to different concentrations. These solutions were soaked onto paper discs. Discs were placed onto agar plates containing bacteria. After 40 hours the clear zones around the discs were measured.

Table 1

Concentration	Mean diameter of clear zone in mm	
	Disinfectant A	Disinfectant B
Low	9	4
Medium	18	17
High	21	20

Table 2

	Disinfectant A	Disinfectant B
Cost in £ per dm³	20	4
Toxicity to people	Medium	Low

Evaluate which **type** and **concentration** of disinfectant the hospital cleaners should use to kill these bacteria. Use information from Table 1 and Table 2.

[6]

08 Adult stem cells can be collected from donated bone marrow. These cells are cultured in the laboratory and used to treat some medical conditions such as blood cell cancer.

08.1 The stem cells are cultured to produce more identical cells. Name this type of cell division.

[1]

08.2 Give **one** risk and **one** benefit of using these stem cells compared to cells from therapeutic cloning.

[2]

09 An experiment was set up as shown. The bag of sugar solution was weighed at the start and after 20 minutes.

09.1 Predict what would happen to the mass of the bag. [1]

09.2 Explain your prediction. [3]

09.3 A student decided to use this equipment to compare the effect of different sugar concentrations.

Explain why temperature and size of the bag should be control variables in this experiment. [4]

09.4 In a different experiment, the bag of solution increased from 10.0 g to 22.0 g in 2 hours.

Calculate the rate of water uptake by osmosis per hour. [2]

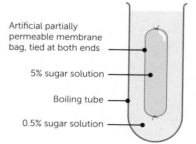

Artificial partially permeable membrane bag, tied at both ends

5% sugar solution

Boiling tube

0.5% sugar solution

PRINCIPLES OF ORGANISATION

The cells of multicellular organisms are organised into levels of tissues, organs and organ systems.

Cells

Cells are the basic building blocks of all living organisms.

Epithelial cell in the small intestine

Tissues

A **tissue** is a group of cells with a similar structure and function.

Epithelial tissue in the small intestine

Organs

Organs are collections of tissues performing specific functions.

The small intestine – an organ

Organ systems

Organs are organised into **organ systems**.

The **digestive system** is an example of an organ system. Several organs work together to carry out the function of digesting and absorbing food.

Different organ systems work together to form the whole living thing – the **organism**.

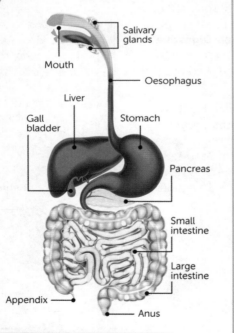

Salivary glands
Mouth
Oesophagus
Liver
Gall bladder
Stomach
Pancreas
Small intestine
Large intestine
Appendix
Anus

You should know the main parts of the digestive system and what they do from your Key Stage 3 studies.

1. In plants, tightly packed cells carry out most of the photosynthesis.
   Name:
   (a) the tissue in which the cells are found. [1]
   (b) the organ in which they are found. [1]
2. Name the animal organ system that neurone cells belong to. [1]
3. The stomach organ contains muscle tissue. Explain the function of this tissue. [3]

   1 (a) *Palisade mesophyll tissue.*[1]
     (b) *Leaf.*[1]
   2. *Nervous system.*[1]
   3. *It contracts*[1], *mixing food with acid and enzymes*[1] *speeding up digestion.*[1]

ENZYMES

Enzymes are biological catalysts. Catalysts speed up the rate of a reaction but are not used up in the reaction.

Lock and key theory

Each type of enzyme catalyses one reaction in a living organism. Enzymes have an area called an **active site** that has the correct shape to bind to one type of molecule. This molecule is called the **substrate**. Enzymes **catalyse** specific reactions according to the shape of their active site. The 'lock and key theory' is a simple model that explains enzyme action.

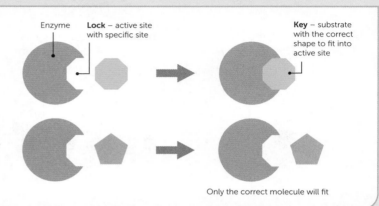

Enzyme

Lock – active site with specific site

Key – substrate with the correct shape to fit into active site

Only the correct molecule will fit

How enzymes work

Without enzymes, all the reactions of our **metabolism** would be too slow for survival. Some enzymes work to break down a substrate into products that are smaller molecules. **Digestive enzymes** work like this. They convert food into small soluble molecules that can be absorbed into the bloodstream.

The products of digestion are used to build new **carbohydrates**, **lipids** and **proteins**. Enzymes are also involved in this process. These enzymes catalyse the reaction to join small molecules together to make larger ones. Some glucose from digestion is also used in respiration - a reaction which is also catalysed by enzymes.

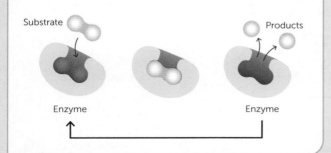

Substrate

Products

Enzyme

Enzyme

Enzymes are proteins

Enzymes are large protein molecules that are folded into a ball-like shape. This shape is held together by chemical bonds. Extremes of temperature and pH can affect this folding and change the enzyme's shape, including the shape of the active site. If the shape of the active site changes too much, the substrate will not fit so the reaction will no longer be catalysed. This means that the enzyme has been **denatured**.

Normal conditions

Enzyme molecule

Substrate molecule

Extreme conditions

Denatured enzyme

Shape of active site has changed so substrate can no longer fit

The effect of temperature on enzyme activity

Enzymes have the highest activity at their **optimum temperature**. This is usually around 40°C for human enzymes but can be different for other enzymes. They work slowly at low temperatures and stop working all together at higher temperatures.

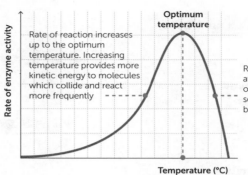

Rate of reaction increases up to the optimum temperature. Increasing temperature provides more kinetic energy to molecules which collide and react more frequently

Optimum temperature

Rate of reaction decreases at temperatures above the optimum. Enzymes denature, so the substrate can no longer bind to the active site

The effect of pH on enzyme activity

Each enzyme has an **optimum pH** at which its activity is highest. The acid or alkaline conditions at other pH levels interfere with the shape of the enzyme and it works less effectively. At extreme pH the enzyme will denature.

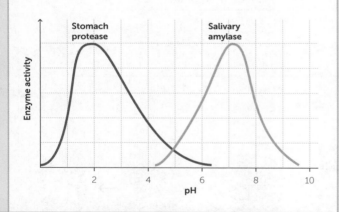

This is also covered in Required Practical Activity 4 on **page 28**.

1. Explain the differences in the optimum pH of salivary amylase and pepsin. Use the graph above. [3]

2. Salivary amylase breaks down starch. Explain why it will not break down proteins. [2]

3. The enzyme catalase breaks down hydrogen peroxide and releases oxygen gas. In an investigation, the enzyme produced 30 cm³ of oxygen in 10 minutes. Calculate the rate of the reaction. [2]

1. *The optimum pH of amylase is 7 and pepsin is 2.[1] Conditions in the stomach are acidic and those in the mouth are neutral[1], so each enzyme has the optimum pH for maximum activity in the area where they are released.[1]*

2. *The active site shape is specific to starch.[1] It is the wrong shape to bind to proteins.[1]*

3. $\frac{30}{10} = 3^{[1]}$ *cm³/min⁻¹.[1]*

Measuring the rate of reactions

A reaction rate describes the amount of product made per unit time. We can calculate the rate of enzyme reactions by calculating the gradient of a graph or by dividing by the time taken for the reaction to take place.

THE HUMAN DIGESTIVE SYSTEM

Human digestive enzymes

Three main types of enzymes break down food in our digestive system. These are the **carbohydrases**, **proteases** and **lipases**.

There are several varieties of these enzymes made in different parts of the digestive system. One important type of carbohydrase is **amylase**. Amylase is a carbohydrase which breaks down **starch** into the simple sugar, glucose.

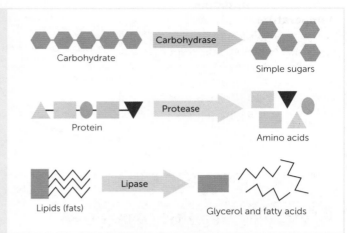

Enzyme	Where it is produced	Where it works
	Salivary glands	Mouth
Amylase	Pancreas	Small intestine (duodenum)
	Wall of small intestine	Small intestine (ileum)
	Wall of stomach	Stomach
Protease	Pancreas	Small intestine (duodenum)
	Wall of small intestine	Small intestine (ileum)
Lipase	Pancreas	Small intestine (duodenum)

Bile emulsifies fats

Bile is made in the **liver** and stored in the **gall bladder**. It is released through the bile duct into the first part of the small intestine. Here, the alkaline bile **neutralises** the hydrochloric acid in the partly digested food coming from the stomach. Bile also **emulsifies** fat to form small droplets which increases the surface area.

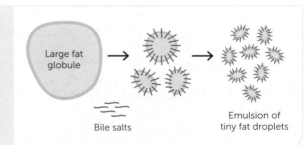

1. Write the word equation for the breakdown of lipids (fats) by lipase. [1]
2. Bile causes alkaline conditions and a large surface area of fat droplets.
   Explain how this increases the rate of fat digestion. [2]

   1. Lipids → Fatty acids + Glycerol (the word 'lipase' may be written above the arrow).[1]

   2. A larger surface area allows lipase enzymes to reach more lipid molecules.[1] This enzyme has an alkaline optimum pH so has a higher activity in alkaline conditions.[1]

REQUIRED PRACTICAL 3

Use qualitative reagents to test for carbohydrates, lipids and proteins

Testing for carbohydrates, lipids and proteins

You can test for different biological molecules by adding reagents that give a colour change. You need to be able to describe and understand results from the tests shown in the table. There are two tests for carbohydrates: the iodine test for starch and the Benedict's test for sugar. Tests can be done in a test tube. The substance to be tested (usually a food) may need to be ground up with water. The iodine test can be done on lumps of food on a spotting tile. If the substance is not present, the colour will not change.

Test for	Reagents	Method	Starting colour	Positive result
Sugar	**Benedict's** solution	Mix the food with a few cm³ of Benedict's solution. **Heat** it	Blue	Orange-red precipitate
Starch	Iodine solution	Add a few drops of iodine solution to the food	Orange-brown	Blue-black solution
Protein	**Biuret** reagent	Mix the food with a few cm³ of Biuret reagent	Blue	Purple
Lipid	Ethanol	Shake the food with ethanol. Pour this into water	Clear solution	An emulsion forms as a cloudy white layer

Using a water bath

Testing for sugar with the Benedict's test requires heating, usually using a water bath. A water bath can simply be a beaker of water heated with a Bunsen burner. Water baths are used to help maintain a constant temperature. It is safer to use an electronic water bath to heat liquids than a beaker of water and Bunsen burner, because there is no naked flame and it is less likely to spill. Water baths with an electric heater have a thermostat so that temperature can easily be set and kept constant.

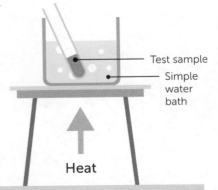

Test sample
Simple water bath

Heat

Safety: To use a water bath safely you must avoid scalding from steam or spilled hot water. Keeping water temperatures below 60°C and standing up while heating helps to reduce the risk of scalding.

REQUIRED PRACTICAL 4

Investigating the effect of pH on the rate of reaction of amylase enzyme

Continuous sampling

The diagram shows how continuous sampling is used to measure the time taken for amylase to digest a starch solution at a range of pH values. Different **buffer solutions** are added which keep the pH at a particular value. The timer is started when the amylase and starch solutions are mixed together. **Iodine** reagent is used to test for starch every 30 seconds. Iodine will turn from orange-brown to blue-black if starch is still present. The time taken for all the starch to be digested (when the iodine first remains orange-brown) is recorded. This approach measures the reaction time to the nearest 30 seconds.

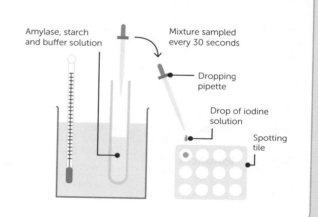

1. Explain why the water bath is used in this investigation. [2]
2. Use the diagram of example results to draw a results table. [3]
3. A student looked at the example results in the diagram. They wrote 'pH 6 is the optimum pH for amylase.' Explain why this may not be a valid conclusion. [2]
3. Use the diagram of example results to draw a results table. [3]

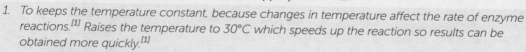

4. The rate of reaction can be calculated using $\frac{1}{\text{Time taken}}$ for the reaction. Calculate the rate of reaction for pH 6. Include appropriate units. [2]

1. *To keeps the temperature constant, because changes in temperature affect the rate of enzyme reactions.[1] Raises the temperature to 30°C which speeds up the reaction so results can be obtained more quickly.[1]*

2. *See table. Times calculated correctly (e.g. pH 4 takes 8 × 30 s to react completely = 240 s) [1]. Table headings correct.[1]*

3. *Not enough different pH levels were tested to make this conclusion.[1] The optimum pH could be anywhere between pH4 and pH8. [1]*

4. *Rate = $\frac{1}{90}$ = 0.11[1] per second, or s^{-1}.[1]*

pH	Time for reaction (s)
4	240
6	90
8	180

THE HEART AND CIRCULATION

Structure of the heart

The **heart** is an organ that pumps blood around the body. It has walls made of cardiac muscle with four chambers inside. The right ventricle pumps blood to the lungs, where **gas exchange** takes place. The left ventricle pumps blood around the rest of the body. The atria collect blood as it returns and pump it into the ventricles. The atria contract together just before the ventricles contract. Blood shown as blue in the diagram has given up its oxygen to body cells – it is **deoxygenated**. The blood shown as red has been **oxygenated** in the lungs. Valves between the atria and ventricles, and in the veins, prevent blood flowing backwards.

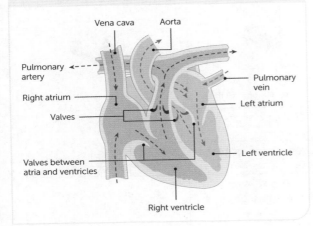

A double circulatory system

Mammals have a double circulation – blood flows through the heart twice during one complete circulation.

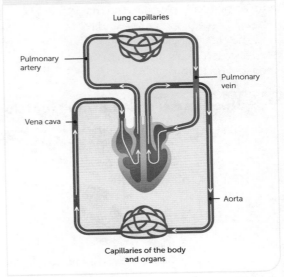

Heart rate

Heart rate is the frequency with which the heart contracts. It is measured in **beats per minute**. The natural resting heart rate is controlled by a group of cells found in the right atrium. They act as a **pacemaker**, producing regular impulses that travel through the heart causing it to contract. **Artificial pacemakers** are electrical devices used to correct irregularities in the heart rate.

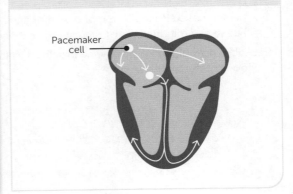

1. The left ventricle has a thicker wall than the right ventricle. Suggest why. [2]
2. Give the benefits of a double circulation. [3]
3. Coronary arteries run down the outside of the heart. State the role of these arteries [1]

   1. *The thicker muscle generates greater force[1] needed to push blood around the body compared to through the lungs.*
   2. *Passing through twice allows a higher pressure to be maintained[1] increasing blood flow to the tissues.[1] Oxygenated and deoxygenated blood do not mix.[1]*
   3. *They supply the cells of the heart / heart muscle with oxygen.[1]*

BLOOD VESSELS

The body contains three different types of blood vessel: **arteries**, **veins** and **capillaries**.

The aorta branches into different arteries that carry blood to the major organs. These branch more and more until they form tiny vessels within tissues called capillaries. Capillaries then join up to form veins.

Note that the muscle in arteries does **NOT** pump blood, it simply adjusts the size of the lumen.

Blood flow

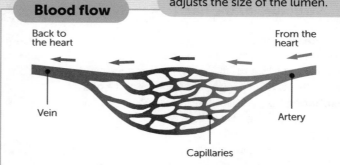

Back to the heart

From the heart

Vein

Artery

Capillaries

Blood vessel structure and function

	Arteries	Capillaries	Veins
	Thick outer wall / Small lumen / Thick layer of muscle and elastic fibre	Very small lumen / Very thin wall, only one cell thick	Thin layer of muscle and elastic fibres / Large lumen / Outer wall is fairly thin
Function	Carry blood at high pressure away from the heart	Exchange of substances with cells	Return blood at low pressure to the heart
Lumen	Narrow to maintain pressure	Very narrow. Keeps red blood cells close to tissue cells	Large, so there is less resistance to blood flow
Wall	Elastic fibres stretch and recoil to maintain pressure. Thick wall resists bursting	Very thin – Short distance to maximise exchange by diffusion	Low pressure so no need for a thick elastic wall
Valve	No – High pressure blood keeps moving	No	Yes – Prevents backflow of low pressure blood

1. A person has a stroke volume of 0.06 dm³ and a heart rate of 65 beats per minute (bpm). Calculate the cardiac output. [2]
2. Explain how blood keeps flowing in veins despite the low blood pressure in these vessels. [3]

   1. *0.06 × 65 = 3.9[1] dm³ per minute[1].*
   2. *Skeletal muscles press on the veins during activity and squeeze blood along.[1] Valves prevent the blood from going in the wrong direction.[1] The lumen is large so there is little resistance to flow.[1]*

Rate of blood flow

The rate of blood flow from the heart into the aorta is called the cardiac output. It is calculated from the stroke volume (the volume pumped with each heartbeat) and the heart rate.

Cardiac output (dm³ per minute) = stroke volume (dm³) × heart rate (beats per minute)

LUNGS AND GAS EXCHANGE

Structure of the lungs

The **trachea** leads from the throat to the lungs. Here it splits into two **bronchi** (*singular*, bronchus). The bronchi divide further and eventually end in small air sacs called **alveoli** (*singular*, alveolus). Muscles in the diaphragm and between the ribs contract and relax again to move air into and out of these airways - a process called **ventilation**.

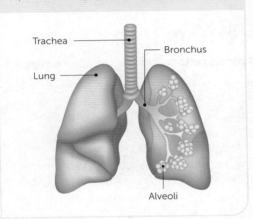

Gas exchange in the alveoli

The alveoli are surrounded by a network of **capillaries**. Capillaries bring blood to them with a high concentration of **carbon dioxide** and a low concentration of **oxygen**. Oxygen moves by **diffusion** from the air in the alveolus into the blood. Carbon dioxide diffuses in the opposite direction. Blood leaving the alveolus has a high concentration of oxygen and low concentration of carbon dioxide.

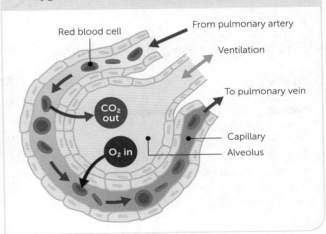

How the lungs are adapted for gas exchange

Lung adaptations increase gas exchange by diffusion. A large **surface area** for exchange is provided by the many alveoli with folded surfaces. There is a **short diffusion pathway** because the alveolar wall and capillary wall are both only one cell thick. The **concentration gradient** for diffusion is maintained by having a good blood supply from many capillaries and by ventilation (by breathing). This maintains significant differences in the concentration of CO_2 and O_2 between alveoli and blood, so that diffusion is maximised.

1. Suggest why trachea and bronchi have rings of stiff tissue called cartilage. [1]
2. Explain the importance of: (a) carbon dioxide removal. [1]
                                 (b) oxygen uptake by the lungs. [1]
3. Describe the path taken by an oxygen molecule as it moves from the air in the alveolus to the blood. [4]

1. *To keep the airways open and allow unobstructed air movement.[1]*
2. *(a) CO_2 from cellular respiration is toxic.[1]   (b) Oxygen is needed for aerobic respiration.[1]*
3. *It diffuses [1] through the single cell layer of the alveolus wall[1] through the single-celled wall of the capillary [1] through the blood plasma[1] then into a red blood cell[1].*

BLOOD

The composition of blood

Blood is a tissue. It has four main components: plasma, red blood cells, white blood cells and platelets.

Plasma is the liquid part of the blood. It is mostly water. All the blood cells and platelets are suspended in this liquid.

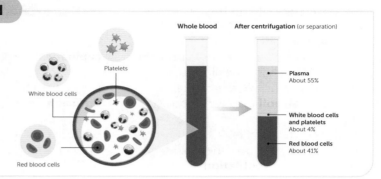

The functions of blood components

Plasma

Plasma is important for **transport** around the body. Useful molecules, such as digested food, are dissolved in the plasma and transported to cells. **Waste** products from cells, such as **carbon dioxide** and **urea**, are transported in solution to the lungs or the kidneys for excretion.

White blood cells

White blood cells are important for **defence against infection**. They can recognise cells that are non-self. White blood cells defend the body in different ways. Some (called phagocytes) are flexible and engulf microorganisms; they then produce enzymes to digest them. Lymphocytes can produce antibodies to kill pathogens. See **section 4.3.1.6 page 50** for more about pathogens.

Phagocyte

Lymphocyte

Red blood cells

Red blood cells are adapted to carry **oxygen**:

- They contain the red pigment haemoglobin, which binds to oxygen in the lungs and releases it in tissues with low oxygen levels.
- The biconcave disc shape increases the surface area for diffusion of oxygen.
- They are small and flexible so they can squeeze through tiny capillaries close to cells.
- They have no nucleus, increasing the space available for carrying oxygen.

1. Suggest why blood is classified as a tissue. [1]

   1. It is a group of similar cells that work together for a specialised function.[1]

Platelets

Platelets are fragments of cells that aid blood **clotting**. They release clotting substances and clump together to help form a clot to plug a hole in damaged blood vessels.

2. Suggest why the formation of blood clots is important. [1]

   2. Clots prevent the loss of blood and entry of pathogens into the blood stream.[1]

CORONARY HEART DISEASE: A NON-COMMUNICABLE DISEASE

Non-communicable diseases cannot pass from person to person. Cardiovascular diseases affect the heart and blood vessels. One type is **coronary heart disease**, this affects the arteries that supply heart tissue with blood.

Causes and treatments

In coronary heart disease, layers of fatty material build up inside the coronary arteries. This makes them narrower, reducing blood flow. Insufficient oxygen reaches the heart muscle, which causes heart attacks and damage to the heart.

Statins are drugs that can be taken to reduce blood cholesterol levels. Lower blood cholesterol levels slow down the deposit of fatty material in arteries.

Stents can be used to keep the coronary arteries open. A stent is a short, wire-mesh tube that is inserted inside the artery to support it.

Faulty heart valves are a common cause of cardiovascular disease. Faulty valves may not open fully, so less blood passes through. Valves may also develop leaks, so that some blood flows backwards. The pumping efficiency of the heart and oxygen transport is reduced, causing problems such as fatigue and chest pain. Faulty heart valves can be replaced using biological or mechanical valves. **Mechanical valves** are made of artificial material. **Biological valves** are made from tissue from cows' hearts or actual pigs' valves.

If the heart is badly damaged, a **donor heart**, or heart and lungs can be transplanted. **Artificial hearts** are sometimes used to keep patients alive whilst waiting for a heart transplant, or to allow the heart to rest as an aid to recovery.

Risks of treatments

Major surgery always presents a risk of infection. Biological treatments, such as organ or tissue transplants, carry a risk of **rejection**, which can be fatal. Patients must take drugs to suppress their immune system.
Mechanical treatments like artificial valves and heart valves present a greater risk of blood clots and can become faulty. Drugs may have side effects. Statins sometimes cause problems with digestion and in rare cases can cause liver damage.

1. Suggest one symptom that might result from a faulty heart valve. [1]
2. Give **two** benefits and **two** risks of a donor heart transplant. [4]

   1. *Symptoms of poor oxygen supply to tissues: fatigue, shortness of breath, chest pain[1]*
   2. *Benefits: prolongs life[1], transplants are more reliable and effective than mechanical hearts[1]. Risks: must take immunosuppressant drugs[1], organ rejection / infection after surgery[1].*

HEALTH ISSUES

Health and disease

Health is the state of a person's **physical** and **mental** well-being.

Diseases are major causes of ill health. Some diseases can be transferred from one person to another. These are known as **communicable** diseases and include those such as influenza, chicken pox and tuberculosis. Such disease-causing microorganisms are called **pathogens**. **Non-communicable** diseases, such as cancer, diabetes and heart disease cannot be passed to other people. Other factors can also affect both physical and mental health. These include diet, stress and life situations.

Diseases may interact

- Immune reactions to a pathogen can trigger allergies, such as skin rashes and asthma. For example, colds and flu are a common trigger for an asthma attack.
- People that have a defective immune system are more likely to suffer from infectious diseases.
- Severe physical ill health can lead to depression and other mental illness.
- Viruses living in cells can be the trigger for cancers. For example, infection with a virus called HPV has been linked to cervical cancer.

Sampling and epidemiology

Epidemiology is the study of patterns of disease in populations. It is usually too difficult and time-consuming for scientists to study the whole population. Instead, they study a sub-set or sample of the population. The sample must be representative of the population – it should avoid bias and cover all types of people found in the population. Random sampling can be used to achieve this. A larger sample size is more likely to be representative. If the effect of one risk factor is to be studied, all other possible risk factors must be controlled or accounted for.

Cancer Research UK

Cancer research UK reports around 850 cervical cancer deaths in the UK every year, and that 99.8% of cases are caused by HPV infections.

A Scottish study of 138,692 women found that vaccination against HPV had led to a 90% reduction in precancerous cells in young women.

1. HPV vaccination may cause side effects, such as temporary pain or fever. Rarely, an allergic response may occur. Use the information provided in the report opposite to evaluate HPV vaccination. [4]

2. AIDS is a disease caused by a virus that damages white blood cells. Suggest why there is a link between the occurrence of AIDS and deaths from the bacterial disease tuberculosis. [2]

1. *HPV causes most cases of cervical cancer[1]. A large study shows HPV vaccination is effective in reducing the signs of cancer[1]. Vaccination could save many lives.[1] Health issues of the vaccine are usually mild compared to cancer. [1] Need to assess balance of risk of severe allergies.[1]*

2. *People with AIDS have a defective immune system because their white blood cells are damaged.[1] They have reduced defence against pathogens such as tuberculosis bacteria.[1]*

LIFESTYLE AND NON-COMMUNICABLE DISEASE

Non-communicable diseases can have **human costs** such as the inability to do enjoyable tasks, damage to mental health, or the loss of loved ones. They can also have **financial costs**, such as an inability to earn money, or the expense of treatments. These costs have effects on individuals, local communities, entire nations or even globally.

Lifestyle risk factors and disease

Lifestyle factors are aspects of the way people live their lives, such as diet, exercise, smoking and alcohol intake. Substances in a person's environment, such as pollution, can also be risk factors.

A link, or correlation, between a factor and a disease does not necessarily mean that the factor causes the disease. The way in which the factor causes the disease must be found. Causal mechanisms have been proven for some risk factors, but not for others.

Obesity is the main risk factor for **type 2 diabetes**. The body cells of overweight people become less responsive to insulin. Blood sugar levels are not controlled, leading to diabetes.

Alcohol intake is a risk factor for many health problems, including poor liver and brain function. Long term alcohol abuse can lead to permanent damage (cirrhosis) of the liver and have effects on the brain, such as memory loss and permanent brain damage.

Smoking and alcohol intake during **pregnancy** can affect the health of unborn babies. Smoking can lead to premature birth, which has risks for the baby, and low birthweight. Excessive alcohol can result in babies with abnormal facial features and problems in the brain and nervous system.

Cardiovascular disease has been linked to a **diet** rich in high energy foods and saturated fat, and also to a lack of **exercise**. This is associated with obesity, build-up of fatty deposits in blood vessels and high blood pressure. Chemicals from **smoking** damage arteries and cause high blood pressure, also leading to cardiovascular disease.

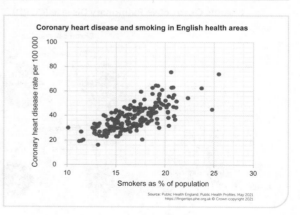

Coronary heart disease and smoking in English health areas

Source: Public Health England, Public Health Profiles, May 2021
https://fingertips.phe.org.uk © Crown copyright 2021

1. What is the meaning of the term 'risk factor' in disease? [1]
2. Describe the relationship shown in the scatter diagram of coronary heart disease and smoking. [2]

   1. *A risk factor is any aspect linked to an increased rate of a disease.[1]*
   2. *There is a positive correlation.[1] As the percentage of smokers in the population increases, so does the rate of coronary heart disease.[1]*

RISK FACTORS AND CANCER

Lifestyle factors and cancer

Cancer is a non-communicable disease. Scientists have identified lifestyle risk factors for many types of cancer. These include factors such as diet, smoking and sunbathing.

The effect of smoking on the lungs

Cigarette smoke contains many substances that can damage the lungs. Some substances from cigarette smoke can be carried in the blood stream, causing damage to cells and triggering cancer elsewhere in the body. Breathing other peoples' cigarette smoke also presents a health risk.

- **Lung cancer** can be caused by the many cancer-causing chemicals (carcinogens) in cigarette smoke. More than 70% of lung cancer cases are caused by smoking. Only 16% of lung cancer sufferers survive for five years or more.

- Smoking is a major risk factor for other forms of **lung disease**. For example, smoking is responsible for 90% of cases of COPD. Toxins in cigarette smoke cause damage to the walls of the airways and alveoli. This causes reduced oxygen exchange and symptoms such as shortness of breath.

Chronic Obstructive Pulmonary Disease (COPD)

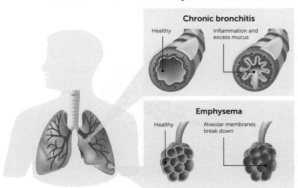

Genetic risk factors

Genetic risk is the contribution our genes play in our chances of developing disease. Inherited mutations of some genes are associated with a higher risk of some cancers. These cancer-risk genes can now be tested for. For example, an inherited mutation in the BRCA1 or BRCA2 gene can cause breast cancer. Up to 10% of breast cancers may be due to genetic factors.

Carcinogens and cancer

A **carcinogen** is a chemical or physical agent that can cause the development of cancer in cells. Carcinogens include **substances** such as asbestos and types of **ionising radiation** such as X-rays. Carcinogens damage the DNA in cells, causing mutations that can lead to cancer. Ultra-violet (UV) light is a type of ionising radiation found in sunlight and tanning lights. Sunscreen can therefore reduce the risk factor of skin cancer.

Melanoma – a skin cancer

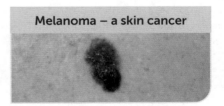

---

1. Male cigarette use in the US. peaked in 1962 but lung cancer peaked in 1990. Suggest why. [1]
2. In emphysema, the alveolar walls break down. Explain why gas exchange is less efficient. [2]

1. It may take years for cancerous cells to develop. / Early cancer may not have symptoms.[1]

2. There is reduced surface area available[1] for diffusion of gases / oxygen and carbon dioxide[1].

CANCER

Uncontrolled division of cells

The cell cycle and division by mitosis are usually carefully controlled. **Cancer** is caused by the uncontrolled division of cells. A group of cells called a **tumour** grows as cancer cells divide. Cancer cells are altered or abnormal cells. Changes in cells can be caused by mutations in their DNA. This can happen by chance or by damage from **carcinogens**.

Malignant tumours

Malignant tumour cells are cancers. This means that they can detach from the tumour and invade neighbouring tissues. This is called **metastasis**. They can also squeeze through capillary walls and move into the blood stream, which carries them to different parts of the body. The cells can then enter other tissues and begin to divide to form growths called **secondary tumours**.

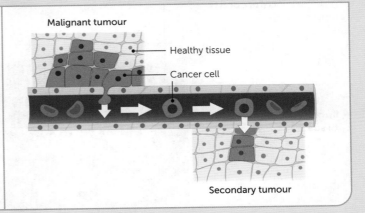

Malignant tumour

Healthy tissue

Cancer cell

Secondary tumour

Benign tumours

Benign tumours are growths of abnormal cells that do not invade other parts of the body. They are not cancer. The tumour is contained in one area, and usually surrounded by a membrane. They may cause problems as they grow if they put pressure on important body structures. They are usually removed by surgery. The health risks are lower than for malignant tumours.

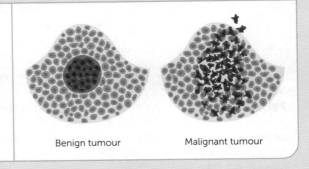

Benign tumour   Malignant tumour

1. State **two** differences between benign and malignant tumours. [2]
2. The cells of malignant tumours are undifferentiated and do not look like the surrounding tissue. Define undifferentiated cells. [1]
3. Explain how a small cancer tumour on the skin can be fatal if not treated. [3]

   1. *Benign tumours are usually enclosed within a membrane, malignant tumours are not.[1] Only cells of malignant tumours can invade other tissues and spread through the body.[1]*
   2. *Undifferentiated cells are not specialised; they do not have the adaptations for a particular function shown by the surrounding tissue cells.[1]*
   3. *Malignant cells can detach from the tumour and enter the blood stream.[1] They may be carried to other tissues and form secondary tumours.[1] If tumours invade key organs they will interfere with body functions that are necessary to live.[1]*

PLANT TISSUES

Tissues in leaves

The leaf is a plant organ that carries out photosynthesis. The structure of some of the tissues that make up the leaf and how they are related to their function is shown in the diagram below.

Palisade mesophyll tissue. Cell shape allows them to be tightly packed. Cells have lots of **chloroplasts** which maximises the amount of light energy that can be captured in photosynthesis.

Epidermal tissue. A single outer layer of flattened cells that protect the leaf. A waterproof waxy surface layer on the upper epidermis prevents water loss. The lower epidermis contains many stomata and guard cells.

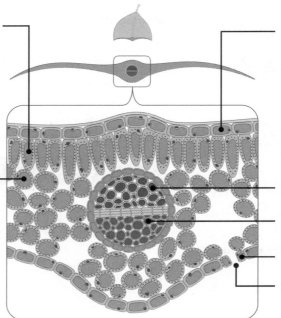

Spongy mesophyll tissue. Cells are loosely packed with a large surface area and lots of air gaps. This maximises the rate of gas exchange for photosynthesis and allows gases to diffuse easily through the leaf to and from the stomata.

Xylem tissue

Phloem tissue

Guard cells open and close the stomata.

Stoma (plural *stomata*) are pores in the leaf that gases can diffuse through.

Transport tissues

Xylem cells transport water and mineral ions through the plant. **Phloem cells** transport sugars made in leaves to the rest of the plant. See diagrams on **page 6, 4.1.1.3**.

Meristem tissue

Meristem tissue is made of unspecialised cells that can divide repeatedly to make other cells. The new cells can then differentiate into other types of plant cell, allowing the plant to grow. See **page 13-14, section 4.1.2.3** for more on meristems.

1. Give **two** adaptations of xylem tissue for transporting water and mineral ions. [2]
2. Describe **two** parts of the plant where meristem tissue is found. [2]
3. Explain why palisade mesophyll is the tissue that carries out most photosynthesis. [3]

   1. *Xylem tissue is made of hollow tubes with no cell contents or end walls for free movement of water.[1] Walls contain lignin that gives strength and stops tubes from collapsing.[1]*
   2. *Meristem tissue is found at the growing tips of shoots[1] and the growing tips of roots[1].*
   3. *Cells here have the largest number of chloroplasts.[1] There are many cells in a small space.[1] It is close to the upper surface of the leaf where more light is available.[1]*

PLANT ORGAN SYSTEM

The roots, stem and leaves form an **organ system** for the **transport** of substances around the plant.

Transport of water and minerals

Root hair cells take up water from the soil by **osmosis**, and mineral ions by **active transport**. For adaptations see section (4.1.1.3) page 6.

Xylem tissue then carries this mineral and water solution from the roots to the stems and leaves. The flow of water up through the plant is called the **transpiration stream**. Water is lost from the leaves of a plant by evaporation and diffusion – a process called **transpiration**. This loss of water from the leaf pulls more water up through the xylem.

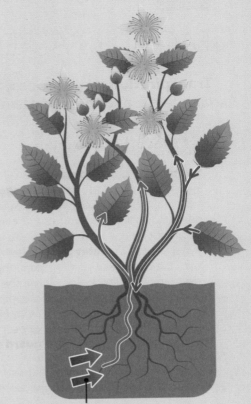

Xylem

Stiffened with ligin

One-way flow

Water and mineral ions

No end walls between cells

Water and mineral ions

Transport of dissolved food

The movement of dissolved food molecules through **phloem tissue** is called translocation.

Food molecules are produced in the leaves by photosynthesis. This food is transported in the phloem to the rest of the plant, mostly as dissolved sugars. This sugar may be used immediately by cells or may be converted into starch for storage.

Food molecules can be transported upwards towards growing shoots or down towards roots and storage organs.

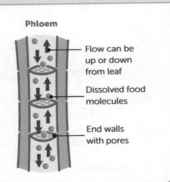

Phloem

Flow can be up or down from leaf

Dissolved food molecules

End walls with pores

1. Describe **two** adaptations of phloem cells for transport of dissolved sugar. [2]
2. Suggest **two** ways in which sugar is used by cells. [2]
3. Give **one** difference and **one** similarity between transpiration and translocation. [2]

1. *Phloem cells are elongated to carry sugar over longer distances.[1] Pores in the end walls allow cell sap containing sugar to pass from one phloem cell to the next.[1]*
2. *Sugar may be used in respiration[1] or used for growth / to make cell walls[1].*
3. *A similarity is that both are important for transport of substances through the plant.[1] A difference is that transpiration drives the transport of water in the xylem, whereas translocation is the movement of dissolved food molecules / sugar in the phloem.[1]*

FACTORS AFFECTING TRANSPIRATION

Transpiration is the process through which plants absorb water from the soil, circulate it and then exhale it as vapour through pores in their leaves.

The role of stomata

Stomata (*singular* **stoma**) are small pores in a leaf. They are vital for **gas exchange**. Without stomata, the supply of carbon dioxide would be too low for photosynthesis, and oxygen could not be released; but they are also a major source of **water loss**. Water evaporates from the surface of spongy mesophyll cells inside the leaf. The water vapour then diffuses through the air gaps in the tissue, passing through the stomata to the air outside. Water loss is controlled by the pairs of cells that surround the stomata, called **guard cells**.

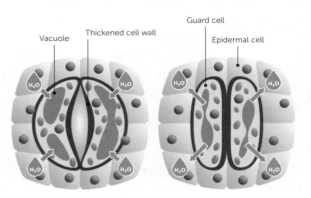

- When water moves into the guard cells by osmosis they swell. The thickened inner cell wall causes them to bend and the stomata to open.

- When water moves out of the guard cells they shrink and become flattened, so the stomata close.

- Stomata open when photosynthesis is taking place and close, reducing water loss, when it is not. If the plant is short of water and beginning to wilt, the stomata will close, reducing water loss.

Environmental factors and the rate of transpiration

- **Humidity** is a measure of the amount of water vapour in the air. **Humid** conditions will **decrease** the rate of transpiration because the concentration gradient for diffusion is reduced.

- **Temperature** increases accelerate the rate of transpiration. Water vapour molecules have more kinetic energy, so evaporation and diffusion increases.

- **Air movement** (such as windy conditions) will increase transpiration rate as air disperses water vapour that builds up outside the stomata. This increases the concentration gradient. Evaporation and diffusion of vapour out of stomata is increased.

- **Light intensity** As light levels increase, so does transpiration. Light stimulates photosynthesis, causing the stomata to open for essential gas exchange. More water vapour is lost through open stomata. In the dark, photosynthesis stops and stomata close. Stomata close at night, conserving water.

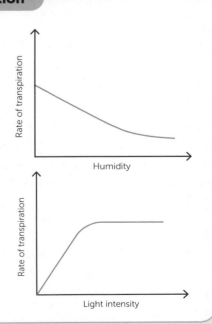

Transpiration rate investigations

You can investigate the effect of any of these environmental factors on transpiration rate. If one factor is changed and investigated, all the others must be kept the same. Transpiration rate can be measured by changes in the mass of a plant as water is lost. Alternatively, you can measure the volume of water taken up to replace the water lost from a cut shoot.

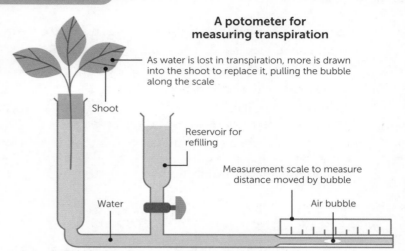

A potometer for measuring transpiration

Shoot

As water is lost in transpiration, more is drawn into the shoot to replace it, pulling the bubble along the scale

Reservoir for refilling

Measurement scale to measure distance moved by bubble

Water

Air bubble

Air movement / fan setting	Transpiration rate in mm³ per minute			
	Test 1	Test 2	Test 3	Mean
Off	3.1	3.2	3.0	3.1
Low	6.2	5.5	5.9	5.9
High	9.3	9.4	9.2	

1. Look at the graph of light intensity and transpiration on the opposite page. Explain why transpiration rate rises then plateaus as light intensity increases. [3]

2. Look at the diagram of a potometer above. State which direction the air bubble will move as the plant transpires. Explain your answer. [3]

3. Look at the results table for the investigation of how air movement affects transpiration.

   (a) Calculate the mean for the high air movement results. [1]

   (b) Describe the results shown in the table. [3]

   (c) Another measurement was taken at a different temperature. The bubble moved 27.5 mm³ in 5 minutes. Calculate the transpiration rate. [1]

1. *As light intensity increases more stomata open / open wider[1] and more water is lost/evaporates through open stomata[1]. Eventually all the stomata will be fully open so there can be no further increase in transpiration.[1]*

2. *Bubbles moves from right to left.[1] The plant takes up water to replace losses in transpiration[1] so the volume of water in the equipment is reduced, drawing the bubble along[1].*

3. *(a) 9.3 mm³ per minute.[1]*

   (b) Transpiration rate increased with increasing air movement.[1] Low air movement almost doubled the transpiration rate compared to no air movement.[1] Increasing the fan/air movement to the highest setting had an even greater effect / tripled the transpiration rate.[1] All repeats showed a similar pattern.[1]

   (c) $\frac{27.5}{5}$ = 5.5 mm³ per minute.[1]

EXAMINATION PRACTICE

01   The small intestine is adapted for the absorption of digested materials.

    01.1   Explain **three** adaptations of this organ for absorption. [3]

    01.2   Lipase enzyme digests lipids in the small intestine. State where this enzyme is produced. [1]

    01.3   Name **one** product of the digestion of lipids. [1]

    01.4   Amino acids absorbed in the small intestine are used by cells to build new molecules. Name the molecules that are made from amino acids. [1]

    01.5   Lipase enzymes are used in some laundry powders to break down fatty stains. Advice on the laundry powder packet says 'For best results wash clothes at 30°C.' Suggest reasons for this advice. [3]

02   A student tested an unknown food powder and noted the following results:

    Identify the substances that were shown to be present in the food. [2]

Biochemical Test	Final colour with food
Biuret reagent	Purple
Benedict's solution	Blue
Iodine solution	Blue-black
Ethanol emulsion test	Clear and colourless

03   The results from an investigation to measure the rate of reaction of amylase enzymes at different pH levels are shown in the table.

pH	Rate of reaction in mg starch per minute
4	0
5	11
6	24
7	27
8	15
9	2

    03.1   Use these results to plot a graph (use graph paper if you have some, if not then sketch the graph on lined paper). [3]

    03.2   Use the graph to estimate the optimum pH of this amylase enzyme. [1]

    03.3   Use the graph to estimate the rate of reaction at a pH of 7.5. [1]

04   The heart is an organ that pumps blood around the body in a double circulatory system.

    04.1   Describe the path followed by blood as it leaves the lungs and carries oxygen to body cells. [4]

    04.2   Explain how the natural resting heart rate is controlled. [2]

    04.3   After a race a runner counted 35 heart beats in 15 seconds. Calculate the heart rate in beats per minute. [1]

    04.4   Name **one** artery that carries deoxygenated blood. [1]

04.5 A person's cardiac output is 2.9 dm³ and heart rate is 70 bpm.
Calculate the stroke volume. [2]
Cardiac output = stroke volume × heart rate.

05 Red blood cells carry oxygen from the lungs to body cells.

05.1 Explain how the lungs are adapted for gas exchange. [6]

05.2 Describe how the structure of red blood cells enables them to carry out their function. [3]

06 Explain **two** treatments for coronary heart disease that can reduce symptoms. [2]

07 Asthma is a condition in which airways can become restricted leading to breathing difficulties.
The graph shows some statistics about deaths from asthma in the UK.

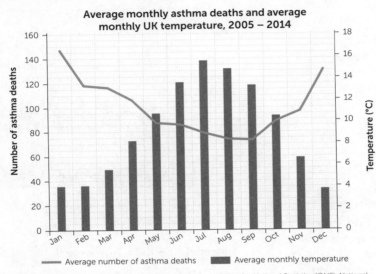

Average monthly asthma deaths and average monthly UK temperature, 2005 – 2014

— Average number of asthma deaths   ▮ Average monthly temperature

Asthma UK analysis of Met Office data (temperature) and Office for National Statistics (ONS), National Records of Scotland and Northern Ireland Statistics and Research agency (NISRA) mortality data.

07.1 Determine the months in which the highest and lowest number of asthmas deaths were reported. [2]

07.2 The charity Asthma UK reports that 74.8% of sufferers report viruses like cold or flu as an asthma trigger while 59.1% report cold weather.
Evaluate the evidence for this shown by the graph. [5]

08 The leaf is an organ that carries out photosynthesis.
Explain how adaptations of the leaf tissues support this function. [6]

09 A student observed stomata on the lower surface of a plant leaf with a microscope. They made this drawing.

09.1 Calculate the number of stomata per mm² of leaf.
Use information from the drawing. [2]

0.5 mm

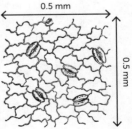
0.5 mm

09.2 The student compared the number of stomata per mm² on the upper and lower surfaces of the leaf. They found very few stomata on the upper surface compared to the lower surface. Suggest the advantage of this to the plant. [2]

COMMUNICABLE DISEASES

Communicable or **infectious** diseases can be passed from one individual to another. They are caused by **pathogens** – microorganisms that cause disease.

Pathogens

Pathogens may be **viruses**, **bacteria**, **protists** or **fungi**. The infected **host** plant or animal provides the ideal conditions and nutrients that the pathogen needs to grow and reproduce.

Pathogens cause damage to tissues and this makes us feel ill. Bacteria may also produce poisons (**toxins**) that damage tissues. Bacteria and viruses can reproduce rapidly inside the body.

How infectious diseases are spread

- **Direct contact** – Touching an infected person, such as by shaking hands or sexual contact.
- **Water** – Pathogens can enter water supplies, especially from human or animal waste. They infect people when they drink the contaminated water or food it is added to.
- **Air** – When an infected person sneezes or coughs they spray out tiny droplets. These can carry pathogens suspended in the air. Another person may breathe them in, or they may land on a surface that a person then touches.

Some diseases, like malaria, may be spread by another organism known as a **vector**.

1. Explain why adding low concentrations of chlorine to drinking water can help to prevent the spread of disease. [2]

2. Doctors advise that a person should sneeze into the crook of their arm if they cannot find a tissue, rather than sneezing into their hands. Explain **two** ways that this may reduce the spread of disease. [4]

   1. *Chlorine kills pathogens in the water at low concentrations[1] but is not harmful to humans at these low concentrations[1].*

   2. *It reduces spread by droplets in the air[1] because droplets are captured on the arm.[1] It reduces spread by direct contact[1] from hands that might be used to touch others or infect other surfaces[1].*

Reducing the spread of disease

Simple precautions can kill bacteria, or reduce or prevent the spread of disease.

Method of spread	How it can be reduced
Contact	Regular hand washing. Disinfecting surfaces. Isolation of sick individuals. Avoiding risky sexual contact or by using condoms.
Water and food	Sterilisation of drinking water using chemicals or UV light. Cooking food and boiling water. Treatment and safe disposal of sewage or wastewater.
Air	Using a tissue when sneezing or coughing. Good ventilation indoors.
All types	Vaccination for a specific disease so that people cannot pass it on.

VIRAL DISEASES

Viruses live and reproduce inside cells, causing cell damage. Unlike bacteria, they are not made of cells themselves.

Structure

Viruses have a variety of structures, but all have genetic material enclosed in a protein coat. They do not carry out the normal processes of living things and can only reproduce with the help of the **host cells** they infect and enter. The viral genetic material is used by the host cell to replicate the virus. This causes damage and death of cells as new viruses burst from the infected cells.

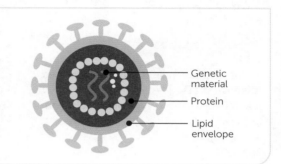

Genetic material

Protein

Lipid envelope

HIV

Infection with **HIV** (Human Immunodeficiency Virus) causes a flu-like illness at first. Early and ongoing treatment with **antiretroviral drugs** can keep the virus under control. The virus enters and destroys the body's **immune cells**. If untreated, HIV infection will lead to **AIDS** (Acquired Immunodeficiency Syndrome). This occurs when the body's immune system becomes so badly damaged that it can no longer defend against other infections or destroy cancer cells. People with AIDS may eventually die from these other diseases. HIV is spread by sexual contact or by exchange of body fluids such as blood. This can happen, for example, when drug users share needles.

Measles

The **measles** virus causes a disease with symptoms of **fever** (high temperature) and a **red skin rash**. Measles is a serious illness that can kill or cause permanent damage in the worst cases. For this reason, young children should be vaccinated against it. The measles virus is spread in droplets when an infected person sneezes or coughs.

1. Plants infected with TMV have leaves with yellow spots. Explain why this leads to stunted growth of the plant. [3]
2. Suggest why viral diseases can be difficult to treat. [1]
3. State **one** way in which the spread of measles could be reduced. [1]

1. *The cells have less chlorophyll in the yellow patches so photosynthesis is reduced.[1] There is less glucose produced[1] needed to provide energy for growth / to make new plant tissue[1].*
2. *Viruses live inside cells, so it may be difficult to destroy them without damaging the patient's own cells.[1]*
3. *Vaccination[1] or using a tissue to catch droplets from sneezes and coughs[1].*

Tobacco Mosaic Virus

TMV is a common plant pathogen. It can infect many species of plants, including tomatoes. Infected plants develop **yellow patches** in a 'mosaic' pattern on the leaves. This significantly reduces photosynthesis, causing **reduced growth** of the plant.

BACTERIAL DISEASES

Bacteria are **prokaryotes**. Their structure is revised on page 3. Most bacteria are not pathogens. Some bacteria can infect the body where they divide rapidly by binary fission. Some thrive in body fluids. Others can invade body cells. Many types produce toxins that cause damage to body cells and illness.

Gonorrhoea

Gonorrhoea is a **sexually transmitted disease (STD)**. The symptoms of infection are a thick green or **yellow discharge** from the vagina or penis and **pain** on urinating. As it is caused by a bacterium, it was once easily treated with the **antibiotic** penicillin. The number of people with Gonorrhoea has increased recently because many antibiotic resistant strains have appeared.

The Gonorrhoea bacterium

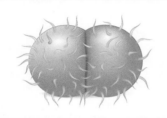

In the UK, all commercial poultry is vaccinated to prevent the spread of Salmonella

Salmonella

Salmonella is a common cause of **food poisoning**. It is usually spread when the bacteria are present in food that is eaten. The source of infection can usually be traced to contamination with faeces from an infected animal or person. Salmonella transmission is more likely when food is prepared in an unhygienic way.

The main symptoms of Salmonella are **fever**, painful **abdominal cramps**, **vomiting** and **diarrhoea**. Symptoms result from the bacteria invading tissues and secreting damaging toxins.

1. Salmonella can be spread when food is prepared in unhygienic conditions. Explain the actions that kitchen workers can take to reduce the spread of Salmonella. [5]

2. Explain why early treatment with antibiotics can help to reduce the spread of Gonorrhoea. [2]

   1. *To kill bacteria and reduce spread from contaminated food.[✓] Wash hands after handling raw meat.[✓] Disinfect surfaces / utensils.[✓] Cook food thoroughly.[✓] Use separate cutting boards for meat and other foods.[✓] To prevent growth of salmonella on food[✓] store food in a refrigerator.[✓] To prevent possible contamination from an infected person[✓], wash hands after using the toilet / before preparing food.[✓] Do not prepare food if ill / have symptoms.[✓]*

   2. *There will be fewer people who can infect others.[1] Early treatment means there is a reduced time period during which an infected person can pass it on.[1]*

FUNGAL DISEASES

Fungi

Fungi are **eukaryotes**. They may be single-celled organisms such as yeasts. Others are multicellular species whose cells form filaments that can grow inside their host. Most fungi are decomposers, but some are pathogens.

Many species can reproduce using microscopic spores. These may be small enough to be carried on the wind. Diseases caused by fungi are very common in plants.

Spores being released from a fungus to be dispersed on the wind.

Read the following extract from a gardening magazine.

"To treat small areas of Black Spot disease on your roses we recommend removing the affected leaves and shoots. These should then be destroyed, for example by burning. Spread a thick layer of clean bark chippings around the base of the plant to stop rain splashes."

Suggest how each of the following helps to control the spread of Rose Black Spot:

(a) Removing infected leaves. [1]

(b) Destroying removed leaves. [1]

(c) Suggest how bark chippings help to reduce spread of the disease. [1]

(a) Removes actively growing fungus that could spread through the plant.[1]

(b) Kills any spores that may survive on discarded plant material.[1]

(c) They prevent rain splashes carrying disease spores onto the plant (from the soil).[1]

Rose black spot

Black spot is a fungal disease that affects roses. The leaves of infected plants develop purple or black spots. The leaves often turn yellow and drop off early. There is less healthy leaf tissue that can carry out photosynthesis, so growth of the plant is reduced.

Rose black spot disease spreads by **water** and **wind** which can carry the reproductive spores to nearby plants.

Rose black spot can be treated by:

- Using **fungicides** – these are chemicals that specifically kill fungi. They are usually sprayed onto the leaves.

- Removing and destroying the affected leaves.

PROTIST DISEASES

Protists are a varied group of eukaryotic organisms. Most are single-celled, but some are multicellular. Some are pathogens but most are not.

Malaria infection

Malaria is caused by a single-celled protist called Plasmodium. The malarial protist has a complicated life cycle. A stage inside a host **mosquito** is an essential part of this life cycle. Malaria is not spread directly from person to person. It is only spread when a mosquito bites an infected person and feeds on blood containing the protist. The mosquito becomes infected and can then infect another person when it bites them. The mosquito is a **vector** for malaria.

The symptoms of malaria are repeated **episodes of fever** with a high temperature and a headache. The disease can be fatal. Globally there are more than four hundred thousand deaths a year from malaria, mostly in young children.

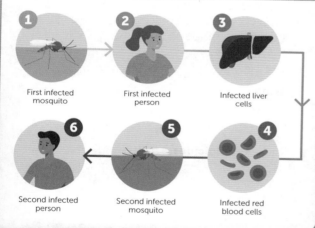

1 First infected mosquito	**2** First infected person	**3** Infected liver cells
6 Second infected person	**5** Second infected mosquito	**4** Infected red blood cells

Control of malaria

The spread of malaria is mainly controlled by preventing the spread of the disease by the vector.

- Mosquitoes lay their eggs in stagnant water where their larvae develop and hatch. Malaria can be controlled if mosquitoes are **prevented from breeding** by draining water from ditches and small ponds, or by using insecticides to kill the larvae.

- **Mosquito nets**. Mosquitoes are most active during the night so this is effective in preventing people being bitten while asleep. Some nets also contain an insecticide.

1. The mosquito is a disease vector. State the meaning of the term 'disease vector'. [1]

2. Describe **two** ways that the spread of malaria is controlled. [2]

3. Populations of carnivorous fish have been introduced to ponds in some malaria regions. Suggest how this may reduce local cases of malaria. [3]

1. *A disease vector is an organism that carries and spreads a disease.[1]*

2. *Mosquito nets prevent bites that transmit the disease.[1] Draining areas of still water and using insecticides prevent mosquitoes from breeding.[1]*

3. *The fish will eat mosquito larvae that live in the ponds.[1] There will be fewer adult mosquitoes to spread malaria.[1] The malaria protist must infect mosquitoes to complete its life cycle.[1]*

NON-SPECIFIC HUMAN DEFENCE SYSTEMS

The human body has a range of defence systems that prevent pathogens from entering the body and causing disease. These are described as **non-specific** because they stop all pathogens – even those that have not been met before.

Skin

The **skin** is tough and waterproof. It forms a **barrier** that most pathogens cannot pass through. Glands in the skin secrete substances that have **antimicrobial** properties which kill or stop the growth of pathogens.

Stomach

The cells of the stomach wall produce hydrochloric acid. The acid in the stomach kills pathogens in food.

Trachea and bronchi

The lining of the **trachea** and **bronchi** have cells that produce a sticky **mucus**. The mucus traps pathogens in inhaled air. The cells lining the bronchi have tiny hair-like projections called **cilia**. These move backwards and forwards in a coordinated wave. This sweeps the mucus and pathogens to the throat to be swallowed. Pathogens then die in the stomach.

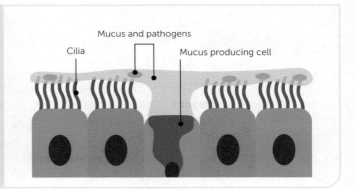

Mucus and pathogens

Cilia

Mucus producing cell

Nose

The passages of the **nose** are lined with **hairs**. The hairs trap pathogens that are suspended in the air or attached to dust particles. This reduces the risk of infection of the lungs and throat.

---

1. Which pH level best represents conditions in the stomach? Tick **one** box. [1]

   ☐ pH 2   ☐ pH 5   ☐ pH 7   ☐ pH 9

2. State **two** differences between hairs in the nose and the cilia that line the bronchi. [2]

3. Name the sticky fluid that coats the inside of the trachea and bronchi. [1]

   1. *pH 2[1]*

   2. *Two from: Nose hairs are much larger[1] and not part of a cell. Cilia are microscopic parts of a cell.[1] Cilia actively move to waft mucus along, nose hairs do not move in this way.[1]*

   3. *Mucus.[1]*

THE HUMAN IMMUNE SYSTEM

The immune system is activated when a pathogen gets through all non-specific defences. The immune system destroys pathogens inside the body and defends against disease. This relies on the actions of different types of **white blood cell** such as **phagocytes** and **lymphocytes**.

Phagocytes

Phagocytes can detect and **engulf** pathogens. The pathogen is surrounded by the cell membrane and taken into the cytoplasm. Enzymes then break down and destroy the pathogen. Phagocytes are not specific to one type of pathogen.

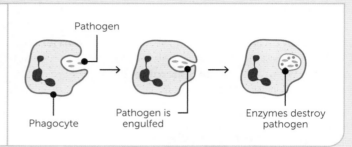

Lymphocytes

Lymphocytes form part of the specific immune response, targeting particular pathogens. Pathogens have proteins on their surface that act as **antigens**. Lymphocytes can detect these non-self antigens. Lymphocytes that can produce the right type of antibody will respond.

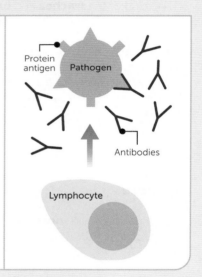

- Lymphocytes produce proteins called **antibodies** that bind to the pathogen, triggering processes to destroy it. Some antibodies make it easier for phagocytes to find and engulf the pathogen.

- Lymphocytes may also produce **antitoxins** that neutralise the toxins made by the pathogen.

- After infection, some of these lymphocytes remain in the body as **memory cells** that can provide immunity to a disease.

---

1. After infection there is a delay before the antibody response can destroy the pathogen. Illness may occur during this time. Suggest a reason for this delay. [2]

2. A gardener is scratched by a thorn and begins to bleed. Describe how immediate body responses help to prevent infection from bacteria. [2]

3. Explain why antibodies for the measles virus are not effective against chickenpox. [1]

1. *Lymphocytes that can make the correct antibody and selected and respond by dividing rapidly.[1] It takes time for enough antibody to be released into the blood stream.[1]*

2. *Platelets trigger blood clotting to stop bacteria from entering the body.[1] Phagocytes engulf and destroy bacteria that enter the cut.[1]*

3. *Antibodies have a specific shape that will only bind to one type of antigen / pathogen.[1]*

VACCINATION

After having a disease we are unlikely to catch exactly the same disease again. The memory cells that remain in our body make us **immune**. Vaccination can provide immunity to a disease in a similar way.

How vaccines prevent illness

This graph shows how a vaccine provides immunity.

A Traditional vaccines introduce small amount of a **dead** or **inactive** form of a pathogen into the body. mRNA vaccines give instructions to cells to make a protein antigen found on the pathogen.

B This stimulates white blood cells (lymphocytes) to produce **antibodies** specific to this pathogen.

Covid-19 caused a pandemic in 2020, so new vaccines were developed and distributed worldwide to prevent continued outbreaks.

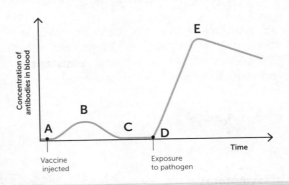

C **Memory cells** for this antibody remain in the body.

D The person is exposed to the same pathogen. It enters the body.

E The white blood cells respond quickly. They produce large amounts of the correct antibodies. These destroy the pathogen quickly, so the person does not become ill.

How vaccines prevent spread of disease

If a large percentage of the population is vaccinated or become immune from infection, it is difficult for the disease to spread. This is known as **herd immunity**.

1. Most people are immune, so they do not become ill, even when exposed to the disease.
2. This means there are fewer people who can pass it on.
3. So there is less chance of unvaccinated individuals being exposed to the pathogen.

1. The measles vaccine contains measles virus that has been weakened so it is inactive. Explain why inactive measles virus is suitable for vaccination. [3]
2. Most vaccines provide immunity for many years. A new vaccine for flu virus is needed every year. Suggest why a new flu vaccine is need each year. [2]

   1. *White blood cells will recognise the virus as measles / active and inactive forms will have the same antigens.[1] Lymphocytes will be stimulated to produce antibodies against measles.[1] An inactive virus will not cause the disease.[1]*
   2. *The flu virus changes over time / mutates / antigens change.[1] White blood cells do not recognise new strains / antibodies will only destroy the old strain.[1]*

ANTIBIOTICS AND PAINKILLERS

Some medicines cure diseases, others just reduce the symptoms.

Antibiotics

Antibiotics are medicines that are used to cure bacterial diseases. They **kill bacteria** inside the body. Antibiotics do not kill viruses.

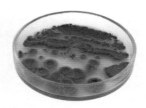

The earliest antibiotics, such as **penicillin**, were collected from microorganisms, but now they are made artificially. Different types are needed to treat specific bacterial diseases.

The use of antibiotics since the 1940s has greatly reduced deaths from infectious bacterial diseases, but this great benefit of antibiotics is at risk of being lost. New strains of bacteria constantly arise that are **resistant** to antibiotics. There is a race to develop new types of antibiotics to kill resistant strains as they develop. **See page 94** for more detail on resistant bacteria.

Painkillers

Painkillers are one type of medicine that will only treat the **symptoms** of disease. They do not kill pathogens and will not cure the disease. Examples of painkillers include aspirin and paracetamol.

1. Antibiotics do not kill viruses. Explain why it is difficult to develop drugs that kill viruses. [1]
2. Antiseptics were used to treat surface wounds before antibiotics were discovered. Suggest **one** advantage of antibiotics over antiseptics [1]

   1. *Viruses live inside cells, so they are difficult to kill without also damaging the body's own tissues.*[1]

   2. *Antibiotics can be used to kill bacteria inside the body. Antiseptics can only be used externally.*[1]

DISCOVERY OF DRUGS

Traditionally, drugs were extracted from plants and microorganisms.

- The drug **digitalis** originates from **foxgloves**. This ancient herbal remedy has been used as a drug by doctors since the 18th century to treat heart conditions.

- The painkiller **aspirin** originates from a chemical extracted from **willow** tree bark.

- The antibiotic **penicillin** was discovered by the Scottish doctor and scientist **Alexander Fleming**. It comes from the mould *Penicillium*. This was a chance discovery when the *Penicillium* fungus contaminated an agar plate, killing the bacteria that Fleming was culturing.

Most new drugs are now made by chemists in the **pharmaceutical** industry. But even today the starting point may still be a substance extracted from a plant.

DEVELOPMENT OF DRUGS

New medical drugs need to be tested and clinically trialled in a standardised way before they can be licensed for use.

Stages in drug testing

Preclinical trials may test toxicity of the drug on cultured human cells and on whole animals such as mice. **Clinical** trials use healthy volunteers at first. Only if the drug is found to be safe is it tested on patients that have the disease.

Patients and doctors might hope that a drug will work and this might influence the results. A **double blind** trial is used to avoid this bias. Some patients are given a false 'drug' with no active ingredient called a **placebo**. Patients are randomly allocated to each group. Both doctors and patients do not know which treatment is being given – they are both 'blind' to whether it is the new drug or the placebo.

Preclinical testing

- Tests on cells, tissues and live animals.
- Tests for toxicity, efficacy and dose.
- Takes place in a laboratory.

Early clinical testing

- Carried out on healthy volunteers.
- Tests for safety and toxicity.
- Uses very low concentrations of the drug.

Further clinical testing

- Carried out on patients with the disease.
- Tests for optimum dosage and efficacy.
- Includes double blind trials in which some patients are given a placebo.

1. After testing a drug, the results of trials are published. This can only be done after results have been peer reviewed. Explain why peer review is necessary. [2]

2. Suggest why the initial clinical tests for toxicity are carried out on healthy volunteers rather than patients. [2]

3. Explain the purpose of tests to determine dose. [2]

1. The methods and data must be checked by other scientists that are experts in the subject. [1] This helps to achieve consensus on the results and prevent false claims about the success of new drugs. [1]

2. In patients, it may be hard to separate drug side effects from the symptoms of illness. [1] Ill people may be more vulnerable to harm from the drug. [1] Healthy individuals are more likely to have a better defence against possible side effects of the drug. [1]

3. Scientists try to find the dose that is most effective in treating the disease [1], but that does not cause serious side effects [1].

EXAMINATION PRACTICE

01 Bacteria and virus pathogens can rapidly increase in number when inside the body.

    01.1 Compare the ways in which they achieve this. [2]

    01.2 AIDS is a condition caused by the Human Immunodeficiency Virus.

        Explain why HIV infection leads to AIDS if proper medical treatment is not received. [3]

    01.3 Name the type of drugs used to control HIV infection. [1]

02 Pathogens cause disease. Name the type of pathogen that causes:

    02.1 Malaria. [1]

    02.2 Tobacco mosaic disease. [1]

    Gonorrhoea is a bacterial disease.

    02.3 Describe how this disease is spread. [1]

    02.4 State the treatment for Gonorrhoea. [1]

03 Malaria is a serious disease that can be fatal. Explain how a person becomes infected with malaria. [3]

04 Smallpox was a serious viral disease that killed one third of those infected. The World Health Organisation coordinated a global vaccination effort in 1966. By 1980 smallpox had been eradicated. The table below shows annual global reported cases of smallpox for selected years from 1955 to 1980.

Year	1955	1960	1965	1970	1975	1980
Annual smallpox cases	91,823	67,127	112,228	33,706	19,278	0

    04.1 Calculate the percentage reduction in reported cases between 1955 and 1975. [2]

    04.2 Evaluate the evidence that the global vaccination programme was responsible for the eradication of smallpox. [2]

    04.3 Suggest why the number of reported cases might be an underestimate. [1]

    04.4 Antibiotics became available for use in the 1940s. Give a reason why this did not lead to a reduction in cases of smallpox. [1]

    04.5 When a vaccinated person was exposed to smallpox they did not become ill. Explain how the immune system of a vaccinated person prevented illness from smallpox. [4]

05 Lung diseases caused by pathogens are a common cause of illness. Explain how non-specific defence mechanisms in the human body help to defend against lung infections. [4]

06 Digitalis is a drug made from foxglove plants. It is used to treat heart conditions. In the late 18th century William Withering began using powdered leaves on patients and recommended its use as a drug.

    06.1 Describe **one** other drug that is extracted from plants. [2]

    06.2 Explain what testing would be done before using digitalis, if it were discovered today. [6]

PHOTOSYNTHETIC REACTION

Photosynthesis is essential to almost all life on earth.

Why photosynthesis is important

Photosynthesis uses sunlight to transfer **energy** from the environment into living things. Plants use photosynthesis to make **glucose** molecules that provide the energy they need. Glucose molecules are also used to build all the other materials that make up a plant – known as the plant biomass. This **biomass** and its store of energy passes into the food chain when animals feed on plants. Photosynthesis therefore supplies energy to the whole food chain.

The process of photosynthesis

Photosynthesis takes place in **chloroplasts**. These contain a green molecule called **chlorophyll** that traps the energy transferred by **light**. This energy allows glucose to be made from water and carbon dioxide. Glucose is a simple sugar. Oxygen is also produced. The products have more energy than the reactants. This is possible because energy is transferred from the environment by light. This is an **endothermic** reaction.

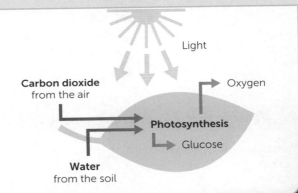

Photosynthesis equations

Photosynthesis can be represented as a word equation:

$$\text{carbon dioxide} + \text{water} \xrightarrow{\text{light}} \text{glucose} + \text{oxygen}$$

Or as a balanced symbol equation:

$$6CO_2 + 6H_2O \xrightarrow{\text{light}} C_6H_{12}O_6 + 6O_2$$

1. For the reaction of photosynthesis, name:
   (a) the reactants. [1]
   (b) the products. [1]
2. Explain why photosynthesis is described as an endothermic reaction. [1]
3. Some leaves are patterned with white patches. Explain why cells within the white patches are not able to carry out photosynthesis. [2]

   1. *(a) Carbon dioxide and water.[1] (b) Glucose and oxygen.[1]*
   2. *Because energy is transferred from the surroundings (to the products / chloroplast).[1]*
   3. *Cells in white patches do not contain any chlorophyll / chloroplasts[1] needed to trap energy from light[1].*

RATE OF PHOTOSYNTHESIS

Rate is something measured per unit of time. Rate can be measured as the amount of a reactant used up, or a product made, in a certain time.

Calculating rates of photosynthesis

Investigations carried out to study photosynthesis often measure the production of oxygen. Calculate the rate by dividing the amount of oxygen produced by the time taken to collect it.

The graph opposite shows the volume of oxygen released by a plant over time. After two hours, 10 cm³ of oxygen had been released. The rate of photosynthesis is $10 \div 2 = 5.0$ cm³/ hour.

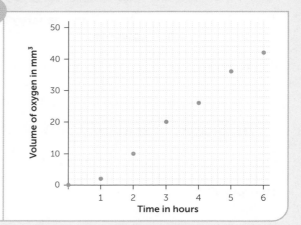

Factors that affect photosynthesis

Various factors affect the rate of photosynthesis. The rate is limited by the factor that is at the least favourable level.

Light intensity - At lower light levels, increasing light intensity increases the rate of photosynthesis. This is because energy transfer is increased. Eventually other factors start to be more limiting, so increasing light gives no further increase in photosynthesis.

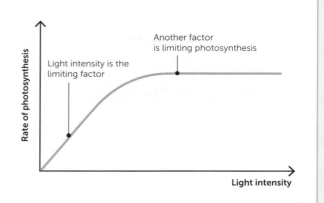

Temperature - Increasing the temperature usually increases the rate of photosynthesis (if nothing else is limiting). This is because some of the reactions of photosynthesis are catalysed by enzymes. At high temperatures the rate decreases because the enzymes become denatured.

Carbon dioxide concentration - If all other factors are favourable, carbon dioxide concentration can become a limiting factor.

Amount of chlorophyll - Chlorophyll is essential to capture the energy from light. If chloroplasts do not contain enough chlorophyll this can limit the rate of photosynthesis.

---

1. Use the graph at the top of the page to calculate the rate of photosynthesis after 5 hours.   [2]
2. Explain why carbon dioxide concentration may be a limiting factor for photosynthesis.   [1]

   1. $36 \div 5 = 7.2$[1] cm³ / hour.[1]

   2. Carbon dioxide is a reactant / needed to make glucose in the photosynthesis reaction.[1]

Economics of limiting factors   Higher Tier

Farmers increase crop growth in **greenhouses** by controlling factors that limit photosynthesis. Artificial lights increase 'daylength' and intensity of light. Heaters increase temperature and carbon dioxide gas can be introduced artificially. These control measures are expensive, but they give higher yields and bring in more income when crops are sold. Farmers must balance the cost of increasing the rate of photosynthesis with the profit.

Interactions between limiting factors   Higher Tier

Any one of the limiting factors may restrict photosynthesis at a particular time. Graphs can show which factor is limiting. On this graph, increasing light intensity increases the rate of photosynthesis up to point **A**. Here, a CO_2 concentration of 0.04% becomes limiting. A higher CO_2 concentration of 0.4% allows the photosynthesis rate to continue to increase. But at point **B**, factors other than light are limiting again.

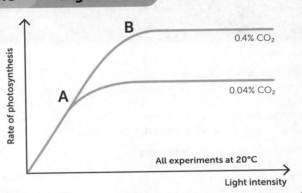

The inverse square law   Higher Tier

Light intensity is often changed by altering the distance between the plant and a lamp. Increasing distance reduces the light intensity, but this is not a linear relationship. Light intensity is **inversely proportional** to the **square** of the distance.

This **inverse square law** is shown as: $l = \dfrac{1}{d^2}$

Where l is light intensity and d is distance from the light source.

This can be used to calculate the effect of distance on light intensity. If distance is doubled, light intensity will be $1 / 2^2$ or ¼ of the original value.

3. A lamp is moved from 20 cm distance from a plant to 10 cm.
   Explain how light intensity changes. Use calculations in your answer.   [3]

4. Gas burners may be used in greenhouses to increase plant growth. Explain how gas burners may increase growth.   [3]

5. Mountain tops are often covered in snow and the concentrations of gases in air are reduced at high altitude. Explain why plants on mountain tops grow very slowly.   [2]

3. *Relative light intensity at 20 cm was $\dfrac{1}{400}$ = 0.0025.[1]*

   Relative light intensity at 10 cm was $\dfrac{1}{100}$ = = 0.001.[1] Distance has halved so the light intensity increased by a factor of four.[1]

4. *Burners increase the temperature which may increase the rate of photosynthesis.[1] Burning gas releases carbon dioxide which may be a limiting factor for photosynthesis.[1] Increased photosynthesis leads to increased plant growth.[1]*

5. *The rate of photosynthesis may be limited by low temperatures[1] and low CO_2 concentrations[1].*

The effect of light intensity on the rate of photosynthesis of pondweed

It is easy to measure the rate of photosynthesis in aquatic plants. The oxygen released from photosynthesis forms bubbles which rise through water and displaces water. The **volume of oxygen** released in a certain time can be measured as the water displaced from a measuring cylinder. This gives the rate of photosynthesis. The number of bubbles released can be counted instead, but this is less accurate. A **stopwatch** is needed to record time.

Differences in light **intensity** can be achieved by changing the distance between a lamp and the pondweed, or by using lamps of different power. This represents the **independent variable**. Other factors that affect photosynthesis must be kept constant. These control variables include temperature, carbon dioxide and the amount of pondweed. Some example results are shown in the table.

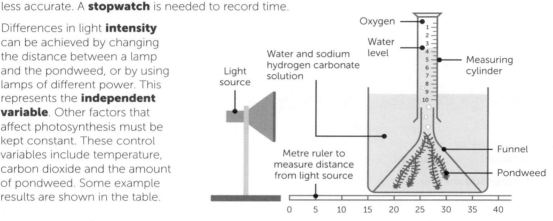

Distance (d) from light source in cm	$\frac{1}{d^2}$	Volume of oxygen produced in 30 minutes in cm³	Rate of photosynthesis in cm³ / hour
5	0.04	4.0	8.0
10	0.01	1.2	2.4
15		0.5	

1. Explain why sodium hydrogen carbonate is added to the water. [2]
2. Complete the results table by calculating the missing figures. [2]
3. A graph was drawn of rate of photosynthesis against $\frac{1}{d^2}$. Explain why $\frac{1}{d^2}$ was plotted instead of distance from the light. [2]

1. *Sodium hydrogen carbonate increases the amount of carbon dioxide dissolved in the water.[1] This makes sure that CO_2 is not a limiting factor for photosynthesis.[1]*

2. $\frac{1}{15^2}$ *is* $\frac{1}{225}$ = 0.004 *(correctly rounded).[1] Rate is 0.5 × 2 = 1.[1]*

3. $\frac{1}{d^2}$ *should be used because this value is proportional to / has a linear relationship with light intensity, while distance (d) does not.[1]*

USES OF GLUCOSE FROM PHOTOSYNTHESIS

The glucose molecules produced in photosynthesis provide the building blocks from which other molecules can be made. In some cases, other elements present in mineral ions taken from the soil are also required.

Energy release and storage

Glucose is essential for energy release and for making storage molecules. Glucose is used:
- directly in **respiration** to release energy.
- to produce **fat** or **oil** molecules for storage.

Glucose molecules can be linked together and converted into insoluble **starch** for storage.

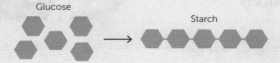

Starch, fat or oil can be stored in cells. When needed by the plant, these molecules are broken down by enzymes and converted back into glucose.

Molecules for growth

Glucose is also used to make molecules that are essential for making new plant tissue. It is used:
- to produce **cellulose**, which strengthens the cell walls.
- to produce **amino acids** that are needed for protein synthesis.

Proteins are essential for growth. **Nitrate ions** are needed to make (synthesise) proteins. Nitrates are absorbed from the soil.

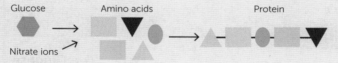

1. Plant seeds may contain a store of lipid (fats and oils).
   Explain how these stores may be used by a growing seedling. [4]
2. Give **one** reason why starch is a good storage molecule. [1]
3. Plants need glucose and nitrate ions to make amino acids for protein synthesis.
   Explain how a plant gets the glucose and nitrate it needs. [3]

   1. *Lipids can be broken down to release glucose[1] which can be used by a seedling in respiration[1] or to make proteins[1] or to make new cell walls.[1]*
   2. *One from: The molecule contains large amounts of glucose.[1] It is insoluble so it stays in cells / does not affect osmosis.[1]*
   3. *Glucose is made through photosynthesis.[1] Nitrate ions are absorbed from the soil by root hair cells[1] through active transport[1].*

AEROBIC AND ANAEROBIC RESPIRATION

What is respiration?

Respiration is a reaction that goes on all the time in living cells. Because it happens inside cells, it is called cellular respiration. It should not be confused with breathing (ventilation).

Respiration is an **exothermic reaction**. It transfers the energy stored in molecules, such as glucose, so that the energy is available for living processes. Note that respiration releases energy, it does not 'produce' it.

Cellular respiration can take place **aerobically** (using oxygen) or **anaerobically** (without oxygen).

Don't forget that all living things respire.

Why organisms need to respire

Respiration provides the **energy** that is needed for:

- **chemical reactions** to build larger molecules from smaller ones, for example, making proteins from amino acids

- **movement**, for example to bring about muscle contraction in animals

- **active transport** where substances are moved against a concentration gradient, such as mineral ion uptake in plants

- **keeping warm** in some animals.

Respiration does not transfer energy efficiently – some is lost as thermal energy. Mammals and birds have high respiration rates. The thermal energy released helps to maintain their warm body temperatures.

Aerobic respiration

When there is enough oxygen available, glucose can be completely **oxidised**. A lot of energy is released. Aerobic respiration takes place in the mitochondria of cells.

Aerobic respiration can be represented by a word equation:

glucose + oxygen → carbon dioxide + water

Or by a balanced symbol equation:

$$6C_6H_{12}O_6 + O_2 \rightarrow 6CO_2 + 6H_2O$$

The equation for respiration is the reverse of the ones for photosynthesis.

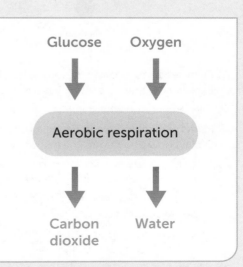

Glucose   Oxygen

Aerobic respiration

Carbon dioxide   Water

Anaerobic respiration

When oxygen levels are low, aerobic respiration stops and **anaerobic** respiration takes over. The products of anaerobic respiration are not the same in every type of organism.

Anaerobic respiration in animals simply converts glucose to **lactic acid**. The **muscles** of animals often respire anaerobically during exercise. This is shown by the equation:

glucose → lactic acid

In anaerobic respiration, much **less energy** is transferred for every molecule of glucose used than in aerobic respiration. This is because the glucose is not completely oxidised.

In **plant** and **yeast** cells, anaerobic respiration converts glucose into **ethanol** (an alcohol) and **carbon dioxide**. This is shown by the equation:

glucose → ethanol + carbon dioxide

Anaerobic respiration in yeast cells is called **fermentation**. It is used in the food and drink industry as an important part of making bread and alcoholic drinks.

Carbon dioxide from fermenting yeast forms bubbles in bread which makes it rise. The ethanol evaporates in the oven.

Fermentation produces (ethanol) the alcohol and carbon dioxide bubbles in alcoholic drinks.

1. Name the products of aerobic respiration. [1]
2. Explain why respiration is described as an exothermic reaction. [1]
3. Explain why protein synthesis in plants requires both photosynthesis and respiration. [3]
4. Anaerobic respiration does not use oxygen, but aerobic respiration does use it.
   Describe **two** other differences between aerobic and anaerobic respiration in mammals. [2]
5. Explain where the glucose for respiration comes from:
   (a) in plants. [1]
   (b) In animals. [1]

1.   *Carbon dioxide and water.[1]*
2.   *One from: Because energy is transferred to the surroundings.[1] Thermal energy is released / the temperature of the organism is increased.[1]*
3.   *Photosynthesis produces glucose molecules that are used to build amino acids.[1] This glucose is also used in respiration.[1] Respiration provides the energy required for protein synthesis.[1]*
4.   *Aerobic respiration transfers more energy than anaerobic.[1] Anaerobic respiration in mammals produces lactic acid, aerobic respiration produces carbon dioxide and water.[1]*
5.   *(a) Photosynthesis.[1]*
     (b) Digested food.[1]

RESPONSE TO EXERCISE

During exercise, our muscles contract more, so there is an increased demand for energy and the rate of aerobic respiration in muscles increases.

Changes during exercise

Muscles need a greater supply of oxygen to meet increased demand. Our body responds:
- by increasing **heart rate** to supply more oxygenated blood to muscles, and
- by increasing **breathing rate** and **breath volume** to increase oxygen supply from the air.

These responses also help to remove the additional carbon dioxide that is released.

Oxygen debt

If we exercise hard, or for a long time, our heart and breathing cannot keep up with the demand for oxygen. Anaerobic respiration begins to take place in muscles, causing a build-up of **lactic acid**. Extra oxygen is needed to remove this lactic acid when exercise stops. This is called an **oxygen debt**. During long periods of intense exercise, muscles become **fatigued** and stop contracting efficiently.

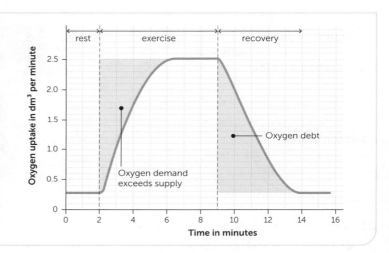

Removing lactic acid   Higher Tier

Blood flowing through the muscles removes lactic acid and takes it to the **liver**. Here, it is converted back into **glucose** in a reaction that needs oxygen. This glucose can be used in respiration or converted to glycogen and stored in cells. Oxygen debt is the amount of extra oxygen the body needs after exercise to react with the lactic acid that has built up and to remove it.

---

1. Explain how changes in breathing help to increase the supply of oxygen to muscles during exercise. [4]
2. Explain why heart rate and breathing rate stay high for a while after exercise has stopped. [2]
3. Look at the oxygen debt graph. Determine the rate of oxygen uptake at rest. [1]

1. *Breathing rate and volume of each breath increases[1] which increases the supply of oxygen from the air.[1] More oxygen can diffuse into the blood through the lungs.[1] Blood then carries the oxygen to muscles.[1]*
2. *Extra oxygen is needed to convert lactic acid into glucose.[1] This is oxygen debt.[1]*
3. *0.25 dm³ per minute.[1]*

METABOLISM

Metabolism is the sum of all the chemical reactions in the cells of an organism.

These reactions can synthesise (build up) new molecules or can break molecules down. They are controlled by **enzymes** and rely on energy released by cellular respiration.

Reactions involving carbohydrates

In plants, reactions join glucose molecules together to make starch for energy storage. In animals, glycogen is made instead. These polymers can be broken down again to release glucose when needed for respiration. Cellulose is also synthesised from glucose to make plant cell walls.

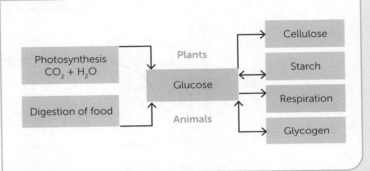

Reactions involving proteins

Proteins are made from chains of amino acids which can be broken down again. Plants make their own amino acids. Animals get proteins from their food, which are digested into amino acids and new proteins are made. In animals, excess proteins are broken down, producing urea which is excreted.

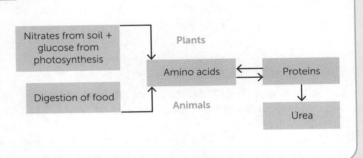

Lipid reactions

Lipid molecules are synthesised from one molecule of glycerol and three fatty acid molecules. Lipase enzymes can break lipids down again. Lipids are made from the products of photosynthesis in plants, and from the products of digestion in animals.

1. Name the metabolic reaction that transfers the energy needed for other reactions. [1]

2. Urea is made in the liver. Explain why animals need to make urea. [2]

3. Protein synthesis is a metabolic reaction.
   Name **two** other metabolic reactions in cells. [2]

   1. *Respiration.[1]*

   2. *Animals may take in more protein than they need in digestion.[1] Excess amino acids can't be stored / ammonia from their breakdown is toxic.[1] They are converted to urea and excreted in urine.[1]*

   3. *Any two from: Synthesis or breakdown of proteins / glycogen / starch, formation of urea, respiration, or photosynthesis (formation of glucose).[2]*

EXAMINATION PRACTICE

01 Plants use the energy from sunlight to make food.

01.1 Complete the word equation for photosynthesis. [1]

Carbon dioxide + _____ → _____ + oxygen

01.2 State where in a cell photosynthesis takes place. [1]

The graph shows the effect of light on the rate photosynthesis at two different temperatures.

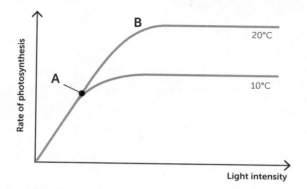

01.3 Look at the line for the temperature of 10 °C. Explain the change in the rate of photosynthesis at light intensities below point A and above point A. [4]

01.4 Explain the reason for the difference between the lines for 10 °C and 20 °C. [2]

01.5 Temperature can be increased in a greenhouse by using a heater and this may increase photosynthesis and crop growth. Explain what factors a farmer would consider when deciding whether to use heaters in a greenhouse.
Use the graph and your own knowledge. [4]

01.6 Increasing light intensity may increase the rate of photosynthesis. Suggest **one** other aspect of lighting that a farmer might change to increase crop growth rates. [1]

02 Students investigated the effect of changing light intensity on the rate of photosynthesis. They used the apparatus shown in the diagram.

Light intensity was changed by using bulbs of different power. The number of bubbles released in one minute were counted.

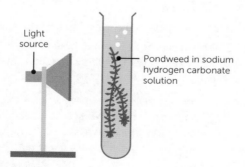

02.1 The bulb in the light source gets hot which may affect the rate of photosynthesis. Suggest how this problem could be reduced while using the same light source. [2]

02.2 Counting bubbles is less accurate than measuring oxygen volume. Suggest **one** reason why counting bubbles is not an accurate measure of oxygen production. [1]

02.3 The amount of pondweed is a control variable. Explain why the amount of pondweed should remain constant. [2]

02.4 Suggest how the amount of pondweed can be kept constant as light intensity is changed. [1]

02.5 Some results from the investigation are shown in the table.

Power of bulb in Watts	Number of bubbles produced per minute
50	14
100	25
150	33
200	35
250	35

Calculate the percentage increase change in the number of bubbles produced as light intensity is increased from 50 to 100 Watts. [2]

02.6 Predict the number of bubbles that would be produced per minute if the power of the bulb is increased to 300 Watts. Give a reason for your answer. [3]

02.7 Photosynthesis allows the pondweed to grow. Explain how the growth of new cell walls relies on the products of photosynthesis. [2]

03 A student observed anaerobic respiration in yeast. They used this apparatus.

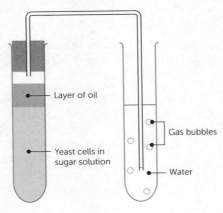

Layer of oil

Gas bubbles

Yeast cells in sugar solution

Water

03.1 Name the gas present in the bubbles. [1]

03.2 The gas is produced by anaerobic respiration. Name one other product of anaerobic respiration in yeast. [1]

03.3 Explain the purpose of the oil layer. [2]

03.4 Describe how this apparatus could be used to measure the rate of anaerobic respiration. Name any additional equipment that is required. [2]

03.5 Yeast can also respire aerobically. Complete the word equation for aerobic respiration in yeast.

_____ + oxygen → water + _____ [1]

04   Students investigated the effect of different activities on heart rate. They measured the heart rate of individuals after each activity and calculated a mean. The bar chart shows their results.

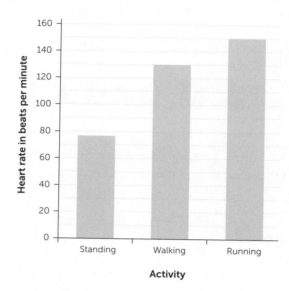

04.1   Compare the heart rates after each of the three activities.   [3]

04.2   The students were asked to do a further investigation to compare the heart rates of athletes and non-athletes after exercise. Describe how this investigation could be carried out to collect valid data. You should include how some variables are controlled.   [5]

04.3   Other students tried lifting a heavy weight as one type of exercise. Individuals could not keep lifting the weight for the full 5 minutes. Explain why.   [2]

04.4   Glucose can be converted and stored in muscle cells for later use in respiration. Name the metabolic reaction that converts glucose for storage in muscle cells.   [1]

Higher Tier only

04.5   Students noticed that individuals had high heart rates and were out of breath for several minutes after they stopped running. Explain this observation.   [2]

BIOLOGY
TOPICS FOR PAPER 2

Information about Paper 2:

Trilogy 8464:

Written exam: 1 hour 15 minutes
Foundation and Higher Tier
70 marks
16.7% of the qualification grade
All questions are mandatory

Specification coverage

The content for this assessment will be drawn from Topics 5–7: Homeostasis and response; Inheritance, variation and evolution; and Ecology

Questions

A mix of calculations, multiple-choice, closed short answer and open response questions assessing knowledge, understanding and skills.

Questions assess skills, knowledge and understanding of Biology.

HOMEOSTASIS

Human body cells function most efficiently in optimum conditions.

How conditions change

External changes, such as missing a meal, sunbathing, not drinking, or exercising can all change the internal conditions in which cells work. Changes in conditions either increase or decrease one or more of the following conditions: water level, blood glucose concentration or body temperature.

1. Give **one** way increased body temperature can affect body cells. [1]

2. Loss of water can cause red blood cells to shrivel. Explain why red blood cells with the correct shape work most efficiently. [3]

3. Exercise can cause a decrease in blood pH. Describe how a fall in blood pH would affect cell function. [2]

   1. *Cell reactions are faster.[1]*

   2. *Biconcave shape[1] gives large surface area[1] for exchange. They are smooth and flexible so they pass easily[1] through capillaries.*

   3. *Enzymes would not be at optimum pH / would be less active[1], slowing some cell reactions[1].*

Effect on cells

If **body temperature** decreases, enzymes are less active. This decreases the rate of many cell reactions, while the diffusion of substances into and out of cells occurs more slowly.

If **water level** increases or decreases, cells gain or lose water by **osmosis**, so they swell or shrink and may change shape.

If **blood glucose concentration** falls, less glucose is available for respiration, so less energy is available for cell activities, such as active transport.

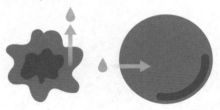

Dehydrated red blood cell

Spherical red blood cell

Normal red blood cell

CONTROL SYSTEMS AND RESPONSE

Maintaining optimal internal conditions is called homeostasis and is achieved **automatically** by a variety of control systems.

Components of control systems

All biological control systems have three components.

Receptors are cells that are able to detect changes in conditions (**stimuli**). Receptor cells can exist singly or in groups.

Effectors carry out **responses** to return conditions to optimal. In the human body, effectors are either **muscles** or **glands**.

In a control system, effectors are linked to receptors via a coordination centre. In the human body, examples of **coordination centres** include the **spinal cord**, the **brain** and the **pancreas**.

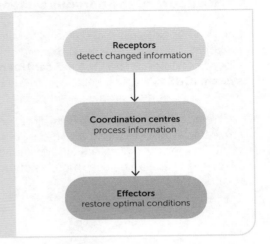

1. Name the group of substances involved in chemical responses. [1]
2. Put the following into the correct sequence: coordinator, stimulus, effector, response, receptor. [1]
3. Draw **one** line between each part of a control system and a function. [2]

Coordination centre | Releases hormones or causes movement
Effector | Detects changed conditions
| Receives and passes on information

1. *Hormones.* [1]
2. *Stimulus, receptor, coordinator, effector, response.* [1]
3. *Line between coordination centre and receives and passes on information.* [1] *Line between effector and releases hormones or causes movement.* [1]

The brain

If the coordination centres are the brain or the spinal cord, information is passed from receptor cells to effectors in the form of electrical impulses.

The pancreas

Some coordination centres are not part of the nervous system. The pancreas uses hormones to pass information from receptor cells to effectors.

THE HUMAN NERVOUS SYSTEM

Humans can respond rapidly to their surroundings because of their nervous system.

Structure of the nervous system

The nervous system is made of long thin cells called neurones. The brain and the spinal cord contain billions of neurones and form the **central nervous system** (**CNS**). Very long neurones extend out to receptors and effectors in different parts of the body.

Information from receptors passes along neurones as electrical impulses into the CNS. Any required reactions to surroundings are coordinated by the CNS which then passes electrical impulses out to effectors (muscles or glands).

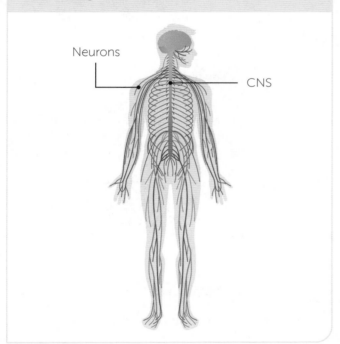

Reflex actions

Blinking when something approaches the eye helps to prevent injury. Rapid, protective movements like these are called reflex actions. They do not require conscious thought because the electrical impulses pass almost directly from receptor to effector along a route called a **reflex arc**.

A reflex arc includes:

- Sensory neurone to carry electrical impulses to the CNS
- Short relay neurone connecting via **synapses** to a motor neurone
- Motor neurone to carry electrical impulses to the effector

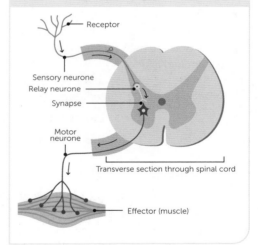

1. Which of these actions is a reflex action? Tick **one** box.

   ☐ Riding a bicycle.   ☐ Seeking shade on a hot day.   ☐ Jumping at a sudden loud noise.   [1]

2. Synapses connect the different neurones in a reflex arc. Give the number of synapses in a reflex arc. [1]

3. Explain how neurones are adapted to their function.   [2]

   1. *Jumping at a sudden loud noise [1].[1]*

   2. *2.[1]*

   3. *They are very long[1] so they are able to connect one place in the body to another[1].*

REQUIRED PRACTICAL 6

Human reaction time

This required practical activity helps to develop your ability to measure and record results carefully, using appropriate apparatus.

Choosing apparatus

Differences between reaction times can be very small, so you need to choose apparatus capable of detecting small differences.

- To measure reaction time, choose a digital timer rather than a clock so you can record milliseconds rather than just seconds.
- Using the ruler drop method, you will need a ruler showing millimetres rather than just centimetres. Carefully measure the distance fallen in exactly the same way each time, for example, to the top of the person's thumb.

| Attempt | Distance fallen (mm) | |
	With right hand	With left hand
1	95	103
2	97	102
3	98	105
Mean	96.7	103.3

Presenting results

Results showing the effect of a categorical factor on reaction time, such as using the right or left hand, should be presented as a bar chart. The bars should show the mean reaction time for each situation.

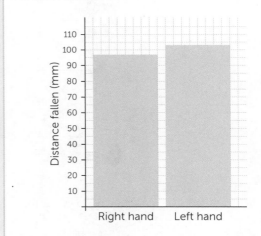

1. Suggest **two** ways you could makes sure the results were valid. [2]
2. Reaction time can be calculated in the table above from the distance fallen. Use the equation:

   Reaction time (in seconds) = $\sqrt{\frac{2d}{g}}$, where d is distance fallen and g is the acceleration due to gravity 9.8 m/s². Calculate the mean reaction time using the right hand. [3]
3. What would the experiment above have concluded using a ruler showing only centimetres? [1]

   1. *Hold the ruler at the same starting height each time.*[1] *Make sure the subject's thumb and finger are the same distance from the ruler each time.*[1]
   2. *Distance fallen = 0.0967m*[1], $t = \sqrt{\frac{2 \times 0.0967}{9.8}}$ [1] *= 0.14 s.*[1]
   3. *There would have been no difference in the reaction time using the right and left hand.*[1]

HUMAN ENDOCRINE SYSTEM

Humans can respond to changes in their internal conditions by releasing hormones from glands in their endocrine system. These responses are slower than those controlled by the nervous system.

Controlling conditions with hormones

Glands release or **secrete** hormones into the bloodstream. They are transported all around the body but only cause responses in specific **target organs**. Hormonal responses are slower than nervous responses.

The **pituitary gland** in the brain releases hormones that cause other glands to release hormones. Because it responds to body conditions and controls some other glands through these hormones, it is called the '**master**' endocrine gland. Other glands, such as the pancreas, detect changes in body conditions themselves and then release hormones.

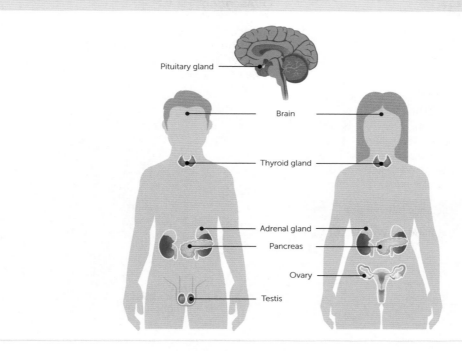

1. Some human endocrine glands exist in pairs. Which glands exist as members of a pair?
   Tick **two** boxes. [1]

   ☐ Pituitary   ☐ Adrenal   ☐ Testis   ☐ Thyroid   ☐ Pancreas

2. Suggest why hormonal responses last longer than nervous responses. [2]

3. Explain why hormonal responses are slower than nervous responses. [2]

   1. *Adrenal[1], testis[1].*
   2. *Hormones remain in the bloodstream[1] for some time[1] after release.*
   3. *Hormones travel in the bloodstream[1] and so take time to reach their target organ[1].*

CONTROL OF BLOOD GLUCOSE CONCENTRATION

The pancreas monitors and controls human blood glucose concentration.

How the control system works

The **pancreas** detects when blood glucose rises above its optimal level and releases (**secretes**) insulin. **Insulin** stimulates increased glucose uptake by all body cells. **Liver** (and muscle) cells store excess glucose as **glycogen**.

Higher Tier only

The pancreas detects when blood glucose falls below the optimal level and releases glucagon. **Glucagon** causes the liver (and muscle) cells to convert stored glycogen to glucose.

Either of these responses returns the blood glucose level to its optimum, so the pancreas releases less of the hormone. This is a called a **negative feedback cycle**.

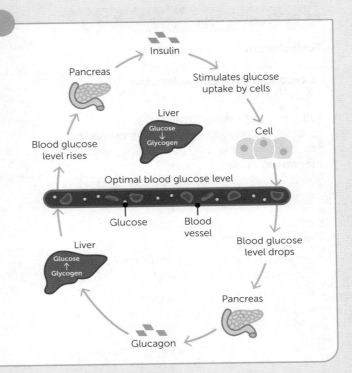

Diabetes

With diabetes, a person is unable to control their blood glucose concentration. There are two types:

Type I	Type II
Pancreas does not produce enough insulin.	Body cells do not respond to insulin.
Blood glucose levels are very erratic and can become extremely high after eating.	Blood glucose levels are too high all the time.

1. Name the coordinator in the control of blood glucose concentration. [1]
2. Explain why obesity is a risk factor in type II diabetes. [2]
3. Describe the treatment for each type of diabetes. [3]

   1. *Pancreas.[1]*
   2. *Excess blood glucose from diet[1] so body cells become resistant to insulin[1].*
   3. *Type I diabetes is treated with insulin injections.[1] Type II diabetes is treated with a carbohydrate controlled diet[1] and sufficient exercise[1].*

HORMONES IN HUMAN REPRODUCTION

Human reproduction is controlled by hormones. Hormones regulate the development of sexual maturity and then control the production of eggs and sperm.

Production of eggs and sperm

Testosterone and **oestrogen**, released by the testes and the ovaries, cause the development of **secondary sex characteristics**. This includes stimulating sperm and egg production.

Following **puberty**, sperm production is continuous but egg production, or **ovulation**, occurs roughly every 28 days as part of the **menstrual cycle**. The menstrual cycle is controlled by four different hormones.

Hormone	Effect
Testosterone	Stimulates sperm production
Follicle stimulating hormone (FSH)	Causes maturation of the egg
Luteinising hormone (LH)	Stimulates the release of an egg
Oestrogen and progesterone	Maintain the lining of the uterus for the egg

- In the first week of the menstrual cycle, FSH causes an egg to begin to mature in an ovary.

- Approximately half way through the cycle, LH stimulates the release of the mature egg.

- Following ovulation, oestrogen and progesterone maintain the thickened lining of the uterus in readiness for an embryo in case fertilisation happens.

- If fertilisation does not happen, oestrogen and progesterone levels fall causing the cycle to start again. The uterine lining breaks down (menstruation) and then FSH causes another egg to mature.

1. Name the hormones that stimulate:
   (a) The release of an egg. [1]
   (b) Sperm production. [1]
2. Describe how hormones cause menstruation. [2]
3. Explain on which day of the menstrual cycle is fertilisation most likely to occur. [2]

1. (a) *Luteinising hormone.*[1]
   (b) *Testosterone.*[1]

2. *Oestrogen and progesterone*[1] *levels decrease*[1].

3. *Day 14.*[1] *Luteinising hormone increases around day 14 and stimulates the release of the mature egg.*[1]

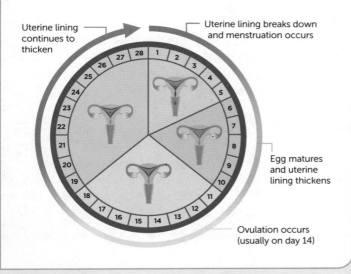

Uterine lining continues to thicken

Uterine lining breaks down and menstruation occurs

Egg matures and uterine lining thickens

Ovulation occurs (usually on day 14)

CONTROLLING THE MENSTRUAL CYCLE

Changing concentrations of the four hormones during the 28 day cycle cause the different events in the cycle.

How the menstrual cycle is controlled

A graph of the levels of the four hormones shows how they interact to regulate:
- the timing of egg maturation and ovulation.
- the thickening and maintenance of the uterine lining and then its breakdown.

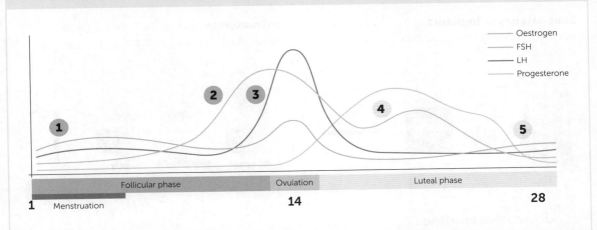

1 A slight increase in the level of FSH during the first week causes egg maturation

2 An increase in the level of oestrogen in the second week causes the thickening of the uterine lining

3 A sudden large increase in LH around the middle of the cycle causes ovulation.

4 High levels of oestrogen and progesterone in the third week prevent the uterine lining from breaking down.

5 Falling levels of oestrogen and progesterone in the fourth week lead to the breakdown of the uterine lining and stimulate the release of FSH, starting the cycle again.

Using the graph above:

(a) Describe how hormones stimulate the release of an egg. [3]

(b) Explain when and how oestrogen prepares the uterine lining for possible fertilisation. [3]

(c) Explain when and how hormones cause the cycle to repeat. [4]

(a) FSH stimulates oestrogen release.[1] Oestrogen inhibits FSH[1] but stimulates the release of LH[1].

(b) Rising oestrogen levels[1] during the week before[1] ovulation thickens[1] the uterine lining.

(c) During the last week[1] of the cycle, if there is no implantation of a fertilised egg[1], falling levels of oestrogen and progesterone[1] stimulate the release of FSH[1].

CONTRACEPTION

Human reproduction is controlled by hormones. Hormones regulate the development of sexual maturity and then control the production of eggs and sperm.

Hormonal

Pills (**oral contraceptives**) contain hormones that prevent the release of FSH so no eggs mature.

Skin patches or **implants** release hormone slowly to stop eggs maturing or being released.

Oral contraceptives

Vaginal ring

Skin patch

Implant

Non-hormonal

Barrier methods prevent sperm reaching eggs or contain **spermicides** that kill or disable sperm.

Intrauterine devices prevent a fertilised egg implanting in the uterus lining.

Female condom
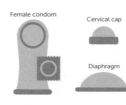
Cervical cap

Male condom
Intrauterine Device (IUD)

Diaphragm

Alternative methods

Alternative ways to prevent pregnancy include avoiding intercourse on days when ovulation is likely, and **sterilisation** (having an operation to physically block the egg or sperm ducts).

Each method of contraception has advantages and disadvantages with different levels of effectiveness. For example, condoms protect against STDs as well as providing contraception.

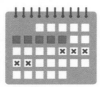
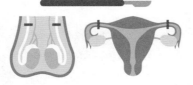

1. Some methods of contraception are called barrier methods. Which are barrier methods? Tick **two** boxes. [2]

   ☐ Intrauterine device
   ☐ Condom
   ☐ Spermicide
   ☐ Diaphragm

2. Name the hormone slowly released by contraceptive implants. [1]

3. Explain why contraceptive implants are more effective than contraceptive pills. [2]

   1. Diaphragms.[1] Condoms.[1]

   2. Progesterone.[1]

   3. Contraceptive implants slowly release progesterone for months or years[1] so the person does not need to remember to take a contraceptive pill.[1] Pills are sometimes forgotten.[1]

THE USE OF HORMONES TO TREAT INFERTILITY

Hormones can be used to stimulate ovulation.

Fertility drugs

FSH (Follicle Stimulating Hormone) and **LH (Luteinising Hormone)** can be used in **fertility drugs** that increase the chances of egg maturation and release. Using fertility drugs can help a person to become pregnant naturally.

Fertility drugs are also used before collecting eggs during **In Vitro Fertilisation (IVF)** treatment.

The four stages of IVF treatment

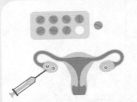

1

2

3

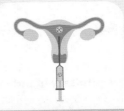

4

FSH and LH are given to encourage egg maturation. Mature eggs are collected from the ovaries.

The eggs are fertilised in the laboratory using sperm collected from the father.

The fertilised eggs are kept in culture until they develop into embryos.

One or two healthy embryos are inserted into the mother's uterus.

Fertility treatment can be very stressful and emotional because it often fails and the disposal of unused embryos raises ethical concerns. Fertility drugs also increase the chances of multiple births (e.g. twins) which increases the risk of problems for both the babies and the mother.

1. State what 'In Vitro Fertilisation' means.   [1]
2. Explain why fertility drugs increase the chances of a multiple birth.   [2]
3. Explain why fertility drugs are used even if treating male infertility with IVF. [2]

   1. *Sperm are added to eggs in the laboratory.*[1]
   2. *Fertility drugs stimulate ovulation*[1] *and often result in the release of several eggs at once.*[1]
   3. *IVF treatment for male and female infertility both require more than one egg*[1] *to be fertilised to maximise the chances of a healthy embryo*[1].

NEGATIVE FEEDBACK SYSTEMS

Once a hormone has caused enough of a response, its secretion from a gland is reduced by **negative feedback**.

Thyroxine

Thyroxine is a hormone secreted by the **thyroid gland**. Thyroxine regulates the rate of metabolism in all body cells. When the thyroid gland secretes more thyroxine, the rate of metabolism (called the **basal metabolic rate**) increases.

Metabolism includes the synthesis of new molecules, including proteins and lipids used for growth. This means that thyroxine plays a part in regulating growth and development. The level of thyroxine in the blood is controlled by a negative feedback cycle.

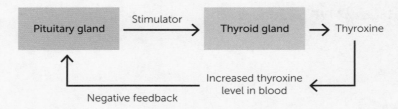

The **pituitary gland** releases a hormone that stimulates the secretion of thyroxine by the thyroid gland. However, as the level of thyroxine increases, it has a negative feedback effect on the pituitary gland so the stimulation of the thyroid gland is reduced again.

Adrenaline

Adrenaline is a hormone secreted by the **adrenal glands**. It is released as a response to fear or stress. One of its effects is to increase heart rate. This increases the rate of blood flow, delivering oxygen to the brain and muscles faster. This is known as the **fight or flight** response.

1. Give the response caused by the hormones below:
   (a) Adrenaline   [1]
   (b) Thyroxine   [1]
2. Explain why the pituitary gland is sometimes called the 'master gland'. [2]
3. Explain why the effects of a fright take a while to wear off. [3]

1. (a) *Adrenaline increase in heart rate.*[1]

   (b) *Thyroxine to increase in metabolic rate.*[1]

2. *The pituitary gland secretes hormones that act on other glands*[1] *stimulating the release of other hormones*[1].

3. *The effects of a fright are caused by release of the hormone adrenaline*[1] *into the blood*[1]. *The adrenaline is carried in the blood for some time.*[1]

EXAMINATION PRACTICE

01 The graph shows some events during the menstrual cycle.

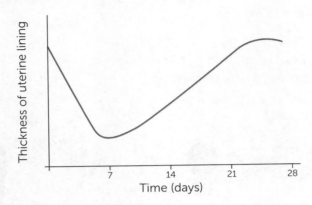

01.1 Describe what is happening between days 1 and 7. [1]

01.2 Explain the purpose of the change between days 7 and 26. [2]

02 The image below shows the human eye in two different conditions, A and B.

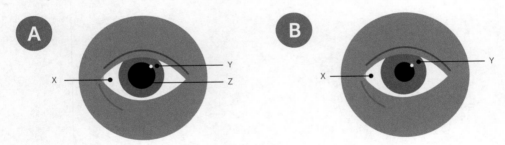

02.1 Name structures X, Y and Z. [3]

02.2 Suggest a change in conditions that might have caused the response seen in B. [1]

02.3 Explain how effectors lead to this response. [2]

03 Explain why it does not matter which way up seeds are planted in the soil. [4]

04 The table shows some information about different methods of contraception.

	Number of unintended pregnancies per 100 women in a year
Diaphragm plus spermicide	8
Contraceptive pill	1
Spermicide	28

04.1 Which method of contraception shown in the table is most effective? [1]

04.2 Suggest **one** reason why the oral pill was not completely effective. [1]

04.3 Give a reason why using a diaphragm with spermicide is more effective than just the spermicide. [1]

05 The drawing shows a human reacting to their surroundings.

Describe how this response is coordinated. [6]

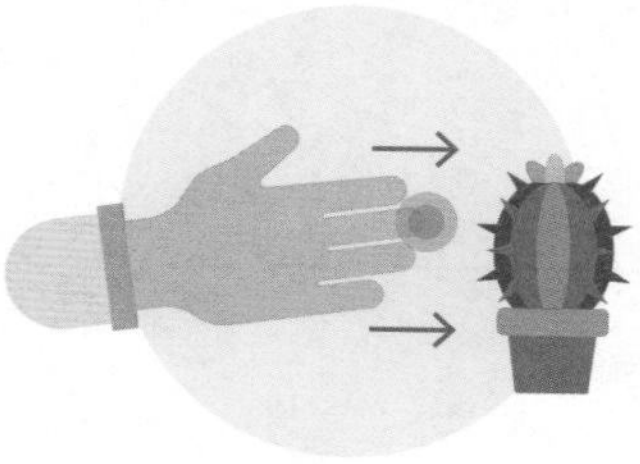

Pain 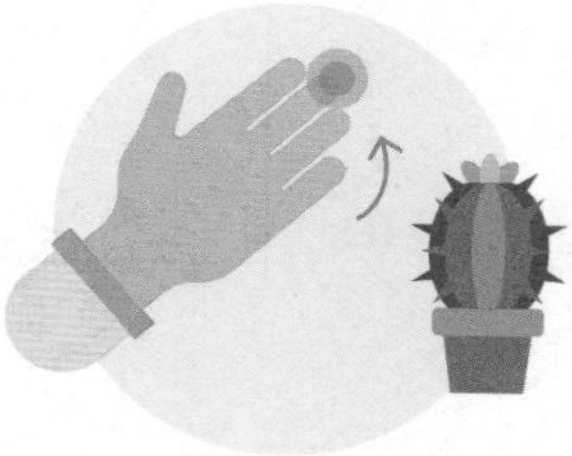   Response

06 The diagram shows the distribution of auxin in some plant roots.

Which distribution would cause the root to bend down? Tick **one** box. [1]

☐ A

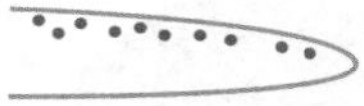

☐ B

☐ C

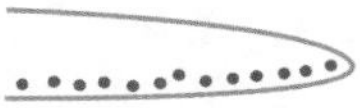

☐ D

07 Electrical impulses travel along neurones at 100 m/s.

07.1 Calculate the time it would take for impulses to travel along a neurone 1 metre long. [1]

07.2 Explain why hormonal responses are slower that nervous responses. [2]

Higher Tier only

08 Explain how insulin and glucagon interact to control blood glucose concentration. [4]

09 Scientists measured the ethene produced by a banana following picking. Their results are shown in the graph.

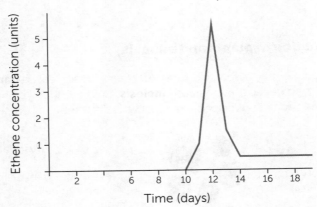

09.1 How many days does it take for the banana to start produce ethene? Tick **one** box. [1]

☐ 10 days ☐ 11 days ☐ 12 days ☐ 13 days

Ethene causes unripe fruit to ripen.

09.2 When would be the best time to put this banana into a container of unripe ones to speed up their ripening? Tick **one** box. [1]

☐ After 8 days ☐ After 10 days ☐ After 12 days ☐ After 14 days

SEXUAL AND ASEXUAL REPRODUCTION

Sexual reproduction results in offspring with mixed genetic information from two parents.

Sexual reproduction in plants and animals

Plants and animals produce specialised cells for sexual reproduction called **gametes**. Gametes are produced by a special type of cell division called **meiosis**. Cells produced by mitosis are genetically identical, but the cells produced by meiosis are genetically different from one another.

Plants	Animals
Gametes made by flowering plants are called pollen and egg cells.	Gametes made by animals are called sperm and egg cells. They are made in the reproductive organs.
Pollen cells are so tiny they can only be seen with a microscope. They are moved from one parent plant to another by wind or by animals.	Sperm cells are able to move in water. In some animals, they are placed inside one parent by the other.
Plant **egg cells** remain inside flowers.	Animal egg cells may leave the parent's body or remain protected inside.

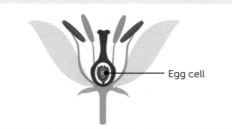

Egg cell

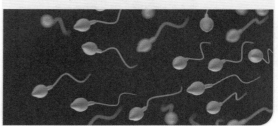

Asexual reproduction

If a single organism reproduces, the reproduction is **asexual**. With only one parent involved there cannot be any mixing of genetic information. The offspring are all genetically identical to one another. Offspring produced like this can be referred to as **clones**.

Asexual reproduction involves small parts of an organism separating from the parent to form a new individual. This means that the new individual is made of cells produced by mitosis.

1. Draw **one** line between each type of cell division and their result. [2]

    Mitosis
        Produces two genetically different cells
        Produces four genetically different cells
    Meiosis
        Produces two genetically identical cells
        Produces four genetically identical cells

2. Suggest why pollen cells have a hard, protective coat. [1]

3. Explain why sexual reproduction leads to variety in the offspring. [2]

    1. *Mitosis to two genetically identical cells.[1] Meiosis to four genetically different cells.[1]*

    2. *To protect them from drying out[1] in the air.*

    3. *Genetic information from two parents[1] is mixed together[1].*

MEIOSIS

Meiosis produces gametes with half the number of chromosomes found in the body cells of the parent. Fusion of two gametes later restores the full number.

What happens in meiosis

Meiosis takes place in cells in the reproductive organs of animals (ovaries and testes) and plants (flowers).

Like **mitosis**, **meiosis** in a cell starts with copying of the genetic information, or DNA. The number of chromosomes is doubled.

Unlike mitosis, the cell then divides twice, ending up with four cells rather than two. Because the doubled number of chromosomes has been halved twice, each of the four cells ends up with half the normal number of chromosomes.

Each time meiosis happens, the mixture of chromosomes ending up in each gamete is different. This means that all gametes are genetically different from one another.

Compare meiosis with mitosis on page 12 (4.1.2.2).

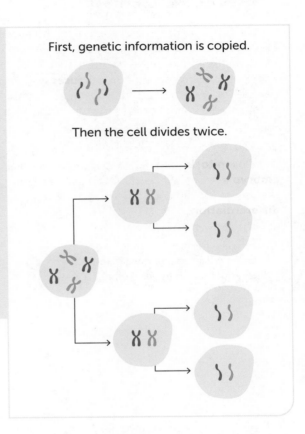

First, genetic information is copied.

Then the cell divides twice.

---

1. Name the cells produced by meiosis. [1]
2. Explain meiosis in plants and why it is important. [4]
3. Explain how meiosis produces cells with half the normal number of chromosomes. [2]

   1. *Gametes.[1]*
   2. *Meiosis occurs in the flowers[1] of plants and produces egg cells and pollen.[1] Pollen and egg cells are genetically different.[1] This is because meiosis results in a different combination of chromosomes[1] in each gamete.*
   3. *Firstly, the chromosomes are copied[1] so the number is doubled. Then the cell divides twice[1], halving the number of chromosomes each time[1].*

FERTILISATION

Sexual reproduction involves the fusion of male and female gametes.

Human chromosomes

Humans have 46 chromosomes in each body cell. Meiosis halves the number of chromosomes in the gametes so sperm and egg cells each have 23 chromosomes. When a sperm cell **fuses** with an egg cell, two single sets of 23 chromosomes are mixed together. Fertilisation restores the normal number of 46 chromosomes.

The new cell formed by fertilisation divides many times by **mitosis**, forming the body cells of the **embryo**. Many of these new body cells become specialised for particular functions. This is called **differentiation**.

You can revise specialisation and differentiation on **pages 6-8. (4.1.1.3-4)**

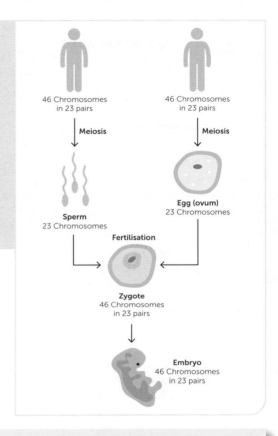

1. Meiosis is a type of cell division. Which one of the following best describes meiosis? Tick **one** box. [1]

   ☐ Produces genetically identical cells.   ☐ Halves the number of chromosomes.
   ☐ Forms the cells in an embryo.   ☐ Doubles the genetic material

2. Human cells contain 46 chromosomes. Fill in the numbers on each of the cells in the diagram. [2]

3. Explain why the cells in an embryo differentiate. [2]

   1. *Halves the number of chromosomes.[1]*
   2. *Egg cell 23 + Sperm cell 23.[1] Fertilised cell 46.[1]*
   3. *The cells need to become specialised[1] to carry out different functions[1].*

DNA AND THE GENOME

The genome is the entire genetic material of an organism.

How genetic material is organised

- The genetic material in the nucleus is composed of a chemical called DNA
- **DNA** is a long polymer made of two strands twisted into a **double helix**
- Long lengths of DNA are tightly coiled into structures called **chromosomes**
- **Genes** are short sections of DNA that code for a specific protein.

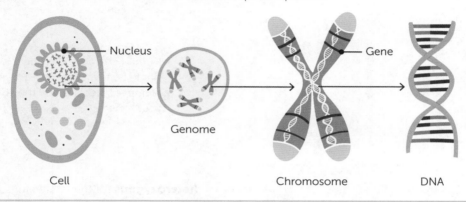

Cell   Nucleus   Genome   Chromosome   Gene   DNA

The human genome

Scientists have completely sequenced the human **genome**. The location of many of the genes on human chromosomes has been discovered and mapped. This information is already extremely useful in medicine and will become increasingly important in the future.

- Scientists continue to search for genes linked to particular diseases.
- Knowing which genes are linked to particular diseases is very helpful for treating people with inherited disorders.
- Mapping chromosomes also allows the tracing of past human migration patterns.

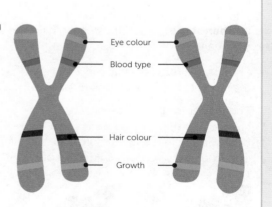

Eye colour
Blood type
Hair colour
Growth

1. Describe what is meant by the genome of an organism.   [1]
2. Suggest how knowledge of the human genome allows some inherited disorders to be identified in a developing human embryo.   [2]

   1. *The genome of an organism is the entire genetic material of that organism.[1]*
   2. *Sequencing the embryo DNA[1] of genes known to be linked to inherited disorders[1].*

GENETIC INHERITANCE

Most features of organisms (**characteristics**) are determined by groups of genes, but some are controlled by a single gene.

Alleles

Genes are carried on **chromosomes**. An organism has two copies, or **alleles**, of each gene in each cell but the **gametes** they produce only carry one copy. When they reproduce, fertilisation results in these single alleles pairing up in the offspring. Different pairs of gametes give different combinations of alleles.

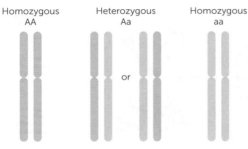

Homozygous AA   Heterozygous Aa   Homozygous aa

or

A = Dominant allele   a = Recessive allele

Human characteristics controlled by a single gene include red-green colour blindness, freckles and the ability to roll your tongue.

Organisms with two different alleles of a gene are called **heterozygous** for that characteristic. Those where the two alleles the same are called **homozygous**.

Where an organism has two different alleles, one allele often overrides the effect of the other. The **dominant** allele determines how the organism appears and the **recessive** allele is hidden.

The alleles an organism has are referred to as its **genotype**, but how the organism appears is called its **phenotype**. Two of these mice have the same phenotype, but different genotypes. Allele B is dominant.

Phenotypes result from alleles exerting their effects in cells at a molecular level.

The recessive allele b can only be expressed if both copies are present. The dominant allele is always expressed, even if only one copy is present in the genotype.

BB   Bb

bb

1. In rabbits, allele W is dominant and gives black fur. Allele w is recessive and gives white fur. Deduce the genotype of a white mouse. [1]

2. Domestic cattle have 60 chromosomes. State how many chromosomes are present in each of these types of cell: [3]

   (a) sperm cell   (b) embryo cell   (c) liver cell

3. Explain why human gametes only carry one allele of a gene. [2]

   1. *ww[1]*
   2. *(a) 30[1] (b) 60[1] (c) 60[1]*
   3. *Gametes are formed by meiosis[1] which reduces 23 pairs of chromosomes to just one of each pair[1].*

Punnett squares

The inheritance patterns of characteristics controlled by a single gene can be predicted. A **Punnett square** predicts the genotypes and phenotypes for offspring produced by two parents.

- The parent's gametes are shown at the top and left.
- The possible alleles in the gametes are at the top of each column and the start of each row.
- Each cell in the square shows the genotype resulting from the fusion of two gametes at fertilisation.
- The phenotype of the offspring is often also shown, sometimes by shading or by diagrams.

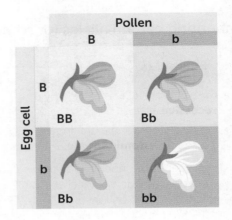

The **probability** that these parents will produce purple flowered offspring is 0.75. The probability that they will produce white flowered offspring is 0.25. The ratio of purple to white flowered offspring is 3:1.

Constructing Punnett squares   Higher Tier

The offspring from crossing a male parent TT with a female parent Tt can be predicted with a Punnett square.

- Draw a grid and put the alleles in the male gametes at the top and the alleles in the female gametes on the left.
- Fill in the pairs of alleles from each possible fertilisation. Then work out the probability of each genotype or phenotype appearing in the offspring.

	T	T
T	TT	TT
t	Tt	Tt

4. The diagram shows a male red-eyed fly crossed with a female white-eyed fly.
   (a) Circle a part of the diagram that shows an allele in a sperm cell. [1]
   The allele for red eyes is dominant.
   (b) Give the genotypes of red-eyed offspring shown in the diagram. [1]
   (c) State the probability that offspring from this cross will be white eyed.
   Tick **one** box. [1]

   ☐ 0.25   ☐ 0.5   ☐ 0.75   ☐ 1.0

   (d) Give the chance that offspring from this cross will be homozygous. [1]

5. Explain why characteristics controlled by a single gene are unusual. [1]

		Male	
		R	r
Female	r	Rr	rr
	r	Rr	rr

4. (a) R or r for male circled[1]   (b) Rr[1]   (c) 0.5[1]   (d) 50%[1]

5. They are unusual because most characteristics are controlled by multiple genes.[1]

INHERITED DISORDERS

Inheriting certain alleles for some genes results in inherited disorders.

Examples

- Cystic fibrosis (CF) is caused by a **recessive** allele.
- Polydactyly (having extra fingers or toes) is caused by a **dominant** allele.

The probability of inheriting disorders caused by recessive and dominant alleles is different. Cystic fibrosis can be inherited by a child of unaffected parents. Polydactyly can only be inherited if one or both parents are affected.

Gene therapy and screening

Gene therapy may reduce the symptoms of people with cystic fibrosis. **Embryo screening** can be used to detect genetic disorders in embryos. But some people have ethical objections to altering the human genome and to abortion should a genetic disorder be detected.

If both parents are heterozygous for cystic fibrosis, the probability of having an affected child is 0.25.

If one parent is affected with polydactyly, the probability of having affected children is 0.5.

If parents think they may have affected children, they can ask for embryo screening to find out the genotype of their baby.

1. The genetic diagram shows parents who are heterozygous for polydactyly.

		Father	
		P	p
Mother	P		
	p		

(a) Complete the genetic diagram. [1]

(b) Circle any children with polydactyly. [1]

(c) What is the probability of them having a child unaffected by polydactyly? [1]

2. Explain the ethical difficulty for parents choosing whether or not to have embryo screening. [2]

3. Explain why cystic fibrosis can be inherited by children of unaffected parents. [3]

1. *(a) PP, Pp, Pp pp[1]*
   (b) PP, Pp, Pp circled[1]   (c) 0.25[1]

2. *The parents have to choose between taking the risk of having a child with a genetic disorder[1] and the possibility that they would need to consider an abortion by knowing the genotype[1].*

3. *If both parents are heterozygous[1] there is a 25% chance that both[1] may pass on a recessive allele[1].*

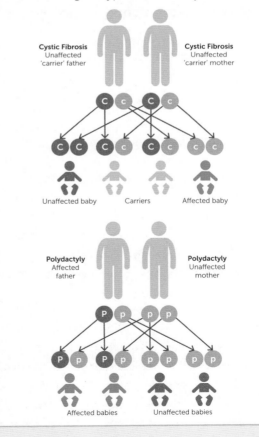

Cystic Fibrosis
Unaffected
'carrier' father

Cystic Fibrosis
Unaffected
'carrier' mother

Unaffected baby   Carriers   Affected baby

Polydactyly
Affected
father

Polydactyly
Unaffected
mother

Affected babies   Unaffected babies

SEX DETERMINATION

One pair of human chromosomes carries the genes that determine sex.

Chromosome pairs

All the cells in a human body (apart from gametes) contain 46 chromosomes. They can be sorted into 23 pairs matched by size. One set of 23 comes from the male parent and one set from the female parent.

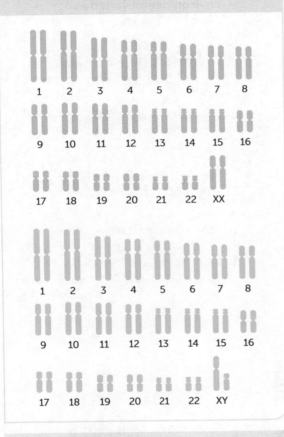

Sex chromosomes

One pair of chromosomes carries the genes that determine the sex of a person. These chromosomes are called the **sex chromosomes**. Females have two X chromosomes in each body cell and males have an X chromosome and a Y chromosome in each body cell.

All egg cells carry an X chromosome. Sperm cells carry either an X chromosome or a Y chromosome.

This means that the probability of having a male child or a female child is always 0.5.

		Male	
		X	Y
Female	**X**	XX	XY
	X	XX	XY

---

1. Explain why the chance of two parents having a girl is always 50%. [2]
2. How many sex chromosomes are found in one human sperm cell? Tick **one** box. [1]

   □ 0.5   □ 1   □ 2   □ 4

3. Describe **one** difference between the X and Y chromosome visible with a microscope. [1]

   1. *Egg cells always contain an X chromosome so the sex of the child is decided by which sex chromosome is contained in the sperm cell[1], 50% of which contain an X chromosome.[1]*

   2. *One.[1]*

   3. *The X chromosome is larger[1] than the Y chromosome[1].*

VARIATION AND EVOLUTION

Differences in the characteristics of all those in a population is called **variation**.

Natural selection and evolution

Mutations occur all the time in populations. Mutations are changes in genes. Most mutations have no effect on the phenotype of the individuals in the population.

Occasionally, a mutation does alter a phenotype and might make it more suited to changing conditions. If this happens, these individuals will be more likely to survive and pass on the mutation to their offspring. This is called **natural selection**.

Changes in inherited characteristics may build up in a population over a very long period of time. This is called **evolution**.

Two separate populations of a species may become so different in phenotype that they cannot interbreed to produce **fertile offspring**. The two populations have then become two distinct species. All species have evolved in this way from simple life forms that first developed more than three billion years ago.

Surviving mice reproduce

Variation

- **Genetic variation** occurs because of differences in the alleles inherited by individuals.

- **Environmental variation** occurs if individuals develop in different conditions.

Often the variation in a population is a mixture of genetic and environmental variation.

Rat tails vary in length. Part of this variation is genetic. It involves multiple genes because if tail length was controlled by a single gene with two alleles there would only be two phenotypes (short or long).

Tail length in rats is also influenced by the climate. Rats reared in warm conditions grow longer tails compared to those reared in cooler conditions. Long tails developed because of warm conditions are not inherited by their offspring.

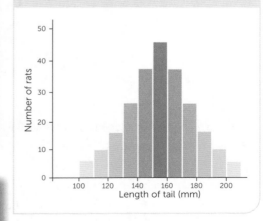

Describe the process of evolution. [4]

Evolution is a change in the inherited characteristics of a population[1] over time[1] due to natural selection[1] which may lead to the formation of a new species[1].

SELECTIVE BREEDING

Humans have been selectively breeding plants and animals for thousands of years.

Artificial selection

The useful characteristics of food crops and farm animals are the result of selective breeding.

Cows have been bred to produce more milk, and potato plants have been bred to produce larger potatoes.

- Parents with the desired characteristics are carefully chosen and bred together.

- The offspring that show the desired characteristics are then selected and bred together.

- This process goes on for many generations until all offspring have the desired characteristics.

Sometimes the desired characteristic is only found in male or female animals. Milk yield is a characteristic of females. Male parents must be chosen whose previous female offspring showed the desired characteristics. This is one reason why animal breeders keep careful records.

Another reason for careful record keeping is to try to prevent **inbreeding** by avoiding crossing closely related animals.

There are also many examples of selective breeding of domestic animals for their appearance or behaviour. Some examples, such as fancy pigeons or goldfish, have taken many years to achieve.

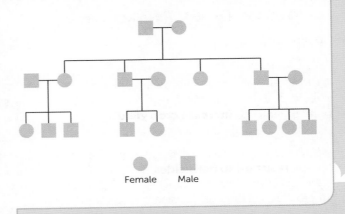

Female   Male

1. Explain why horse breeders keep careful stud records showing which animals were used for breeding. [2]
2. Give **one** problem associated with inbreeding. [1]
3. Describe how scientists could produce goats that produce large volumes of milk. [4]

   1. *To avoid inbreeding[1] and to record the characteristics of offspring to identify suitable future parents[1].*

   2. *One from: Offspring have a higher probability of inheriting recessive genetic disorders[1] / becoming more prone to disease[1].*

   3. *Select females with high milk yield.[1] Select males whose female offspring have shown high milk yield.[1] Cross these animals[1] then select those with highest milk yield from the offspring and continue for a number of generations[1].*

Plants	Animals
Food crops with disease resistance	Animals that produce more meat or milk
Plants with large or unusual flowers	Domestic dogs with a gentle nature

GENETIC ENGINEERING

Genetic engineering is used to give organisms useful characteristics. Genetic engineering alters the genome of one organism by adding a gene from another organism.

Genetic engineering in agriculture

Some plants have been genetically modified to be resistant to diseases or to have larger fruit.

Crops such as cotton have been genetically modified to be resistant to insect **pests**. Crops with these characteristics have the potential to **increase crop yield** in tropical countries where pests cause huge damage to plants.

Some crops such as soya beans have been genetically modified to be **resistant to herbicides**. This means that competitive weeds can be killed by spraying the crop with herbicide. The herbicide will only kill the weeds and not the crop.

Genes are 'cut out' from the chromosomes of human cells or other organisms. They are then added to the genome of another organism which becomes a **genetically modified (GM)** organism.

1. Describe **two** ways that crop plants have been genetically modified. [2]

2. Explain why GM crops show increased yields. [2]

   1. *Any two from: To be resistant to herbicides[1] and insect pests[1], to grow larger (better) fruit[1], to be resistant to disease[1].*

   2. *If crops are not affected by pests or disease[1] or outcompeted by weeds[1], they grow larger[1].*

Genetic engineering in medicine

One of the most successful uses of genetic engineering in medicine has been the **production of human insulin** by genetically modified bacteria.

The gene for human insulin was 'cut out' from the chromosomes of human cells and added to the genome of bacterial cells. These bacteria could then make insulin. If large quantities of bacteria are grown in an industrial process, large amounts of insulin can be collected from them.

This has increased the supply of affordable insulin for people with diabetes.

How genetic engineering works   Higher Tier

There are three key stages in genetic engineering:

1 Enzymes used to 'cut out' or isolate required gene → **2** Required gene inserted into a vector → **3** Vector inserts gene into cells of target organism

Enzymes are needed to cut and join DNA and **vectors**, such as viruses or bacterial plasmids, are needed get the gene into target cells.

Concerns with genetic engineering

There are concerns over the safety of GM crops and their environmental effects. Some people are worried that consuming food products made from GM crops may have harmful effects on humans, although none have yet been identified.

There are also concerns that growing genetically modified crops has had a negative effect on biodiversity.

Investigations show that populations of some wild flowers decrease around GM crops. This may be because more herbicide tends to be used on them and some herbicide may drift onto surrounding wild plants, killing them too.

Investigations also show that populations of some insects decline in and around GM crops. This may be because more pesticide tends to be used on GM crops. It may also be because fewer wild flowers means less food for insects. Either way, there seems to be some environmental cost for growing some GM crops.

Current research is looking at the potential of treating genetic disorders like cystic fibrosis with **gene therapy**. Gene therapy involves the genetic modification of some of the cells in a person with a genetic disorder but some people have ethical objections to altering the human genome.

3. Vectors are used in genetic engineering. Give an example of a vector. [1]

4. Explain the concept of gene therapy. [1]

5. Evaluate the use of genetic engineering in agriculture. [4]

6. **Higher only:** Explain why genes should be transferred to organisms at an early stage of their development. [2]

3. *A virus / bacterial plasmid.[1]*

4. *Genes are inserted into some of a person's cells[1] to reduce the symptoms of a genetic disorder.[1]*

5. *GM crops give bigger yields if larger or resistant to pests or disease.[1] This is particularly beneficial in tropical countries where pests can inflict huge losses.[1] Increased nutritional content may improve health.[1] GM crops may reduce biodiversity[1] and significant numbers of consumers object to GM products / are concerned about health risks as more research is needed[1].*

6. *So that the genetic modification can be transferred to more cells by mitosis[1] during growth and so the GM organism has time[1] to develop the desired characteristic.*

EVIDENCE FOR EVOLUTION

Since Darwin and Wallace proposed the theory of evolution, supporting evidence has built up and the theory of evolution by natural selection has become widely accepted.

Antibiotic resistance

It takes such a long time for natural selection to result in changed phenotypes that there are few examples of scientists being able to follow the process in real time. One example is the adaptation of some species of bacteria to resist antibiotics.

Bacteria reproduce very quickly by asexual reproduction. A mutation that occurs in one cell can be rapidly copied to other bacteria.

A mutation giving **resistance to antibiotics** can occur randomly in a bacterial cell in a person being treated with antibiotics.

That bacterial cell continues to reproduce, leading to many bacteria with antibiotic resistance. This is evidence for **natural selection**.

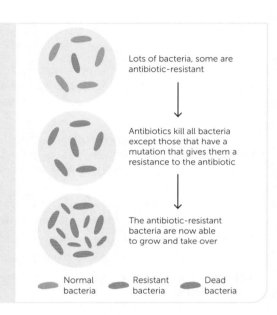

Lots of bacteria, some are antibiotic-resistant

Antibiotics kill all bacteria except those that have a mutation that gives them a resistance to the antibiotic

The antibiotic-resistant bacteria are now able to grow and take over

Normal bacteria	Resistant bacteria	Dead bacteria

Fossil record

Fossils are found in layers of sedimentary rock. The oldest fossils are found in the oldest or lowest layers of rock. The fossils of some species formed in successive layers show changes in their phenotype over time. This is also evidence for **natural selection**.

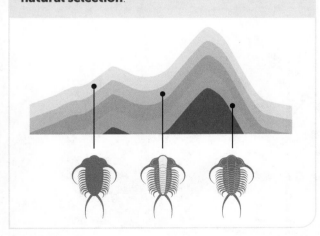

1. Explain why bacteria evolve rapidly. [2]
2. Explain why the fossil record is evidence for natural selection. [2]
3. Use your knowledge of the human genome to describe evidence now available for evolution in addition to antibiotic resistance and the fossil record. [2]

1. *Because they reproduce asexually[1], they can reproduce at a very fast rate.[1]*
2. *Changes can be seen in the characteristics of a species[1] over a long period of time[1].*
3. *Scientists have identified a number of human genes[1] and shown how inherited disorders are passed on to offspring[1].*

FOSSILS

Fossils are found in rocks. They are the 'remains' of organisms from millions of years ago.

How fossils are formed

There are three ways that dead animals or plants can form fossils:

1	2	3
Some fossils form when parts of animals or plants are prevented from decaying. This is because they die in a place where one or more of the conditions needed for decay are absent.	Most fossils form when parts of the animal or plant are slowly replaced by minerals as they decay.	Some fossils are the preserved traces of animals and plants, such as footprints, burrows and rootlet traces.

There are several reasons why the fossil record is incomplete. Animals or plants that died in water were likely to form fossils if they sank in mud or became covered with sediment. However, animals that died on land were more likely to be eaten by scavengers. This means many organisms did not form fossils. Also, many fossils were destroyed by later geological activity.

There is a particular gap in the fossil record of early life on earth. Many of the organisms at that time were soft bodied. Hard parts of organisms are more likely to form fossils whereas soft bodied organisms leave little trace. This is why little is known about how life on earth began.

Evolutionary trees

Groups of dated fossils can be used to make **evolutionary trees** which show how each species on the tree is related to the others. Species on the same branch of an evolutionary tree are more closely related to one another than they are to other species on the tree. Find out more about evolutionary trees in **4.6.4 Classification page 98.**

1. The photograph shows a fossil shell. Explain how the fossil may have formed. [3]

2. Describe **one** piece of information that can be learned from fossils of a species found in different layers of rock. [2]

3. Explain what is shown by an evolutionary tree. [1]

   1. *The animal sank into mud[1] where the soft parts decayed rapidly.[1] The shell was then slowly replaced by minerals[1] as it decayed.*

   2. *How much or how little the species has changed[1] over a period of time[1].*

   3. *How closely related species are to one another.[1]*

EXTINCTION

If there are no individuals of a species remaining alive, it is described as being **extinct**. Species can become extinct in one area, such as a country, or they can become globally extinct.

Causes of extinction

Individual species my become extinct for a number of reasons. These include:

- The arrival of new predators in an area.
- The arrival of new disease causing pathogens in an area.
- Being out competed by another species for food.
- Loss of habitat, such as deforestation.
- A change in habitat such as a change in the climate due to global warming.

The fossil record also shows evidence of some 'mass extinction' events where very large numbers of species became extinct during the same period of time. Scientists think these were caused by large scale catastrophic events, including:

- Asteroids or meteors colliding with the earth
- Ice ages
- Widespread drought
- Volcanic activity

Scientists believe the world is in the middle of another mass extinction caused by human activity.

The splendid poison frog is recently believed extinct owing to a loss of habitat.

1. This drawing shows a tree that has become extinct in the wild. The Franklin tree was found in Georgia in the US. Scientists think it was infected by cotton pathogens when cotton was introduced as an agricultural crop.

   (a) Suggest what is meant by 'extinct in the wild'. [1]
   (b) Explain what is meant by a pathogen. [1]
   (c) Suggest how scientists could increase the number of trees to conserve the species. [1]

2. Suggest why plants are more at risk of extinction due to climate change than animals. [2]

   1. (a) There are no wild individuals of the species left alive[1], but there may be some cultured in laboratories or botanic gardens.

   (b) A microorganism that causes an infectious disease.[1]

   (c) Produce more individuals by using plant tissue culture.[1]

   2. Animals can migrate to places with better climatic conditions[1], whereas plants cannot move[1].

RESISTANT BACTERIA

MRSA is a strain of bacteria resistant to antibiotics.

A mutation giving resistance to an antibiotic can occur in a bacterial pathogen in a person being treated with antibiotics. If this bacterial cell survives and reproduces, a new **strain** of bacteria has evolved.

MRSA

MRSA is resistant to a number of antibiotics. Many people carry MRSA harmlessly on their skin but it can cause serious illness if it gets inside the body. MRSA infections mainly occur in hospitals. This is because hospital patients often have surgical wounds. Once the bacteria are inside the body, the infection can be very difficult to treat.

Patients going into hospital for a planned procedure, such as an operation, are screened for MRSA. If they are carriers, they are treated beforehand. This helps reduce the risk of the patient getting an MRSA infection or passing MRSA on to another patient.

Scientists are working to develop new kinds of antibiotics to kill resistant strains of bacteria but finding new antibiotics is slow and expensive. Resistant strains could evolve faster than new antibiotics can be developed.

To slow down the evolution of antibiotic resistant strains, patients must take their full course of antibiotics so all the bacteria are killed and none can survive to mutate.

Doctors should avoid giving people antibiotics for minor bacterial infections or for infections caused by viruses. Antibiotics are not effective against viruses.

1. This image (right) shows the result of a test for antibiotic resistance. Bacteria have been grown on the agar gel. Each disc contains a different antibiotic.

   (a) Which antibiotics are the bacteria resistant to?
      Tick **one** box. [1]

      ☐ 1 and 4   ☐ 4 and 10   ☐ 2 and 3   ☐ 4 and 8

   (b) Describe MRSA. [2]

2. Explain the risk if farm animals are given antibiotics in their food to keep them productive. [4]

   1. (a) 4 and 10.[1]

      (b) MRSA is a strain[1] of bacteria resistant to antibiotics[1].

   2. There is a chance that a mutation providing antibiotic resistance[1] could occur in bacteria in an animal being fed antibiotics[1] and this resistant strain could then spread to humans[1] and there would be no effective treatment available[1].

CLASSIFICATION OF LIVING ORGANISMS

The way that living things are sorted into groups, or **classified**, is always changing.

Linnaean system

Carl Linnaeus developed an early classification system based on the visible characteristics of organisms. He used the structure of animals and plants to put them into groups that shared the same characteristics.

Linnaeus created seven kinds of group. Closely related species are placed into the same genus. Genera with shared characteristics are placed in the same family and so on.

Linnaeus also gave each kind of organism a unique two-part name consisting of its genus and its species. His **binomial system** is still in use and provides an internationally recognised 'scientific' name for every recorded species on earth. Humans, for example, are named **Homo sapiens**.

| Species |
| Genus |
| Family |
| Order |
| Class |
| Phylum |
| Kingdom |

New models of classification

Classification now uses 'hidden' characteristics as well as visible ones. Microscopes allowed features of the cells of organisms to be used. More recently, similarities in DNA have been used.

Chemical analysis has resulted in **Carl Woese** placing living things into one of three **domains**:

- **Archaea** (primitive bacteria usually living in extreme environments)
- **Bacteria** (true bacteria)
- **Eukaryota** (all other living things which includes protists, fungi, plants and animals)

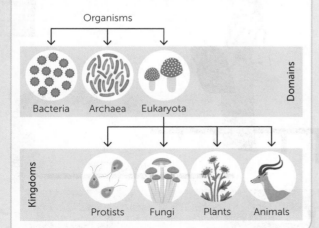

1. Name the scientist who developed the 'three-domain' system. Tick **one** box. [1]
   ☐ Gregor Mendel   ☐ Carl Linnaeus
   ☐ Alfred Russel Wallace   ☐ Carl Woese

2. Suggest which biological structure is found in the cells of all organisms in the domain Eukaryota. [1]

3. The diagram shows an evolutionary tree based on classification data.

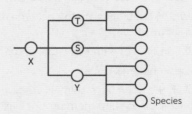

(a) Name the groups at X and Y. [2]
(b) Explain what the diagram indicates about the species in T and S. [2]

1. *Carl Woese.[1]*
2. *Nucleus.[1]*
3. *(a) Family.[1] Genus.[1]*
   (b) The species in T are more closely related to each other[1] than that in S[1].

EXAMINATION PRACTICE

01 The image shows an insect in fossilised tree resin called amber which is 46 million years old.

01.1 Explain why the remains of the insect are preserved. [1]

Fossilised leaves from trees are sometimes found but fossilised tree trunks are more likely to be found.

01.2 Explain why tree trunks are more likely to form fossils than leaves. [2]

01.3 Explain what a fossilised rootlet trace is. [2]

Carl Linnaeus   created the three-domain system

                                 created the binomial system

Carl Woese   created the species system

02 Many scientists have worked to develop our classification system.

02.1 Draw **one** line from each scientist to the description of their work. [2]

02.2 Describe the binomial system. [2]

The table shows part of the classification of the meadow buttercup.

Classification group	Name
	Plantae
	Angiospermae
Class	Dicotyledonae
Order	Ranunculales
	Ranunculaceae
	Ranunculus
Species	Acris

02.3 Complete the table to give the names of the missing classification groups.   [2]

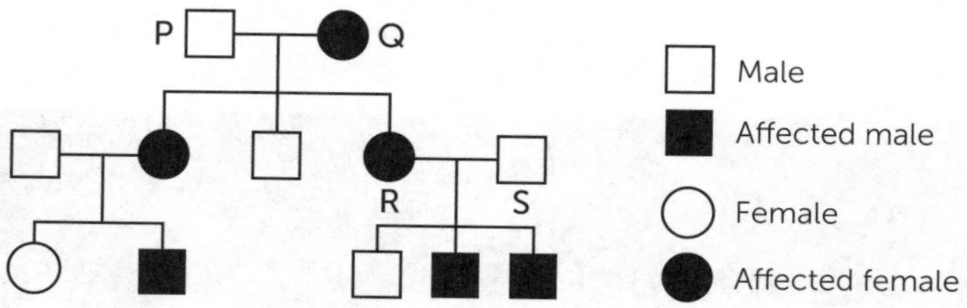

P ☐—● Q

☐	Male
■	Affected male
○	Female
●	Affected female

R S

03   The diagram shows the inheritance of the genetic disorder polydactyly in a family.

03.1   Explain what causes polydactyly.   [1]

03.2   If the symbol for the gene is D, what is the genotype of
       individual P? Tick **one** box.   [1]

☐ dd   ☐ DD   ☐ dD   ☐ Dd

	D	d
d		
d		

03.3   Complete the Punnet square for R and S.   [1]

03.4   What is the probability of R and S having another affected child?   [1]

04   Garden peas usually have green pods. Pod colour is controlled by a single gene. A mutation causes
     a recessive allele that gives yellow pods. The Punnet square shows the inheritance of pod colour
     when two plants with green pods are crossed.

	G	g
G	GG	Gg
g	Gg	gg

04.1   Circle a part of the diagram that shows a gamete carrying a recessive allele.   [1]

04.2   Which offspring are heterozygous? Tick **one** box.   [1]

☐ GG   ☐ Gg   ☐ gg

04.3   What is the chance that seeds collected from this cross will develop into plants with
       yellow pods?   [1]

05   Draw a line between each process and a suitable description.   [3]

Fertilisation   Doubles the normal number of chromosomes

                             Preserves the normal number of chromosomes

Mitosis

                             Halves the normal number of chromosomes

Meiosis

                             Restores the normal number of chromosomes

06   The diagram below shows an evolutionary tree for some fossils.

     Give **one** piece of information indicated by the diagram about fossil species B, C and D.   [1]

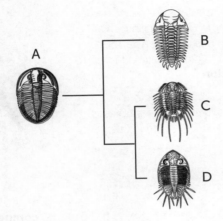

COMMUNITIES

A community is a group of **interdependent** species of plants, animals and microorganisms.

Interdependence

In a community, each species depends on other species in a variety of ways. If one species is removed it can have a knock-on effect on the whole community. This is called interdependence.

Hummingbirds are a good example of interdependence. They eat a mixture of small insects and nectar. While feeding on nectar, they get pollen stuck to their beak, which is then transferred to flowers on another plant. Some flowering plants have adapted to depend on hummingbirds for **pollination**.

Other examples of interdependence include animals or birds **dispersing** the seeds of plants to new places or using holes in trees as a place to **shelter**.

Competition

In any community, if organisms depend on the same thing there will be **competition**.

Plants growing beside each other compete for light and space, and for water and mineral ions in the surrounding soil.

Animals compete for food, for mates and for **territories** in which to raise offspring. In a coral reef community, competition can be intense.

A community is **stable** when all the species and the environmental factors are in balance so that although some population sizes may fluctuate in cycles, most stay fairly constant.

1. Describe **two** ways in which the species in a community may depend on one another. [2]

2. The image shows two male animals in a community competing. They only compete in this way at a certain time of the year.

   (a) Suggest what they are competing for. [1]

   (b) Suggest the advantage of this sort of competition. [2]

3. The table below shows some data for a species of bird that eats insects.

Population size	Number of breeding pairs	Size of territory (m²)
8	4	3100
12	6	3100
20	10	1600
54	26	1350
62	26	1351

   (a) The data indicates that there might be a maximum territory size for this species of bird. Describe how. [1]

   (b) Suggest why there might be a maximum territory size for this species of bird. [2]

   (c) Describe and explain what happens to the territory size and number of breeding pairs as the population size increases. [6]

1. *Any **two** from: For pollination[1] / for seed dispersal[1] / for food[1] / for shelter.[1]*

2. *(a) A territory / a mate.[1]*

   (b) The fittest male will gain the mate/territory[1], so his (advantageous) characteristics are passed on to his offspring.[1]

3. *(a) The territory size stays the same once the number of breeding pairs has fallen past a certain point.[1]*

   (b) There is a maximum area that the male[1] bird is able to defend.[1]

   (c) Indicative content: As the population size increases the territory size decreases[✓], but only to a certain point.[✓] This is because there is a minimum size of territory required to supply enough insect food[✓] for a pair of birds and their offspring.[✓] As the population size increases, some birds are unable to breed[✓] because they do not compete successfully enough[✓] to claim a territory.

ECOSYSTEMS

Ecosystems consist of a community and its physical environment.

Food supply

Organisms need a supply of materials to **survive** and **reproduce**. Plants take materials from the air and the soil. Animals get materials by feeding on other organisms. Respiration by plants and animals, and decomposition by microorganisms both release materials back to the air and the soil.

The interaction between a community and its physical environment forms an **ecosystem**.

Organisation of ecosystems

- The **biotic** part of an ecosystem is made up of all the living organisms. All the individual organisms of one species are adapted to the conditions of their particular **habitat** in the ecosystem.

- To help study ecosystems, scientists organise the individual **organisms** into levels. The first level above individual organisms contains **populations**. A population is all the individual organisms of one species in an ecosystem. An ecosystem contains a number of populations.

- All the populations put together form a **community**.

- The interaction between the community and its **abiotic** or non-living surroundings forms the top level, or **ecosystem**.

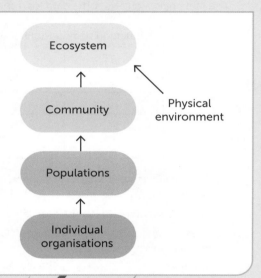

1. The image shows a pond ecosystem.

   Give the part of the ecosystem represented by the following components of the pond: [3]

   (a) the fish

   (b) all of the plants, the fish and other organisms in the pond

   (c) the water and mud

2. Describe a population. [2]

3. The graph shows the total number of individuals over a period of ten years for two communities containing different numbers of species.

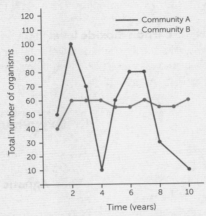

   (a) Which community is the most stable? Give a reason for your answer. [2]

   (b) Suggest why community A probably contains fewer species than community B. [2]

   1. (a) A population.[1]

      (b) The community / biotic part.[1]

      (c) Abiotic part.[1]

   2. A population is all the individuals[1] of one species[1] in an ecosystem.

   3. (a) Community B[1] because the organism numbers vary very little, whereas the number of organisms in community A varies from 10 to 100[1].

      (b) Community A shows large changes in population sizes[1] which means there is less interdependence between fewer species[1].

When answering question on interpretation of graphs, remember to give actual data from the graph as evidence to support your point.

ABIOTIC FACTORS

Communities are affected by the chemical and physical conditions of their surroundings. Abiotic factors often determine where species can live.

Abiotic conditions and photosynthesis

The rate of photosynthesis is affected by **light intensity** and by temperature. Faster photosynthesis means that plants and algae grow faster, which supplies more food to animals.

The rate of photosynthesis affects the **carbon dioxide level** in the atmosphere. In conditions where photosynthesis occurs more slowly, the carbon dioxide level may increase. Experimental evidence shows that rising global CO_2 levels could increase photosynthesis, but the effect would be short-lived because other factors soon limit plant growth.

Temperature

The **temperature** affects all organisms because of its effect on enzymes. Different species are adapted to survive in different temperature ranges. Temperature affects the rate of decay in ecosystems. It also affects the rate of transpiration in plants.

Oxygen in aquatic habitats

The oxygen level in the atmosphere is constant, but it can vary a lot in streams or ponds. **Aquatic** animals live in water and absorb oxygen for respiration through their body surface or through gills. **Low oxygen levels** can kill them.

Abiotic factors in the soil

Moisture level and **mineral content** in the soil affects plant growth. Moisture is required for decay to occur in the soil. Decay releases mineral ions to the soil. Different plant species are adapted to different **soil pH**.

Abiotic factors in the air

Humidity, **wind speed** and **wind direction** affect the rate of transpiration in plants. This can limit the distribution of plant species.

---

1. A rose bush has poor growth and few flowers. Give **two** abiotic factors that could be responsible. [2]
2. Explain why the carbon dioxide level in the atmosphere varies seasonally. [4]
3. Suggest how a windy habitat with dry soil might prevent a plant species from surviving. [2]

1. *Any two of temperature[1], soil pH[1], light intensity[1], mineral content of the soil[1], soil moisture level[1].*

2. *Higher temperatures in the spring and summer mean faster photosynthesis[1] so carbon dioxide uptake by plants is higher[1] and the carbon dioxide level in the atmosphere falls. When the temperature falls in autumn and winter, photosynthesis slows[1] but respiration continues in all organisms[1], so the level of carbon dioxide in the atmosphere rises.*

3. *Unless adapted to the conditions, the plants would lose too much water by transpiration[1] but would be unable to take up more from the dry soil[1] so would wilt and die.*

BIOTIC FACTORS

Communities are affected by the activities of the living organisms within them. Biotic factors often determine the size of the populations in a community.

Factors

Some biotic factors which affect communities include:

- Availability of **food**
- New **predators**
- New **pathogens**
- One species **outcompeting** another, so their numbers are no longer sufficient to breed

The availability of food affects the size of the population eating it. If there are too many individuals for the amount of food available, there will be competition and some individuals may not have enough to survive. If food supply increases, the population size is able to increase.

New predators or pathogens can kill individuals and reduce the size of a population. Find out more about predator-prey relationships in **4.7.2.1 levels of organisation on page 109.**

If two populations compete for the same food, one species may outcompete the other.

The graph shows the number of deer in the populations of two species. The two species both ate the same plants.

When Species B was introduced to the community, it outcompeted Species A.

Eventually, with insufficient individuals in the population of Species A to breed successfully, they died out.

1. Some fish feed on shrimps. A disease kills many of the fish. Explain the likely impact on the shrimp population. [2]

2. Describe **two** sorts of new arrival into a community which could reduce its stability. [4]

3. Look at the graph below showing the population size of two deer species. Suggest how humans could have controlled Species B to enable both species to survive. [1]

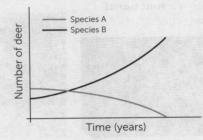

1. *Fewer shrimps are eaten[1] by fish so the shrimp population increases[1].*

2. *A new predator[1] that eats one or more of the species[1] in the community. A new pathogen[1] causing disease in a species[1] in the community.*

3. *Any one of: Introduce a pathogen which affects species A but not species B; give species A 'contraception' (limit reproduction); Feed species B; cull species A.*

ADAPTATIONS

Organisms show features that enable them to survive the conditions of their particular habitat in an ecosystem.

Examples

Octopuses

Some species of octopus can change colour to match their background on the seabed. This is because of changes in special pigment-containing cells in their skin. Colour changing camouflage is a **functional adaptation** to predation.

Mangrove trees

Mangrove trees grow in soft mud, which does not provide as much support as soil.

Some species have roots which act as props to help them remain upright. These features are **structural adaptations** to the soft mud they grow in.

Bats

Some species of bat hibernate in the winter when there is less insect food available. They often cluster together in a huddle so each bat loses less heat to the environment.

Hibernation and huddling are **behavioural adaptations** to colder winter conditions.

Extreme environments

Organisms adapted to extreme environments which are very salty, very hot or very cold, where it is especially difficult to survive, are called **extremophiles**. Bacteria living in deep sea vents are examples of extremophiles.

You do not have to learn these examples. You do have to be able to identify adaptations in pictures and information, and you should be able to spot the difference between structural, behavioural or functional adaptations.

1. This image shows a mole. Moles live underground in the soil community. Suggest **one** way in which the mole is adapted to the conditions where it lives. [2]

2. Brine shrimps can survive in very salty water. What sort of organism is a brine shrimp? Tick **one** box. [1]

   ☐ Thermophile   ☐ Aquatic   ☐ Extremophile   ☐ Halophile

3. Wood frogs that live in Alaska are able to survive the winter because they have very high glucose levels in their blood, preventing freezing. Name this type of adaptation. [1]

   1. *The mole has very large claws[1] on its front feet for digging[1] in the soil.*
   2. *Extremophile.[1]*
   3. *A functional adaptation.[1]*

LEVELS OF ORGANISATION

Feeding relationships within a community can be represented by food chains.

Producers

All food chains start with **producers**. Producers are organisms with the ability to make their own food molecules.

Most producers are plants or algae. They are **photosynthetic organisms** that use photosynthesis to make glucose molecules.

On land, plants are the main producers. In water, including the sea, algae are the main producers. Microscopic algae are the producers starting the food chains, which support large aquatic animals such as seals and whales.

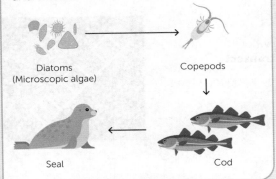

Diatoms
(Microscopic algae)

Copepods

Seal

Cod

Consumers

Organisms that get their food molecules by eating other organisms are called consumers. **Primary consumers** eat plants or algae. **Secondary consumers** and **tertiary consumers** eat animals.

Animals that kill and eat other animals are called **predators**. If they are the ones being eaten, they are called **prey**.

In a stable community, the population sizes of a predator and its prey often rise and fall in cycles.

As predators increase, they eat more prey so competition for food increases. Some predators do not get enough food so they die. There is then less predation, so the number of prey increases and in turn the number of predators can increase.

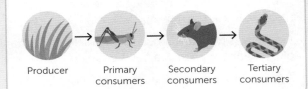

Producer | Primary consumers | Secondary consumers | Tertiary consumers

1. Name **two** kinds of producer that require sunlight to survive. [2]
2. The graph shows the population size of a predator and its prey over several years.
   (a) Which line represents the prey population? [1]
   (b) Explain why the population of prey rises and falls. [3]
   (c) Explain the impact of another predator that eats the same prey when it is introduced to the community. [2]
3. Explain why the animals in a food chain are called consumers. [2]

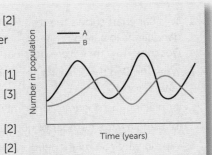

1. *Plants[1] and algae[1].*
2. *(a) A[1]*

   (b) *When there is less predation, the number of prey increases.[1] More food for predators[1] means that their number increases[1], so predation increases and the number of prey falls.*

   (c) *There will be a population decrease of prey.[1] One predator may outcompete the other.[1]*
3. *They get their food molecules[1] by eating (consuming) other organisms[1].*

SAMPLING METHODS

Ecologists can use quadrats and transects to find the size (**abundance**) and **distribution** of populations.

Finding the size of a population

It is not usually practical to count all the individuals in a population. Instead, a sampling method is used.

A **quadrat** is used to mark out a small area. The number of individuals in the quadrat is counted. The count is repeated for a large number of randomly placed quadrats. A mean count per quadrat is calculated. Random placement removes any bias caused by choosing the areas. It also increases the accuracy of the mean count.

The mean count is then multiplied by the number of times the quadrat would fit into the whole area. This gives the estimated population size.

Finding the distribution of a population

Finding the distribution of a population means finding where the individuals live in the ecosystem. This is done using a **transect**.

A transect is a line across the area where the population is found. A tape measure is used to mark the line.

A quadrat is placed at regular intervals along the transect and the number of individuals in each quadrat is counted.

A bar chart showing the number of individuals at each point along the line can then be drawn.

1. This bar chart shows the results of a transect survey of dandelions in a field.
   Describe how they were obtained. [3]

2. When finding the size of a population, what is the reason for:
   (a) placing the quadrats randomly? [1]
   (b) using a large number of quadrats? [1]

3. Explain why you might need a tape measure to find the size of a population. [1]

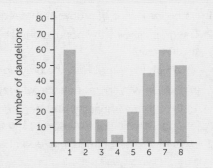

1. *Place a tape measure[1] across the field. Place quadrats at regular intervals[1] and count the dandelions[1] in each quadrat.*

2. *(a) To avoid bias.[1]*

   (b) Repeating the count in a large number of quadrats gives a more precise mean value.[1]

3. *To measure the size of the whole area[1] to find how many times to multiply the mean value[1].*

DEALING WITH SAMPLING DATA

There are particular ways to process and present data from sampling populations.

The final estimate of a population should be given as a whole number because it does not make sense to have a fraction of an organism.

Statistical analysis

After sampling with a quadrat to estimate the abundance of organisms in an area, the **mean** number of organisms per quadrat is calculated. This is called the arithmetic mean. The mean is calculated by adding together all the quadrat counts and dividing the total by the number of quadrats. It is a way of summarising the data from all of the quadrats.

It is important to give mean values a suitable number of **significant figures**. When organisms are counted in quadrats, the data values are all whole numbers. Mean values used for estimating the abundance of organisms should be given an extra significant figure. This is because the accuracy improves when it is multiplied to give an estimate of the population size.

The top table opposite shows the number of dandelion plants counted in each of five quadrats. If the mean value was rounded to 11, the estimate of the population size would be higher and might not reflect the whole range of values so well.

There are other ways of summarising abundance data. The **median** value is the middle value, found by putting all the values in order, as in the bottom table. The count for quadrat 1 (11) is the median for this set of data. If there was an even number of quadrats, the median would be the mean of the middle two values. The advantage of a median is that it is not affected so much by outlying values (like the value 3 in this set of data) and is quicker to find but it might not give such an accurate estimate of population size.

The **mode** is the most common value in a set of data. On the bar chart of transect data on the opposite page, the mode is 4 dandelions. A bar chart is the most appropriate type of graph for showing the abundance of an organism at different points along a transect (**see RP7 on page 112**).

Quadrat	Count
1	11
2	14
3	8
4	17
5	3
Mean	**10.6**

Quadrat	Count
5	3
3	8
1	11
2	14
4	17
Median	**11**

1. (a) Explain why a bar chart is the most appropriate way to represent the abundance of an organism at different points along a transect. [2]

   (b) Suggest a suitable axis label for the x-axis of the bar chart opposite. [1]

2. Suggest how the bar chart opposite should be modified if the quadrats had been placed at different distance intervals along the transect. [2]

   1. (a) *Each quadrat provides a separate/discrete value[1] and the sampling provides no information about dandelion abundance in between each quadrat[1].*

      (b) *Distance in metres.*

   2. *The x-axis should be changed to distance in metres and each bar should be placed at the correct distance, so they would no longer be spaced evenly.[1]*

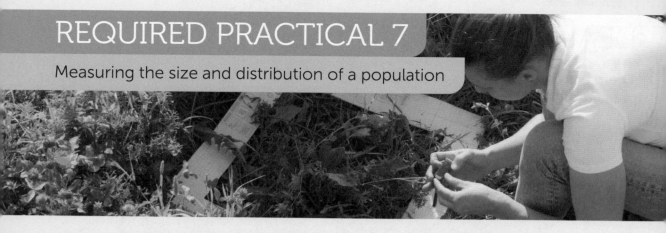

REQUIRED PRACTICAL 7

Measuring the size and distribution of a population

This required practical activity helps to develop your ability to use the right sampling techniques to investigate the distribution or abundance of organisms.

Positioning quadrats randomly

If you are using a quadrat to find the size of a population, it is important to use a method of randomly positioning the quadrat. Throwing a quadrat is not effective because it can introduce bias.

One method is use **coordinates**. A random number table can be used to pick pairs of numbers. Use a pencil to touch the paper with your eyes closed. Pace from your current position to the new position and place the quadrat.

If the chosen coordinates extend beyond the edge of the sampling area, simply pick another pair of numbers.

62686   44711
50033   14021
46176   42391
87989   72248

3 paces
3 paces
Place quadrat

Setting up a transect

Lay a tape measure across the area where you want to record the species distribution.

Decide how often to place a quadrat. This depends on the length of the transect and how patchy the species distribution is. You do not want to count too many quadrats but neither do you want to miss small patches of the species you are sampling.

Place the quadrat alongside the tape measure at regular intervals.

Count the number of each species in each quadrat.

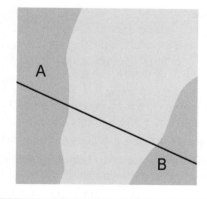

A

B

Investigating the effect of a factor on distribution

A transect can be used to investigate the effect of a factor on the distribution of a population.

The factor could be light intensity, soil moisture level or a biotic factor.

To do this, you need a way of measuring the factor. Some abiotic factors have to be measured with apparatus such as a light meter.

To carry out the investigation you need to:

- Make sure the transect runs across areas where the factor varies.
- Measure the factor at each quadrat position as well as counting the organisms.
- Present the data. For the data in the table, you could plot a double bar chart showing the number of daisies and the light intensity at each quadrat or a scatter diagram.
- Compare the numbers of daisies at different light intensities and look for any **correlation**.

Quadrat	Number of daisies	Light intensity
1	7	2
2	10	3
3	23	5
4	45	10
5	40	10

1. When calculating an estimate of a population size, which would you use? Tick **one** box. [1]

   ☐ Modal number per quadrat.   ☐ Median number per quadrat.   ☐ Mean number per quadrat.

2. Suggest an explanation for the transect results shown in the table above. [3]

3. How could a student with results for a transect ensure their conclusion was more valid? [1]

   1. *Mean.[1]*
   2. *There are fewer daisies[1] growing in quadrats with more shade[1] compared to the areas with greater light intensity[1].*
   3. *Carry out another transect to check the results[1] or compare their results to those from another student's transect.[1] Consider other factors that might have changed along the transect.[1]*

HOW MATERIALS ARE CYCLED

Materials continuously cycle between the biotic and abiotic parts of an ecosystem. All materials in the living world are recycled for future organisms.

The carbon cycle

Photosynthesis uses carbon in the form of carbon dioxide from the atmosphere and builds it into food molecules.

Carbon in the form of food molecules is transferred from plants to animals by feeding.

Plant and animal respiration returns carbon in the form of carbon dioxide to the atmosphere.

Microorganisms are involved in the **decay** of dead organisms and their waste materials which returns carbon in the form of carbon dioxide to the atmosphere.

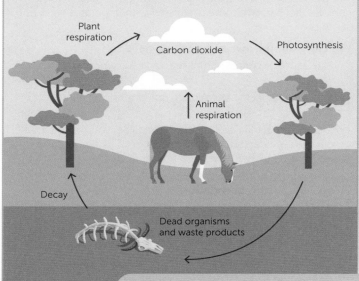

Decay by microorganisms also releases mineral ions back into the soil.

1. Draw a line to connect each process to its role in the carbon cycle. [3]

Decay

Photosynthesis

Respiration

transfers carbon from plants to animals

transfers carbon from the atmosphere to living organisms

transfers carbon to the atmosphere from living organisms

transfers carbon from dead materials to the atmosphere

2. (a) Explain why the carbon cycle depends on microorganisms. [2]

(b) Give **two** ways that plants benefit from the activity of microorganisms in the carbon cycle. [2]

1. *Decay → transfers carbon from dead materials to the atmosphere.[1] Photosynthesis → transfers carbon from the atmosphere to living organisms.[1] Respiration → transfers carbon to the atmosphere from living organisms.[1]*

2. (a) *Microorganisms cause decay.[1] Decay recycles the carbon in dead materials[1] to the atmosphere.*

   (b) *Microorganisms release carbon dioxide[1] and mineral ions[1] for plants to re-use.*

The water cycle

Water is vital for the survival of living organisms on land. **Condensation** and **precipitation** provide a supply of fresh water for plants and animals on land.

Water is absorbed from the soil by plant roots and by animals drinking from surface water.

Water drains (or percolates) from the land into rivers and returns to the sea.

Transpiration by plants and evaporation returns water as water vapour to the atmosphere.

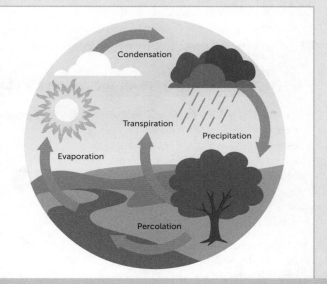

Along with carbon and water, other materials in the living world are also recycled to provide building blocks for the growth of future organisms.

3. The diagram shows the water cycle. Draw a line to connect each letter in the diagram to a process in the water cycle. [3]

A   evaporation

B   transpiration

            condensation

C   precipitation

4. Describe **two** ways that water enters the biotic parts of an ecosystem. [2]

5. Describe **two** ways that the sun affects the water cycle. [2]

> 3. A → transpiration.[1] B → evaporation.[1] C → precipitation.[1]
>
> 4. Water is taken up by plant roots.[1] Animals drink water.[1]
>
> 5. Any two from: Heat from the sun increases the rate of transpiration.[1] Heat from the sun increases the rate of evaporation.[1] Light causes plants to open stomata so transpiration can occur.[1]

BIODIVERSITY

The variety of species in an ecosystem, or on earth, is called biodiversity.

High biodiversity gives ecosystems stability

Having many species in a community means there are more complex feeding relationships. This reduces the dependence of one species on another for food. If one prey species becomes extinct or becomes scarce, predators have others they can rely on for their own survival.

Having many species in a community also means a wider range of habitats where animals can live and shelter. In a forest with a variety of trees and bushes, there are more places to shelter than in open grassland.

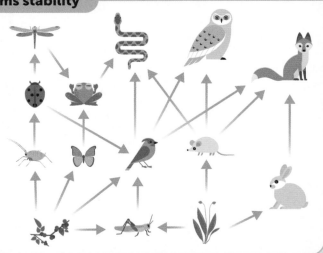

Human survival depends on high biodiversity

The future of the human species relies on maintaining sufficient biodiversity on earth. This is because of our dependence on a wide range of species for food, for recycling of materials and for maintaining our physical environment such as the composition of the atmosphere.

1. There are a very large number of species on earth. What is the variety of species on earth called? Tick **one** box. [1]

   ☐ Genetic variation   ☐ Speciation

   ☐ Biodiversity   ☐ Interdependence

2. Explain how a high biodiversity contributes to ecosystem stability. [3]

3. Suggest **one** reason why human survival depends on maintaining a high biodiversity in the soil. [4]

   1. *Biodiversity.[1]*

   2. *It reduces the number of species depending on just one other species[1] for food or shelter[1] so there are less likely to be large knock on effects[1] if one species is lost from the ecosystem.*

   3. *Humans depend on having many species of microorganisms[1] in the soil to recycle mineral ions[1] for plants to re-use during growth[1] and humans depend on sufficient plant growth for an adequate supply of food[1].*

WASTE MANAGEMENT

The human population is growing rapidly which requires more resources and produces more polluting waste.

You can revise atmospheric pollutants in the Chemistry specification.

Pollution

Pollution kills organisms, resulting in reduced biodiversity. The poor management of waste and chemical materials causes pollution in different ways:

- **In water**, including sewage, fertiliser or toxic chemicals.
- **In air**, including smoke and acidic gases.
- **On land**, including landfill and toxic chemicals.

Fertiliser

Fertilisers become harmful when they are washed from land into rivers when it rains. Fertilisers encourage the growth of aquatic algae and plants. They grow so much that they outcompete each other and some die.

Bacteria then decay the dead plants and use up the dissolved oxygen in the water as they respire. The reduced oxygen level can kill aquatic animals such as fish.

Acidic gases

Acidic gases become harmful when they dissolve in rain and in the sea. Carbon dioxide levels in the atmosphere have increased due to human activities, including the burning and decay of biological materials such as peat. This has caused more carbon dioxide to dissolve in the sea, reducing the pH of the water. Increased acidity in the water dissolves the shells and skeletons of marine animals such as crabs and coral, and can kill them.

Landfill

Solid waste is dumped in landfill sites. If these are poorly managed, toxic chemicals from the waste can leak into the surrounding soil causing contamination that can kill plants and animals.

1. Give **two** examples of pollution that kill organisms by changing the abiotic conditions in aquatic ecosystems. [2]
2. Explain why untreated sewage and fertilisers both encourage the reproduction of bacteria if they enter rivers. [4]
3. Suggest how plants and animals could be protected from landfill site pollution. [3]

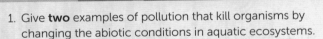

1. *Any two from: Sewage[1], fertilisers[1] and acidic gases.[1]*
2. *Both lead to a large increase of waste biological material in the water.[1] Fertiliser leads increased growth and then the death of aquatic plants[1] which also forms waste biological material. Bacteria decay[1] waste biological materials.*
3. *Prevent toxic materials leaking[1] from waste and contaminating soil[1] by lining the waste site with an impermeable material[1].*

LAND USE

Human activities reduce the amount of land for other animals and plants.

The demand for land for building, dumping waste, quarrying and farming is increasing as the human population increases. Natural habitats are lost and the animals and plants dependent on them move away or die.

Loss of peatlands

Peat bogs are rare habitats. Peat is the remains of plants that have not decayed because conditions are anaerobic.

Many of the plant species that live there require the wet and acidic conditions of peat bogs. Rare wetland insects depend on these plant species for food.

Peatlands are dug up to extract peat. Peat-based compost is a cheap natural fertiliser for plants, but this demand conflicts with the need to conserve biodiversity.

Digging up the peat destroys the habitat, including the plants that live there. If insects are unable to migrate, they die. This reduces biodiversity.

When the vegetation is stripped and the peat is dug up, the remaining exposed peat decays. It also burns easily so can set on fire. The decay and burning of peat releases huge amounts of stored carbon dioxide back into the atmosphere.

1. Explain why humans are destroying the rare habitats in peat bogs. [2]
2. Suggest why peat extraction causes increased carbon dioxide emissions. [4]
3. Lowland heath is a rare habitat supporting rare species such as sand lizards. The main threat to lowland heath is building. The table shows data for the area of lowland heath in Dorset.

Year	Total area of heath (ha)	Number of areas of heath
1765	40 000	10
1811	30 000	21
1896	23 000	39
1934	18 000	70
1960	10 000	104
1973	6100	120

(a) Calculate the percentage of lowland heath lost between 1765 and 1973. [1]
(b) What does the data indicate about the areas of lowland heath left in 1973 compared to 1765? [2]
(c) Sand lizards are unable to migrate far. Suggest how the pattern of heath loss affected the sand lizard population in Dorset. [3]

1. *To extract peat[1] for making compost[1] to improve plant productivity.*
2. *Peat is exposed to oxygen in air[1] so microorganisms are then able to decay[1] the plant remains. Decay releases carbon dioxide.[1] Exposed peat may also catch fire and release carbon dioxide.[1]*
3. *(a) 85%[1]*
   (b) Many much smaller areas in 1973[1] compared to fewer large areas in 1765.
   (c) Sand lizards in a small area of heath cleared for building are less likely to be able to move to suitable habitat[1] and more likely to go locally extinct[1] reducing[1] the size of the population[1].

DEFORESTATION

Large-scale deforestation in tropical areas has reduced biodiversity. Tropical forests contain a large variety of tree species providing food and habitats for animal and plant species that cannot live anywhere else.

Reasons for deforestation

Some tropical forests have been cleared and replaced by crops that are used to make **biofuels**. Biofuels are seen as more sustainable than fossil fuels.

Tropical forest has also been cleared to provide space for the rearing of cattle.

This provides a supply of cheap meat for the world's growing population.

Replacing trees with cattle grazing increases soil erosion. In intensive farming, cattle are often crowded together in pens and fed processed food to reduce their movement. Their waste is poorly managed and pollutes local rivers.

Forests have also been cleared to grow rice. Large areas of one species of plant provide a restricted range of food and habitats for animals compared to the large number of tree species they replace.

Rice is a wetland plant species and is grown in fields flooded with water. The mud in these fields becomes anaerobic. When the mud is disturbed, methane is released which adds to greenhouse emissions.

1. Give **two** reasons for large-scale deforestation in tropical areas. [2]

2. The picture shows palm trees planted on land cleared of tropical forest. Biodiesel uses oil harvested from palm trees.

   (a) Suggest why the palm forest supports fewer animal and plant species compared to the tropical forest it replaced. [2]

   (b) Evaluate the production of biofuel on land cleared of tropical forest. [4]

3. Other than biodiversity loss, describe the environmental impacts of deforestation on the soil and the atmosphere. [3]

   1. *Any two from: Growing rice[1], growing crops for biofuel[1], rearing cattle for meat[1].*

   2. *(a) There is only one tree species[1] so the number of habitats / food for other species is limited.[1]*

      (b) Although using biofuel means less fossil fuel is burned[1] the cleared forest is burnt adding to carbon dioxide emissions[1] and the loss of habitat[1] leads to loss of biodiversity which is not sustainable.[1]

   3. *Loss of trees reduces uptake of carbon dioxide from the atmosphere by photosynthesis.[1] Loss of trees reduces uptake of water from soil; so reduces the capture of heavy rainfall altering the water cycle;[1] so water drains from the land more quickly washing mineral ions required by plants out of the soil[1] into rivers[1].*

GLOBAL WARMING

Global warming is having an impact on living organisms.

Impact of human activity

Human activities are contributing to increased levels of **carbon dioxide** and **methane** in the atmosphere.

- Carbon dioxide is released by the burning of fuels and biological materials including wood and peat. It also results from the decay of peat in damaged peatlands.

- Owing to deforestation, less carbon dioxide is being absorbed by plants in photosynthesis.

- Methane is released from cattle and from flooded fields used to grow rice.

Carbon dioxide and methane are greenhouse gases. They trap heat from the sun in the earth's atmosphere, increasing the mean air temperature.

There is now general agreement amongst scientists, based on thousands of **peer reviewed** publications, that this natural warming process is being accelerated by human activities.

Biological consequences of global warming

Global warming is causing **climate change**, **sea level rise** and **desertification**. These processes are having a variety of effects on living organisms.

- **Loss of habitat**: A rise in sea level is flooding coastal marshes.

- **Changes in distribution**: Some animals are now able to live in different places because of higher mean temperatures but some pathogens may also spread to new areas.

- **Extinctions**: Animals and plants unable to migrate are becoming extinct in some areas as conditions change.

Some plants are growing faster. Higher carbon dioxide concentration and higher temperature means that photosynthesis happens faster. This may result in higher crop yields in some areas.

---

1. Explain how deforestation contributes to global warming. [3]
2. State how increased methane in the atmosphere causes an increase in air temperature. [1]
3. Give **two** possible effects of increased mean air temperature on living organisms. [2]

1. *Cleared forest is burned releasing carbon dioxide.[1] Dead biological material decays releasing more carbon dioxide[1] and there are fewer trees to absorb carbon dioxide[1] for photosynthesis.*

2. *It traps more heat[1] from the sun in the atmosphere.*

3. *Any two from: Higher crop yield[1], change in distribution[1], extinction[1], loss of habitat[1].*

MAINTAINING BIODIVERSITY

Humans can have positive as well as negative effects on ecosystems.

Protection and regeneration of rare habitats

Creating protected areas such as national parks and nature reserves helps to stop habitat destruction. Unfortunately, the growing human population requires food, fuelwood and access to land for farming and grazing. This means it is controversial for some countries to set aside large, protected areas. Even when they do, there is often illegal grazing or poaching.

Recycling to reduce landfill

Recycling waste materials reduces the volume of waste taken to landfill. This reduces the amount of land that needs to be used for new landfill sites, so it reduces local habitat loss. It also reduces the need to obtain replacement resources by activities such as mining.

Breeding programmes for endangered species

Where species are at risk of extinction, breeding of captive animals and cultivating wild plants can increase their numbers for potential release into the wild.

Reintroduction of field margins and hedgerows

In large areas of a single crop there is less food and fewer habitats for wild plants and animals. A variety of plant species provides more food and a wider range of habitats to support insects, birds and bats not supported by the crop.

Reduction of deforestation and CO_2 emissions

Some governments are actively trying to reduce the rate at which their tropical forests are cleared. Many countries have agreed to try to reduce their carbon dioxide emissions.

Farmers plant strips of wildflowers and hedges around the edges of their crops.
The graph shows the results of sampling insects in a field margin over a period of time.

(a) Explain how field margins increase the biodiversity of insects in agricultural areas. [2]

The graph indicates the abundance of two types of insect.

(b) Explain which type of insect is most beneficial to a farmer growing a crop alongside the field margin. [2]

(c) Describe what the graph indicates about how the farmer should manage the field margin to benefit both the field margin ecosystem and the crop. [3]

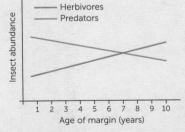

(a) *The variety of plant species[1] in the hedgerows provides more food and habitats for animals[1].*

(b) *Insect predators[1] benefit the farmer more because they eat other insects that feed on the crop[1].*

(c) *The farmer should renew the wildflowers in the field margin every 6 years[1] because this gives the best balance between a stable insect community in the field margin to support biodiversity[1] and the natural control be predators of insects[1] that eat the crop[1].*

EXAMINATION PRACTICE

01   The diagram show an aquatic food chain.

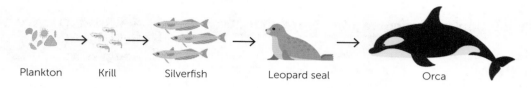

Plankton   Krill   Silverfish   Leopard seal   Orca

01.1 Draw **one** line from each term to the correct organism in the food chain.   [3]

Tertiary consumer   Silverfish

Primary consumer   Krill

Apex predator   Leopard seal

                                              Orca

01.2 Complete the sentences using words from the following list.   [3]

Ingested   **Digested**   **Transpired**   **Egested**   **Respired**   **Excreted**

When silverfish eat krill, some of this food is not absorbed so it is _____.
Materials absorbed from the gut may enter the body cells of the silverfish. Some of these
materials are _____ which breaks them down into carbon dioxide and water.
The carbon dioxide and other waste materials from the body cells are _____.

02   Some students investigated the distribution of plants on some sand dunes. The students placed
quadrats at measured distances along a line and counted the number of three plant species in
each quadrat.

02.1 Which sampling technique was used by the students? Tick **one** box.   [1]

☐ Regular quadrats.   ☐ Transect.   ☐ Random quadrats.   ☐ Survey line.

The student plotted their results on a bar chart.

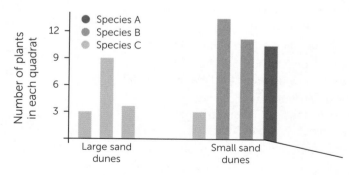

02.2 Compare the distribution of Species B with that of Species C.   [2]

02.3 Suggest which species is most at risk of losing its habitat due to sea level rise caused by
global warming.   [2]

03 Explain the role of microorganisms in cycling carbon through an ecosystem. [6]

04 Explain why using peat to make compost is being banned in some areas. [4]

05 A student investigated the rate of decay of dead leaves in a wood. The student weighed samples of dead leaves and placed them into mesh bags which were buried and left for a year. One bag was buried in dead leaves beneath trees and another was buried in dead leaves in an open clearing. The student made the following hypothesis:

The dead leaves beneath trees will decay slower because they are in the shade.

05.1 Give **one** variable the student should control in the investigation. [1]

05.2 The dead leaves beneath the trees had an initial mass of 21.5 g and after a year had a mass of 7.4g. There are 365 days in a year. Calculate the rate of decay in g per day. Give your answer in standard form. [2]

05.3 Suggest why the student made the hypothesis about the rate of decay. [2]

06 The table shows the population sizes of deer and their predators, wolves, in a community over a number of years.

Year	Wolf population	Deer population	Deer offspring
2003	10	2040	810
2004	12	2280	930
2005	16	2550	1000
2006	22	2360	912
2007	28	2090	765
2008	24	1920	730
2009	18	1951	805
2010	12	1968	840

06.1 Calculate the percentage decrease in the population size of the deer between 2006 and 2008. [2]

06.2 Explain the relationship between the population size of the deer and the population size of the wolves. [6]

07 Explain how pollution by acidic gases can kill animals. [3]

08 Earthworms live in the soil. A student designed an investigation to find the number of earthworms in a lawn. The student dug out the soil to a depth of 10 cm from ten 20 cm × 20 cm quadrats. They sieved the soil and counted all the earthworms found in each quadrat. The mean number of earthworms per quadrat was 7. The lawn was 10 m × 10 m.

08.1 Suggest how the student should have decided where to place each quadrat. [2]

08.2 Calculate an estimate of the number of earthworms in the lawn. [3]

08.3 How could the student have made the estimate more valid? Tick **one** box. [1]
☐ Dug out the quadrats more deeply. ☐ Surveyed another lawn.
☐ Used more quadrats. ☐ Used bigger quadrats.

08.4   Earthworm distribution is affected by soil temperature. Suggest **two** other abiotic factors
that might affect earthworm distribution.   [2]

09   A student measured the oxygen level in a stream every 50 metres for 500 metres and plotted a
graph.

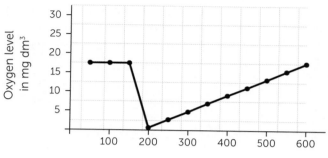

Distance along stream in m

09.1   Give **two** ways that the student could have improved their graph.   [2]

09.2   Give the lowest oxygen level that the student measured.   [2]

09.3   Use data from the graph to suggest **one** possible reason for the pattern of oxygen levels
shown on the graph.   [4]

10.   Discuss how governments can encourage the sustainable production of protein-rich,
plant-based foods.   [4]

CHEMISTRY
TOPICS FOR PAPER 1

Information about Paper 1:

Trilogy 8464:

Written exam: 1 hour 15 minutes
Foundation and Higher Tier
70 marks
16.7% of the qualification grade
All questions are mandatory

Specification coverage

The content for this assessment will be drawn from topics on: Atomic structure and the periodic table; Bonding, structure, and the properties of matter; Quantitative chemistry; Chemical changes; and Energy changes.

Questions

Multiple-choice, structured, closed short answer and open response questions. They may include calculations.

Questions assess skills, knowledge and understanding of Chemistry.

ATOMS, ELEMENTS AND COMPOUNDS

All substances are made from **atoms**. An atom is the smallest part of an **element**.

Elements and compounds

About 100 different elements are known. Each element is represented by a chemical symbol. These consist of a capital letter, which may be followed by one other letter in lower case. For example, C represents carbon and Cr represents chromium. The full **periodic table** (on **page 384**) shows all the known elements but no compounds.

A **compound** is a substance that consists of two or more different elements. The elements in a compound are chemically combined in fixed proportions as a result of a chemical reaction. For example, carbon dioxide always contains one carbon atom for every two oxygen atoms. A **chemical formula** shows the symbol and number of each element in a unit of a compound. The chemical formula of carbon dioxide is always CO_2.

Oxygen is an element, but carbon dioxide and water are compounds

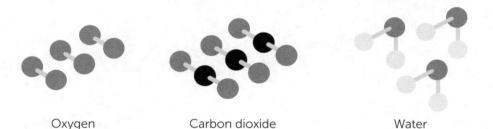

Oxygen   Carbon dioxide   Water

Chemical reactions

Chemical reactions always produce one or more new substances. They involve energy changes that can often be detected. A chemical reaction is the only way to produce a compound from elements, and to separate a compound back into the elements it contains. You can show what happens in a chemical reaction using a **word equation**. This shows the names of the **reactants** (the substances that react together) and the **products** (the substances made in the reaction).

1. Here is the word equation for a chemical reaction.

    silver nitrate + sodium chloride → silver chloride + sodium nitrate

    Describe what this equation shows about the reaction.   [2]

2. The chemical formula for water is H_2O. Explain what this shows   [4]

    1. Silver nitrate and sodium chloride react[1] to produce silver chloride and sodium nitrate[1].

    2. Water contains hydrogen and oxygen[1] atoms[1] chemically bonded[1] in the ratio 2 : 1[1]

CHEMICAL EQUATIONS

Chemical reactions can be represented using **balanced equations**. These use chemical symbols and formulae rather than words.

Formulae with brackets

The formula for aluminium hydroxide, $Al(OH)_3$, represents 1 aluminium, $(3 \times 1) = 3$ oxygen and $(3 \times 1) = 3$ hydrogen.

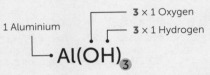

1 Aluminium
3×1 Oxygen
3×1 Hydrogen
$Al(OH)_3$

> Do not change chemical formulae in order to balance an equation. For example, H_2O is water but H_2O_2 is hydrogen peroxide.

Ammonium sulfate is commonly used as a soil fertiliser. Its formula $(NH_4)_2SO_4$ represents:

$(2 \times 1) = 2$ nitrogen $(2 \times 4) = 8$ hydrogen   1 sulfur   4 oxygen

Balancing equations

A chemical equation is balanced when the numbers of atoms of each element in the reactants and products are the same. It is important to write the correct symbol or formula for each substance, and to use balancing numbers where necessary.

Methane burns in oxygen to produce carbon dioxide and water:

$$CH_4 + 2O_2 \rightarrow CO_2 + 2H_2O$$

The balancing number 2 in front of the formula for water means that there are:

$(2 \times 2) = 4$ hydrogen atoms and $(2 \times 1) = 2$ oxygen atoms

Check: On each side of the arrow: 1 carbon atom, 4 hydrogen atoms, and $(2 + 2) = 4$ oxygen atoms.

1. Explain why the following chemical equation is balanced: [3]
$$Ca + 2H_2O \rightarrow Ca(OH)_2 + H_2$$

2. Sodium reacts with oxygen. Complete the equation for this reaction.   [2]
$$\_\_\_Na + \_\_\_O_2 \rightarrow \_\_\_Na_2O$$

3. Potassium reacts with water to produce hydrogen and potassium hydroxide:
$$2K(s) + 2H_2O(l) \rightarrow H_2(g) + 2KOH(aq)$$
Describe what the state symbols show about this reaction.   [4]

State symbols

State symbols show the physical state of a substance in a reaction, or whether it is in aqueous solution (dissolved in water).

1. The 2 in $Ca(OH)_2$ shows that there are 2 oxygen atoms[1] and 2 hydrogen atoms in the formula[1]. On both sides: 1 calcium atom, 4 hydrogen atoms, and 2 oxygen atoms[1].

2. **4**Na + O_2 [1] → **2**Na_2O[1]

3. Potassium is in the solid state[1], water is in the liquid state[1], hydrogen is in the gas state[1] and potassium hydroxide is in aqueous solution.[1]

HALF EQUATIONS AND IONIC EQUATIONS

Ions are charged particles formed when atoms, or groups of atoms, lose or gain **electrons**. Half equations and ionic equations are used in reactions involving ions.

Half equations

A **half equation** can show how a substance loses electrons to form ions. An electron is shown as e^- in half equations. The superscript negative sign shows that it carries a single negative charge.

It is important to balance the charges in equations so that there are equal numbers of positive charges and negative charges, as well as equal numbers of atoms and ions.

Sodium atoms lose one electron when sodium reacts with non-metals:

$$Na \rightarrow Na^+ + e^-$$

Calcium atoms lose two electrons when calcium reacts with non-metals:

$$Ca \rightarrow Ca^{2+} + 2e^-$$

Half equations can also show how a substance gains electrons to form ions. For example, chlorine atoms gain electrons when chlorine reacts with metals:

$$Cl_2 + 2e^- \rightarrow 2Cl^-$$

You can also use half equations to show what happens at each electrode during **electrolysis**. During the electrolysis of molten calcium chloride, calcium ions gain electrons to form calcium, and chloride ions lose electrons to form chlorine. The half equations for these reactions are:

$$Ca^{2+} + 2e^- \rightarrow Ca \qquad 2Cl^- \rightarrow Cl_2 + 2e^-$$

---

1. Aluminium reacts with oxygen to produce aluminium oxide. Complete each half equation.
   (a) $Al \rightarrow Al^{3+} +$ _____ [1]
   (b) $O_2 +$ _____ $\rightarrow$ _____ O^{2-} [1]
2. Oxygen forms at the positive electrode during the electrolysis of acidified water.
   Complete the ionic equation for this reaction: _____ $OH^- \rightarrow O_2 +$ _____ $H_2O +$ _____ e^- [2]

   1. (a) $Al \rightarrow Al^{3+} + $**3e**$^-$ [1]   (b) $O_2 + $**4e**$^- \rightarrow $**2**$O^{2-}$ [1]
   2. **4**$OH^- \rightarrow O_2 + $**2**$H_2O + $**4**$e^-$ [2]

Ionic equations

An ionic equation shows how two ions combine to produce a substance. For example, copper(II) ions react with chloride ions to produce copper(II) chloride:

$$Cu^{2+} + 2Cl^- \rightarrow CuCl_2$$

Just like half equations, it is important to balance the charges.

3. Complete this ionic equation: $Pb^{2+}(aq) +$ _____ $Br^-(aq) \rightarrow PbBr_2(s)$ [2]

   3. $Pb^{2+}(aq) + $**2**$Br^-(aq) \rightarrow PbBr_2(s)$ [1]

Eight elements exist as diatomic molecules (containing two atoms):
H_2, N_2, O_2, F_2, Cl_2, Br_2, I_2, At_2.

SEPARATING MIXTURES

The substances in a mixture can be separated using processes such as **filtration**.

Mixtures

A **mixture** contains two or more elements or compounds. Unlike compounds, the individual substances in mixtures are not chemically combined. This means that:
- the chemical properties of each substance stay the same
- the substances can be separated from one another by physical processes.

No chemical reactions happen when a mixture is separated, so no new substances form.

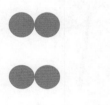
Pure element

Pure compound

Mixture of elements

Mixture of element and compound

Mixture of compounds

Filtration

A **soluble** substance will dissolve in a **solvent** such as water to form a **solution**. An **insoluble** substance will not dissolve. Filtration is a separation method that can separate an insoluble substance from a liquid, gas or solution.

Using filtration to separate sand from a mixture of sand and water

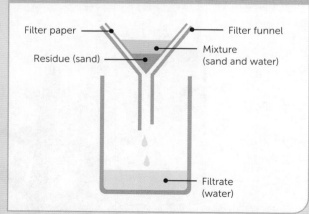

Filter paper — Filter funnel
Residue (sand) — Mixture (sand and water)
Filtrate (water)

Explain why sand can be filtered from a mixture of sand and salt solution. [4]

Sand does not dissolve in water[1]. Its particles are too large to pass through the microscopic holes in the filter paper[1]. The dissolved salt particles are small enough to pass through[1] together with the water particles.[1]

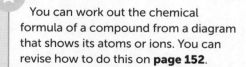

You can work out the chemical formula of a compound from a diagram that shows its atoms or ions. You can revise how to do this on **page 152**.

CRYSTALLISATION

Crystallisation separates a solid solute from a solution.

Solubility

When no more solute will dissolve in a given volume of solvent, a solution is described as being a **saturated** solution. **Solubility** is the mass of solute in a saturated solution at a given temperature.

The solubility of most substances in the solid state increases as the temperature increases. For example, the solubility of copper(II) sulfate in 100 cm³ of water is 32 g at 20 °C, but 84 g at 80 °C. If you cool 100 cm³ of a saturated solution from 80 °C to 20 °C, (84 − 32) = 52 g of copper(II) sulfate cannot stay dissolved and will return to the solid state. This is how crystallisation works.

Crystallising copper(II) sulfate over a boiling water bath

Evaporating basin — Copper(II) sulfate solution — Beaker — Boiling water — Gauze mat — Tripod — **Heat** — Heat-resistant mat

Crystallisation method

(a) Put the solution into an evaporating basin and heat over a boiling water bath. The solution becomes more concentrated as water in the solution evaporates.

(b) Stop heating before all the water has evaporated. Crystals form as the solution cools down.

(c) Leave the evaporating basin with its contents aside for a few days, such as on a windowsill. Remove the crystals and gently pat them dry with a paper towel or filter paper.

1. Explain why a boiling water bath is used instead of heating directly with a Bunsen burner. [3]

2. Suggest **one** way to dry the crystals other than leaving the evaporating basin for a few days. [1]

   1. *A boiling water bath heats the solution more gently.[1] This reduces the chance of hot solids or liquids jumping out of the evaporating basin[1], which would be unsafe[1].*

   2. *Place the evaporating basin in a warm oven.[1]*

CHROMATOGRAPHY

Chromatography separates a mixture of coloured solutes in a solution.

Paper chromatography

Chromatography relies on two 'phases':
- a **stationary phase** that does not move - usually a porous solid.
- a **mobile phase** that moves through the stationary phase.

In paper chromatography, the stationary phase is contained in the paper, and the mobile phase is a solvent such as water or propanone.

Different solutes in a solution form chemical bonds with both phases. The relative strengths of these bonds determine how far a solute travels up the paper with the solvent. The more strongly a solute bonds to the mobile phase, the further it travels up the paper.

Paper chromatography of a sample of ink

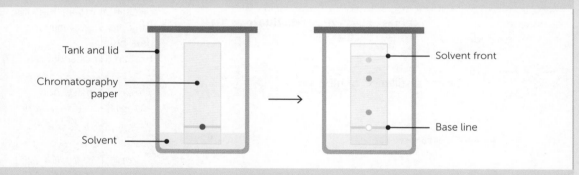

Method

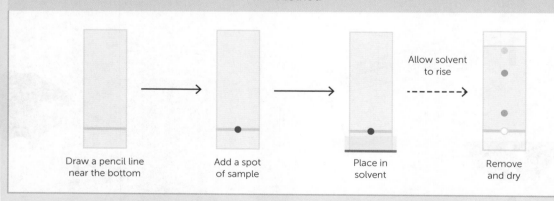

1. Explain why the sample must be higher than the solvent in paper chromatography. [2]
2. Give **two** ways to determine whether two coloured spots are the same substance. [2]

   1. To stop it dissolving in the solvent[1] so it does not leave the chromatography paper[1].

   2. They will be same colour[1] and travel the same distance on the same chromatogram[1].

SIMPLE DISTILLATION

Simple distillation separates the solvent from a solution.

Boiling points

Distillation relies on the different **boiling points** of the components in a solution. In a solution made by dissolving a solid in a liquid, the **solvent** boils at a lower temperature than the **solute**. This means that when the solution is heated enough:

- solvent evaporates and escapes the solution
- the solute is left behind, and the solution gradually becomes more concentrated.

State changes

In simple distillation, solvent evaporates and travels as a vapour into a **condenser**. This is a glass tube with cold water surrounding it. The vapour cools and condenses inside the condenser. Pure liquid solvent is collected as a **distillate** as it leaves the condenser.

1. Give **one** practical use of simple distillation. [1]

2. During the simple distillation of pale blue ink, the colour gradually turns dark blue.
   Explain this observation. [2]

   1. *Drinking water can be made by the simple distillation of sea water[1].*

   2. *Blue pigment stays in the ink as the solvent leaves[1] so the ink gets more concentrated[1].*

Simple distillation of blue ink

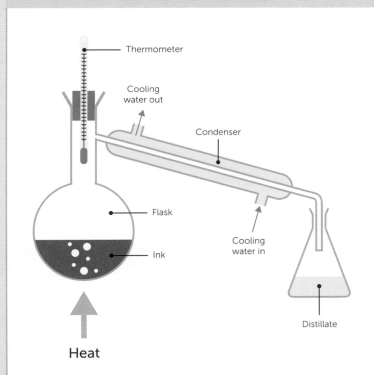

- Thermometer
- Cooling water out
- Condenser
- Flask
- Ink
- Cooling water in
- Distillate

Heat

FRACTIONAL DISTILLATION

Fractional distillation separates a liquid from a mixture of liquids.

How it works

Fractional distillation works in a similar way to simple distillation, but the apparatus includes a **fractionating column**. This is placed between the flask and the condenser.

During fractional distillation, a **temperature gradient** forms:

- The bottom of the fractionating column becomes hotter than the top.
- When the mixture of liquids is heated, the liquid with the lowest boiling point evaporates first. Its vapour travels up the fractionating column and into the condenser, where it cools and condenses.
- The liquid which leaves the condenser is called a **fraction** because it is only a part of the original mixture of liquids.
- With continued heating, liquids with higher boiling points may be collected, one after the other.

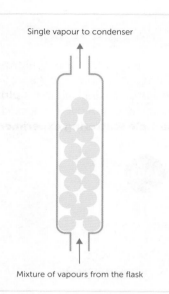

Single vapour to condenser

Mixture of vapours from the flask

Choosing separation and purification methods

Different separation methods are suitable for different mixtures. They can be used on their own, or in a combination of two or more different methods.

Method	Separates
Filtration	Insoluble solid from a liquid
Crystallisation	Solid solute from a solution
Simple distillation	Solvent from a solution
Fractional distillation	Liquid from a mixture of liquids
Chromatography	Different coloured solutes in a solution

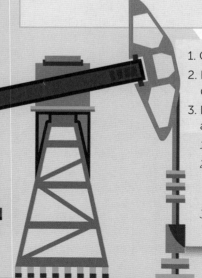

1. Give **two** mixtures that can be separated using fractional distillation. [2]
2. Describe the function of the fractionating column in fractional distillation. [3]
3. Describe how to separate sand and salt from a mixture of sand and salt solution. [3]

1. Crude oil[1] and the mixture formed by the fermentation of sugar[1].
2. The column provides a large surface area[1] for vapours to cool, condense and then evaporate again[1]. This improves the separation of the different liquids in a mixture[1].
3. Filter to separate the sand[1] then use crystallisation to produce salt[1] from the filtrate[1].

DEVELOPING THE ATOMIC MODEL

The **atomic model** has changed over time because of new experimental evidence.

Atomic theories

In the early 19th century, atoms were imagined as tiny, solid spheres. The discovery of the **electron** by J.J. Thomson in 1897 led to his **plum pudding model**. This model was disproved by a series of results from the **alpha particle scattering experiment**.

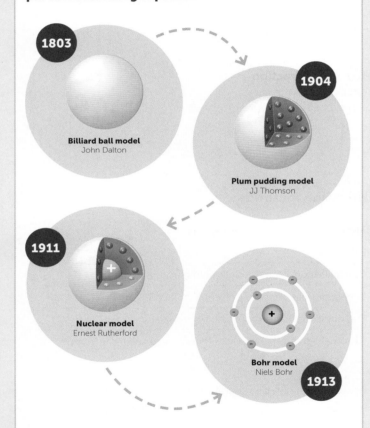

1803

Billiard ball model
John Dalton

1904

Plum pudding model
JJ Thomson

1911

Nuclear model
Ernest Rutherford

Bohr model
Niels Bohr

1913

In this experiment, beams of alpha particles (tiny positively charged particles) were aimed at thin gold foil. The results led to the **nuclear model**. Shortly afterwards, Niels Bohr carried out theoretical calculations showing that electrons orbit the nucleus at set distances. Observations from experiments supported his **electron shell model** and also showed the existence of positively charged **protons**. About 20 years later, James Chadwick demonstrated the existence of **neutrons**.

1. Compare the plum pudding and nuclear models of the atom [3]

2. The diagram shows paths taken by alpha particles through gold foil.

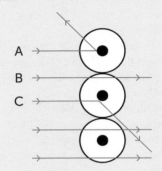

A →
B →
C →

(a) Give a reason why most particles followed the path labelled B. [1]

(b) Explain why some particles followed path C. [2]

(c) Explain why a very small number of particles followed path A. [2]

1. *Both models have negatively charged electrons[1]. These are embedded in a sphere of positive charge in the plum pudding model[1] but surround a positively charged nucleus in the nuclear model[1].*

2. *(a) Atoms are mostly empty space[1] so particles passed straight through.*

(b) The nucleus was positively charged[1] so it repelled the positively charged alpha particles[1].

(c) The nucleus was very small[1] but it had a relatively high mass[1] and high charge[1].

SUBATOMIC PARTICLES

Atoms are very small, but they are made from even smaller **subatomic particles**.

Protons, neutrons and electrons

The nuclei of all atoms contain protons, and almost all contain neutrons. Electrons are arranged around the nucleus. These three subatomic particles have different masses and electrical charges. Instead of giving the actual values, mass and charge are compared to the mass and charge of a proton. This gives relative values.

Name of particle	Relative mass	Relative charge
Proton	1	+1
Neutron	1	0
Electron	Very small	−1

Protons and neutrons have a lot more mass than electrons. This means that most of an atom's mass is in its nucleus. An atom is neutral overall because it contains the same number of protons and electrons.

Atomic number

An **atomic number** is the number of protons in an atom:

- the atoms of an element have the same atomic number
- the atoms of different elements have different atomic numbers.

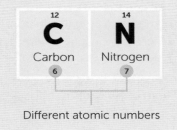

Different atomic numbers

Atomic radius

Atoms vary in size but all of them are very small. Their typical radius is about 0.1 nm. Remember that 1 nm is 10^{-9} m in standard form, so this is 1×10^{-10} m. The radius of a nucleus is about 1×10^{-14} m, around 10 000 times smaller than the atom itself.

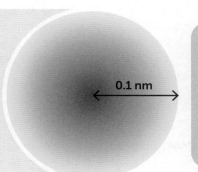

0.1 nm

The actual mass and charge of a proton are incredibly small, just 1.67×10^{-27} kg and $+1.60 \times 10^{-19}$ C. This is why relative values are used.

The mass of the electron is 1836 times less than the mass of proton.

1. A bacterium is 2×10^{-6} m long. Calculate how many times larger the bacterium is compared to an atom. [2]
2. Compare the charges of protons, neutrons and electrons. [2]

1. Diameter of atom = $2 \times 1 \times 10^{-10}$ m[1] Number of times larger = $\dfrac{2 \times 10^{-6} \text{ m}}{2 \times 10^{-6} \text{ m}}$ = 10,000 times[1].

2. Protons and electrons are charged but neutrons are not[1]. Protons and electrons have opposite charges[1].

ISOTOPES

Isotopes are atoms of the same element with different atomic mass numbers.

Representing atoms

The **mass number** of an atom is the total number of protons and neutrons in its nucleus. Atoms are represented using their mass number, atomic number and chemical symbol.

Mass number $\longrightarrow$ 39 K

Atomic number $\longrightarrow$ 19

The isotopes of an element have identical chemical properties because their atoms have the same number of electrons. You can revise the relationship between electrons and properties on **page 141**.

Numbers of particles in atoms

The numbers of neutrons, protons and electrons in an atom are calculated from its mass number and atomic number:

- number of neutrons = mass number − atomic number
- number of protons = atomic number

Remember that atoms have equal numbers of protons and electrons. This means that an atom of $^{39}_{19}K$ contains (39 − 19) = 20 neutrons, 19 protons and therefore 19 electrons.

Numbers of particles in ions

An **ion** is an atom or molecule with a net electrical charge greater or less than 0. It forms when an atom loses or gains electrons. Positively charged ions contain fewer electrons than protons, and negatively charged ions contain more electrons than protons. The number of charges on an ion, and whether it is positively charged or negatively charged, is shown using a superscript:

- K^+ is an ion carrying a single positive charge
- O^{2-} is an ion carrying two negative charges.

1. Explain why $^{23}_{11}X$ and $^{23}_{12}X$ are **not** isotopes of the same element. [2]
2. Explain, in terms of subatomic particles, why $^{12}_{6}C$ and $^{14}_{6}C$ are isotopes of an element. [3]
3. Give the numbers of protons, neutrons and electrons in the following ions. [6]
   (a) $^{39}_{19}K^+$ (b) $^{18}_{8}O^{2-}$

   1. *All atoms of an element have the same atomic number[1] but these have different atomic numbers[1].*
   2. *Both contain 6 protons[1] but different numbers of neutrons[1]. ^{12}C atoms have (12 − 6) = 6 neutrons and ^{14}C atoms have (14 − 6) = 8 neutrons[1].*
   3. *(a) 19 protons[1], (39 − 19) = 20 neutrons[1], (19 − 1) = 18 electrons[1]*
      (b) 8 protons[1], (18 − 8) = 10 neutrons[1], (8 + 2) = 10 electrons[1]

RELATIVE ATOMIC MASS

Periodic tables often show the **relative atomic mass** of each element.

A weighted average

Mass number and relative atomic mass are not the same thing, even though they may have the same value for some elements. Mass number refers to atoms and relative atomic mass refers to elements:

- mass numbers are always whole numbers
- relative atomic masses are rarely whole numbers, unless rounded up or down.

Relative atomic mass (symbol A_r) is an average value that takes into account the relative **abundances** of all the isotopes in a sample of an element.

Calculating relative atomic masses

In a typical sample of chlorine, 75% of the atoms are ^{35}Cl and 25% are ^{37}Cl:

$$\text{Relative atomic mass} = \frac{(75 \times 35) + (25 \times 37)}{(75 + 25)} = \frac{2625 + 925}{100} = \frac{3550}{100}$$

35.5

Cl

Chlorine

17

Relative atomic masses have no units. Where a periodic table shows the relative atomic mass of each element, take care not to confuse these values with mass numbers. The box for chlorine in the periodic table shows 35.5 rather than 35 or 37.

A_r values are given relative to the mass of a ^{12}C atom, which is taken as 12 exactly.

1. Give **one** reason why 35.5 cannot be the mass number of chlorine. Answer in terms of subatomic particles. [2]

2. The table shows the percentage abundance of three zinc isotopes. Calculate the relative atomic mass of zinc. Give your answer to **1 decimal place**. [3]

Isotope	Percentage abundance (%)
^{64}Zn	51
^{66}Zn	29
^{68}Zn	20

1. *If it was 35.5 it would mean that chlorine atoms could not have whole numbers of protons or neutrons[1].*

2. *An answer of 65.4 scores **3** marks.*

$$\text{relative atomic mass} = \frac{(51 \times 64) + (29 \times 66) + (20 \times 68)}{(51+29+20)} \text{[1]}$$

$$= \frac{3264 + 1914 + 1360}{100} = \frac{6538}{100} \text{[1]}$$

$$= 65.4 \text{ to 1 decimal place[1].}$$

ELECTRONIC STRUCTURE

The **electronic structure** of an atom shows how its electrons are arranged around its nucleus.

Energy levels

Electrons can occupy **energy levels** in atoms. The energy of each energy level increases as the distance from the nucleus increases. Each energy level can hold different numbers of electrons. The table shows the maximum numbers of electrons in each energy level of the first 20 elements in the periodic table.

Energy level	Maximum number of electrons
1	2
2	8
3	8
4	18

Energy levels may also be described as electron **shells**. Electrons in an atom occupy and fill the lowest available energy levels, or the innermost available shells, before any outer shells.

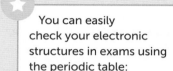

You can easily check your electronic structures in exams using the periodic table:

- total number of electrons = atomic number
- number of levels = period number
- number of electrons in outer level = group number (except for Group 0)

Deducing electronic structures

Electronic structures can be represented by numbers and by diagrams. The atomic number of calcium is 20, so its atoms have 20 electrons. The electronic structure of calcium therefore, is 2,8,8,2. This shows that:

- 2 electrons occupy the first energy level
- 8 electrons occupy the second and third energy levels, leaving
- 2 electrons to occupy the fourth energy level.

The diagram for this electronic structure has four concentric circles, one for each energy level. Each electron is shown as a cross. You can spread the crosses evenly around each circle, but they are easier to count if you show pairs of crosses (as here). This also helps you to show molecules.

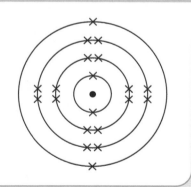

An electron diagram for calcium

Using the Period table, write the electronic structures of the following atoms. [2]

(a) Lithium	(b) Carbon	(c) Aluminium	(d) Sulfur	(e) Argon
(a) 2,1[1]	(b) 2,6[1]	(c) 2,8,3[1]	(d) 2,8,6[1]	(e) 2,8,8[1]

THE PERIODIC TABLE

The periodic table shows the elements in order of increasing atomic number.

Groups

The rows in the periodic table are called periods. The columns in the periodic table are called groups. The elements in a group have these properties:

• similar chemical reactions, and
• the same number of electrons in their highest occupied energy level (outer shell).

The number of outer electrons is the same as the group number. For example, all the elements in Group 1 have 1 outer electron and all the elements in Group 7 have 7 outer electrons. All the elements in Group 0 have full outer shells.

The number of outer electrons determine the chemical properties of an element, which is why elements in a group have similar chemical properties.

Elements and their positions

An element's chemical properties include its **reactivity** and reactions. You predict an element's typical chemical properties from its atomic number shown in the periodic table. There is a copy of the periodic table on **page 384**.

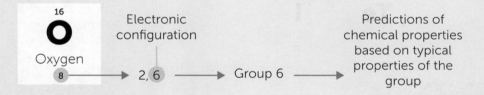

An electron diagram for oxygen

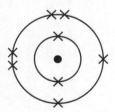

The electronic configuration and electron diagram for oxygen shows that it has two occupied energy levels. This means that oxygen is placed in period 2 of the periodic table.

1. Nitrogen is placed in period 2, group 5, of the periodic table. Give the electronic structure of nitrogen and its atomic number.   [2]
2. The atomic number of sodium is 11. Explain the position of sodium in the periodic table. [3]

   1. 2,5[1] atomic number (2 + 5) = 7[1].

   2. The electronic structure is 2,8,1[1], so sodium is in period 3[1] Group 1[1].

You can revise the general properties of metals and non-metals, and their positions on the periodic table, on **page 143**.

The complete table is printed on page 384.

THE DEVELOPMENT OF THE PERIODIC TABLE

New scientific discoveries and ideas have led to the modern periodic table.

Mendeleev's work

Mendeleev first placed the elements in order of increasing atomic weight, as other scientists had done. However, he left gaps in his periodic table for elements that he thought would be discovered later. This meant that:

- elements with similar chemical properties were placed in groups
- he could make predictions about the **physical properties** and **chemical properties** of the unknown elements.

Mendeleev sometimes changed the positions of some elements to match their properties better. However, he was unable to adequately explain why this should work.

The positions of tellurium and iodine are an example of one of Mendeleev's 'pair reversals'

128	127
Te	**I**
Tellurium	Iodine
52	53

1. Give **one** reason why Mendeleev's predictions were important to the success of his table. [1]

2. Explain how scientists today can explain why Mendeleev was correct when he changed the order of some elements in his periodic table. [4]

1. *When the missing elements were discovered, their properties were found to be similar to his predictions[1].*

2. *Protons, neutrons and electrons had not been discovered then[1] and Mendeleev did not know about isotopes[1]. The modern periodic table places elements in order of atomic number[1] but the existence of isotopes means that some elements have higher relative atomic masses than expected from their atomic number alone[1].*

Atomic weights

Early attempts to produce periodic tables relied on **atomic weights**. These were first determined using attempts at working out chemical formulae but were often inaccurate. They differed in value from modern relative atomic masses, and were often too high or too low.

Early periodic tables were incomplete because a lot of elements were unknown at the time. When a strict order of increasing atomic weight was used, some elements were located in the wrong groups. Dmitri Mendeleev devised periodic tables that overcame some of the problems with other early periodic tables.

The periodic table given to you in exams shows three other pair reversals, including argon and potassium. See if you can find the other two on **page 384**.

METALS AND NON-METALS

Most elements are metals rather than non-metals.

Chemical properties

An element is a **metal** if it forms positively charged **ions**, and a **non-metal** if it does **not** form positively charged ions. Metals and non-metals are found in different places on the periodic table.

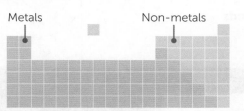

Metals   Non-metals

Chemical bonding

Metals and non-metals form different types of chemical **bonds**:

- metals have metallic bonding
- non-metal elements have covalent bonding
- compounds of metals and non-metals have ionic bonding.

See bonds on **page 150**.

The elements in groups 1 and 2, and between groups 2 and 3, are all metals.

Physical properties

In general, metals and non-metals have opposite physical properties. For example, metals are malleable – they can be hammered into shape without shattering. Non-metals in the solid state are brittle – they shatter when hammered. The table compares some other typical properties.

Property	Metals	Non-metals
Appearance	Shiny	Dull
Melting and boiling points	High	Low
Density	High	Low
Ability to conduct electricity and thermal energy	Good	Poor

1. Describe the positions of the metals and non-metals in the periodic table. [1]

2. The element mercury is in the liquid state at room temperature. Give a reason why this is unusual. [1]

3. Hydrogen is a non-metal. It can form H^+ ions and H^- ions. Explain why this is unusual. [3]

1. Metals are found towards the bottom and left, and non-metals are found towards the top and right[1].

2. Metals are usually in the solid state at room temperature[1].

3. Elements that do not form positive ions are non-metals[1] but hydrogen can form positive ions just like metals do[1].

Chemical properties

The Group 0 elements are unreactive non-metals. The highest occupied energy levels of their atoms are completely filled. These stable arrangements mean that the noble gases:

- have little tendency to lose or gain electrons in chemical reactions, so they do not easily form **ionic compounds**.
- have little tendency to share electrons, so they do not easily form **molecules**.

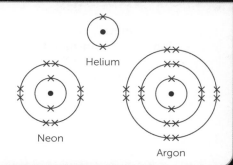

Helium

Neon

Argon

Physical properties

The noble gases have very low boiling points, so they are all in the gas state at room temperature. There is a gradual change or **trend** in their boiling points going down the group.

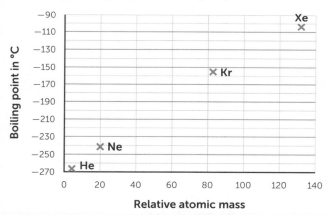

1. Compare the electronic configuration of helium with the electronic configurations of the other elements in Group 0. [1]

2. (a) Describe the relationship in Group 0 between boiling point and relative atomic mass. [1]

   (b) The relative atomic mass of argon is 40. Predict the boiling point of argon. [1]

   1. Helium has 2 outer electrons but the other elements have 8 outer electrons[1].

   2. (a) Boiling point increases as the relative atomic mass increases[1].

   (b) Between −230 °C and −210 °C [1].

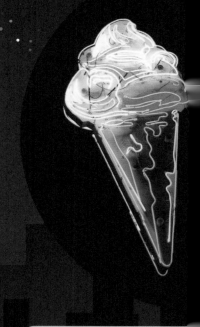

Group 0 atoms do lose electrons to form ions when high voltages are applied to them. They give off coloured light when the electrons return to the ions. This is how neon lights work.

GROUP 1

The Group 1 elements are called the **alkali metals**.
They have one outer electron.

Reactions with chlorine

The alkali metals react vigorously with chlorine when heated. They burn with coloured flames to produce metal chlorides. In general, where M stands for the alkali metal:

$$2M(s) + Cl_2(g) \rightarrow 2MCl(s)$$

Lithium burns with a red flame, sodium burns with an orange flame, and potassium burns with a lilac flame. The reaction of potassium with chlorine is the most vigorous.

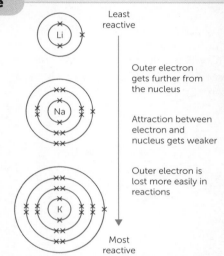

Least reactive

Li

Outer electron gets further from the nucleus

Na

Attraction between electron and nucleus gets weaker

Outer electron is lost more easily in reactions

K

Most reactive

Lithium-ion batteries are essential components of devices such as smartphones. They are lightweight, rechargeable and store a lot of energy.

Reactivity

Alkali metals lose their outer electrons in reactions with non-metals. They become more reactive going down the group.

Reactions with water

The alkali metals react with water to produce metal hydroxides and hydrogen. In general, where M stands for the alkali metal:

$$2M(s) + 2H_2O(l) \rightarrow 2MOH(aq) + H_2(g)$$

The first three elements all float on water and disappear as they react:

- lithium fizzes steadily
- sodium melts to form a silvery ball, fizzes quickly, and may burn with an orange flame.
- potassium burns with a lilac flame very quickly, then pops or explodes.

The metal hydroxides produced in the reaction are all soluble. They dissolve in water to form **alkaline solutions** that turn **universal indicator** blue or purple.

1. (a) Describe the reactions of lithium, sodium and potassium with oxygen. [4]

   (b) Balance the equation for the reaction of lithium with oxygen. Include state symbols.

   ___Li(___) + O$_2$(___) → ___Li$_2$O(___) [2]

2. Rubidium is placed below potassium in Group 1. Predict **two** observations you would see when rubidium is added to water. [2]

1. (a) *Lithium burns with a red flame[1], sodium with an orange flame[1] and potassium with a lilac flame[1]. The reactions become more vigorous from lithium to potassium[1].*

   (b) **4**Li(**s**) + O$_2$(g) → **2**Li$_2$O(**s**) *Correctly balanced[1], correct state symbols[1].*

2. *Two from: a violent or explosive reaction[1], a coloured flame[1], the metal disappears faster than potassium does[1], the water turns universal indicator blue or purple[1].*

GROUP 7

The Group 7 elements are called the **halogens**. They have seven outer electrons.

Physical properties

The melting points and boiling points of the halogens increase going down the group. Fluorine and chlorine are in the gas state at room temperature, bromine is in the liquid state and iodine is in the solid state. Bromine and iodine slowly produce coloured vapours at room temperature.

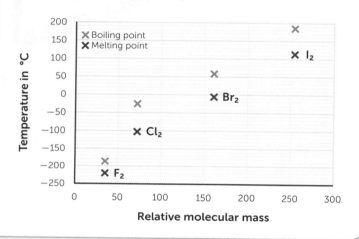

Reactions with metals and non-metals

Halogen molecules each consist of two atoms joined by a **covalent bond**. Halogens react with metals to form **ionic compounds**. For example, chlorine reacts with sodium to produce sodium chloride:

$$2Na(s) + Cl_2(g) \rightarrow 2NaCl(s)$$

Depending on the conditions, halogens can react with some non-metals to form **covalent compounds**. They dissolve in water and react with it to form **acidic solutions**.

Trend in reactivity

Halogens gain electrons in chemical reactions with metals. They become less reactive down the group.

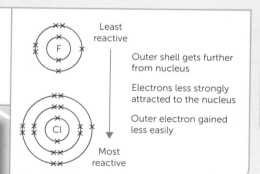

The reaction between chlorine and water creates a solution that is used to kill harmful microorganisms in swimming pools.

Astatine, At, is placed below iodine in Group 7.

(a) Predict the state of astatine at room temperature. Give a reason for your answer. [2]

(b) Predict the formula of astatine. [1]

(a) Solid[1] because iodine is solid and melting points increase down group 7[1]

(b) At$_2$.[1]

HALOGEN DISPLACEMENT REACTIONS

A **displacement reaction** happens when an atom or ion replaces an existing atom or ion in a compound.

Representing displacement reactivity

A more reactive halogen can **displace** a less reactive halogen from its compounds, particularly salts in aqueous solution. For example, chlorine is more reactive than iodine. Chlorine can displace iodine from potassium iodide solution:

$$Cl_2(aq) + 2KI(aq) \rightarrow 2KCl(aq) + I_2(aq)$$

Higher Tier only

Two **half equations** represent this reaction:
- chlorine is **reduced** to chloride ions: $Cl_2(aq) + 2e^- \rightarrow 2Cl^-(aq)$
- iodide ions are **oxidised** to iodine: $2I^-(aq) \rightarrow I_2(aq) + 2e^-$

An **ionic equation** also represents it: $Cl_2(aq) + 2I^-(aq) \rightarrow 2Cl^-(aq) + I_2(aq)$

The potassium ions in the reaction mixture are **spectator ions**. They do not take part in the reaction and are unchanged, which is why they do not appear in the ionic equation.

Determining reactivity

Chlorine, bromine and iodine dissolve in water to form aqueous solutions. These solutions can be mixed with aqueous solutions of **halide** salts, such as sodium chloride solution. A colour change to a darker colour after mixing indicates that a reaction has happened.

A student carries out an investigation to determine the reactivity of three halogens. Nine different mixtures of halogen solution and salt solution are prepared. The table shows the results. A tick shows that a visible reaction occurred.

	Potassium Chloride	Potassium Bromide	Potassium Iodide
Chlorine	✗	✓	✓
Bromine	✗	✗	✓
Iodine	✗	✗	✗

(a) Determine the order of reactivity, starting with the most reactive halogen. [1]

(b) Explain your answer to part (a). [3]

(a) Chlorine, bromine, iodine[1].

(b) Chlorine displaces bromine and iodine from their salts[1], bromine displaces iodine but not chlorine[1] and iodine does not displace chlorine or bromine[1].

EXAMINATION PRACTICE

01   Give **two** differences between a compound and a mixture. [2]

02   Iron(III) oxide reacts with carbon to make iron and carbon dioxide.

Balance the equation for this reaction. [1]

$$\underline{}Fe_2O_3 + \underline{}C \rightarrow \underline{}Fe + \underline{}CO_2$$

03   This question is about atoms and ions.

03.1   Explain how the discovery of the electron led to an improvement in the model of the atom. [2]

03.2   Complete the following table. [3]

Name of subatomic particle	Relative charge	Relative mass
		1
		Very small
	+1	

03.3   A nitride ion is represented as $^{15}_{7}N^{3-}$.
Give the numbers of protons, neutrons and electrons in this ion. [3]

03.4   The diameter of a gold atom is 0.28 nm. Give this value in m in standard form. [1]

04   A sample of copper contains two isotopes, 69% $^{63}_{29}Cu$ and 31% $^{65}_{29}Cu$.

04.1   Calculate the relative atomic mass of copper in this sample. Give your answer to 3 significant figures. [3]

04.2   Explain why these isotopes have the same chemical properties. [2]

05   This question is about the periodic table.

05.1   Describe **two** ways in which Mendeleev overcame some of the problems of earlier periodic tables. [2]

05.2   The electronic configuration of an element X is 2,8,5.

Determine the position of element X in the periodic table. [2]

05.3   Neodymium is an element placed between groups 2 and 3. Predict whether neodymium forms negatively charged ion or positively charged ions in reactions. Explain your answer. [2]

06   A dark green ink consists of insoluble carbon particles, mixed with three different coloured substances dissolved in a mixture of two liquids, water and propanol.

Design an experiment to separate all six substances in the ink. Include essential steps and safety precautions in your answer. [6]

07   This question is about Group 0 elements.

07.1   Explain why the elements in Group 0 exist as single atoms, rather than as molecules.   [2]

07.2   Describe the trend in boiling points of the Group 0 elements.   [1]

The table shows the densities of some Group 0 elements.

Element	He	Ne	Ar	Kr	Xe	Rn
Relative atomic mass	4	20	40	84	131	222
Density in g/dm³	0.179	0.900	1.784		5.89	9.73

07.3   Predict the density of krypton. Explain your answer.   [3]

07.4   Oganesson is a Group 0 element that was discovered in 2016. Its relative atomic mass is 294.
       Suggest a reason why its predicted density is 5000 g/dm³.   [1]

08   This question is about Group 1 reactions.

08.1   Write a balanced equation for the reaction of sodium with water. Include state symbols.   [3]

08.2   Explain why potassium is more reactive than lithium.   [3]

09   A student carries out an investigation into the reactions of halogens with their salts. The salts
      solutions are colourless. The table show the student's results.

	Sodium chloride	Potassium bromide	Potassium iodide
Chlorine	Not done	Changes to orange	Changes to brown
Bromine	No visible change	Not done	Changes to brown
Iodine	No visible change	No visible change	Not done

09.1   Give **one** reason why the student did not investigate three of the possible mixtures.   [1]

09.2   Explain what the student's results show about the reactivity of the halogens.   [3]

09.3   Write an ionic equation for the reaction between bromine and potassium iodide.   [2]

CHEMICAL BONDS

There are three types of strong **chemical bonds**: ionic, covalent and metallic.

Metals and non-metals

You can predict the type of bonding in an element or compound if you know whether it contains metals, non-metals or both:

- **Ionic bonding** nearly always involves a compound of a metal and a non-metal.
- **Covalent bonding** exists in most non-metal elements and compounds of non-metals.
- **Metallic bonding** exists in metal elements and their alloys.

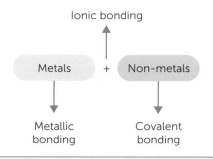

Electrostatic forces of attraction

All chemical bonds involve **electrostatic** forces of attraction between opposite charges. These forces arise in different ways depending on the type of bonding.

Type of bond	Forces of attraction between:
Ionic	Oppositely charged ions
Covalent	Nuclei of two atoms and a shared pair of electrons
Metallic	Nuclei of atoms and delocalised electrons

You can revise these three types of bond in detail: ionic bonding on **page 151**, covalent bonding on **page 153**, and metallic bonding on **page 155**.

1. (a) Predict the type of bonding present in sodium chloride. [1]

   (b) Explain why ammonium nitrate has ionic bonding, even though it does not contain metals. [2]

2. Predict the type of bonding present in sulfur dioxide, SO_2. Give a reason for your answer. [2]

   1. (a) Ionic bonding.[1]

      (b) Ammonium nitrate contains NH_4^+ ions[1] and NO_3^- ions[1].

   2. Covalent bonding[1] because it is a compound of two non-metal elements[1].

Transferring and sharing electrons

Chemical bonds form when electrons are transferred or shared.

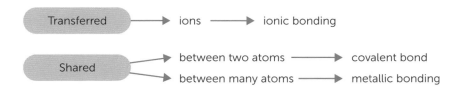

IONIC BONDING

Ions involved in ionic bonding can be represented by **dot and cross diagrams**.

Predicting the charges on ions

In a reaction between a metal and a non-metal, the outer electrons from metal atoms are transferred to the outer shells of non-metal atoms. Metals form positively charged ions and non-metals form negatively charged ions.

The charge on ions formed by atoms in groups 1, 2, 6 and 7 is related to the group number.

	Metals		Non-metals	
Group number	1	2	6	7
Charge on ions	+1	+2	−2	−1

1. Predict the charge on the ions formed by:
(a) sodium
(b) calcium
(c) oxygen
(d) chlorine   [4]

1. (a) +1[1]
   (b) +2[1]
   (c) −2[1]
   (d) −1[1]

Dot and cross diagrams

Dot and cross diagrams represent the electrons in different atoms and ions. The diagrams show the electronic structures for sodium atoms and sodium ions. Each dot represents an electron. It is usual to use shortened electronic structures in dot and cross diagrams for ionic bonding.

Dots represent electrons in metal atoms and metal ions. Crosses represent electrons in non-metal atoms and non-metal ions (except for electrons that have come from metal atoms).

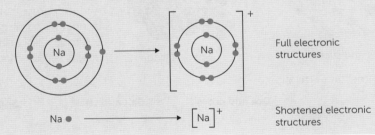

Full electronic structures

Shortened electronic structures

2. Draw a dot and cross diagram to represent the formation of sodium chloride.   [3]
3. (a) Write the electronic structures of (i) a sodium ion, and (ii) a chloride ion.   [2]
   (b) Suggest what your answers to part (a) show about these ions.   [3]

2. Correct structures of Na and Cl[1], correct structures of ions[1], correct charges[1].

$$Na\bullet + {}_{\times}^{\times\times}Cl_{\times}^{\times} \longrightarrow \left[Na \right]^+ \left[{}_{\times}^{\times\times}\overset{\bullet}{Cl}_{\times}^{\times} \right]^-$$

3. (a) (i) 2,8[1]   (ii) 2,8,8[1]

   (b) They have complete outer shells which gives them the same electronic structures as Group 0 elements (noble gases)[1]. Na^+ has the same electronic structure as neon[1], and Cl^- has the same electronic structure as argon[1].

IONIC COMPOUNDS

The ions in **ionic compounds** are held together by **ionic bonding**.

Giant ionic lattice

Ionic compounds have a **giant ionic lattice** structure:
- Lattice – A regular structure.
- Ionic – The structure consists of ions with ionic bonding.
- Giant – The regular structure is repeated very many times.

Ionic bonding acts in all directions in the lattice. It is the strong electrostatic force of attraction between oppositely charged ions.

> ⭐ You should know the structure of sodium chloride, but not the structures of other ionic compounds

Representing ionic structures

Ionic structures extend in three dimensions. You can represent these structures using plastic molecular modelling kits. Each ball represents an ion and each stick represents the bonding.

You can also show ionic structures in two dimensions.

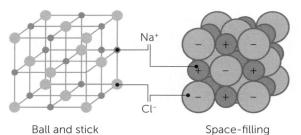

Na⁺

Cl⁻

Ball and stick   Space-filling

A third representation of bonding in ionic compounds is through dot and cross diagrams. (See **page 154**.)

The different diagrams have limitations.

	Dot and cross	Ball and stick	Space-filling
Shows bonding	In detail	As lines	✗
Shows relative sizes of atoms	✗	Inaccurately	✓
Shows shape of lattice	✗	✓	✓

The diagram shows the structure of silver bromide.

(a) Explain why the diagram shows that silver bromide is an ionic compound. [2]

(b) Define the term **empirical formula**. [2]

(c) Determine the empirical formula of silver bromide. [2]

Key
- ⚪ Br⁻
- ⚫ Ag⁺

(a) It shows oppositely charged ions[1] from more than one element[1].

(b) The simplest whole number ratio[1] of ions or atoms in a substance[1].

(c) The diagram shows 9 Ag⁺ ions and 9 Br⁻ ions[1], the simplest whole number ratio is 1 : 1 so the empirical formula is AgBr[1].

COVALENT BONDING

Molecules of non-metal atoms are held together by **covalent bonds** created by sharing pairs of electrons.

Types of molecules

Substances with covalent bonds exist as molecules. There are three types of molecule:

- Small molecules
- Very large molecules
- Giant covalent structures

> A covalent bond is a shared pair of electrons.

Representing molecules

You can represent molecules in two dimensional diagrams. A **displayed structural formula** is a common type of these diagrams. Each atom is shown by its chemical symbol, and each covalent bond is shown as a single line.

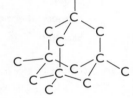

Methane
(Simple / small molecule)

Poly(ethene)
(Very large molecule)

Diamond
(Giant structure)

Polymers such as poly(ethene) have a repeating structure. Their molecules consist of very many atoms of two or more elements. You could not sensibly draw every bond and atom label in these molecules. Instead, you show the **repeating unit** in brackets. The letter n stands for a very large number, which could be many hundreds or thousands.

Methane molecules and most other molecules extend in three dimensions. You can represent them using plastic molecular modelling kits. Each ball represents an atom and each stick represents a covalent bond.

These models are often drawn as ball and stick diagrams.

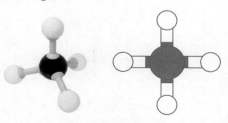

This diagram shows the full displayed structural formula of ethane.

$$H-C-C-H$$

(a) Define the term **molecular formula**. [1]

(b) Determine the molecular formula of ethane. [1]

(a) The actual number of atoms of each element in a molecule, shown as chemical symbols and numbers. [1]

(b) C_2H_6 [1]

DOT AND CROSS DIAGRAMS

Dot and cross diagrams can represent the bonding in small molecules.

Representing covalent bonds

The diagram is a dot and cross diagram for a chlorine molecule, Cl_2. In this type of diagram:

- only the outer shell of each atom is shown
- dots or crosses represent electrons in each atom
- a dot and cross represents a single covalent bond
- lone pairs of electrons are the ones that are not in bonds.

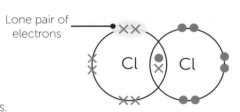

Lone pair of electrons

Drawing a dot and cross diagram

Hydrogen atoms can only form one covalent bond each. For other non-metal elements:

Group	4	5	6	7
Number of bonds	4	3	2	1
Example	C	N	O	Cl

This helps you work out how to complete a dot and cross diagram when you know the chemical formula for a substance. For example, the chemical formula for ammonia is NH_3.

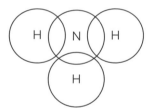

Draw overlapping circles, one for each atom's outer shell

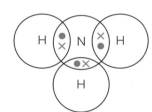

Add a dot and cross for each covalent bond

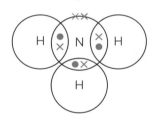

Make sure that all outer shells are full

---

1. The structural formula of oxygen is O=O.

    (a) Describe what the double line in the formula represents.   [1]

    (b) Draw a dot and cross diagram for oxygen.   [2]

2. Draw a dot and cross diagram for nitrogen, N_2.   [2]

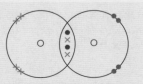

1. (a) A double covalent bond.[1]

   (b) Two shared pairs[1] and correct outer electrons[1].

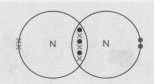

2. Three shared pairs[1] and correct outer electrons[1].

METALLIC BONDING

Metals and alloys are held together by **metallic bonding**.

Giant structures

In solid metals, the metal atoms are arranged in regular patterns. Two-dimensional diagrams often show just one layer of atoms. However, there will be many layers of atoms which form a **giant structure** in three dimensions.

Graphite is a non-metal that contains delocalised electrons. You can revise the structure and properties of this form of carbon on **page 162**.

Delocalised electrons

An atom consists of a central nucleus surrounded by **electrons** arranged in energy levels or **shells**. Electrons usually remain with individual atoms. However, outer electrons can become **delocalised** in some chemical structures – they leave individual atoms and become free to move through some or all of the structure.

Positively charged **ions** form when electrons permanently leave atoms. However, delocalised electrons do not permanently leave atoms. They are just free to move from atom to atom instead. Overall, the number of electrons in each atom remains equal to the number of **protons**. The outer electrons in metals are delocalised, and this gives rise to metallic bonding.

Representing metallic bonding

Metallic bonding is the electrostatic force of attraction between the positively charged centres of metal atoms and a 'sea' of delocalised electrons.

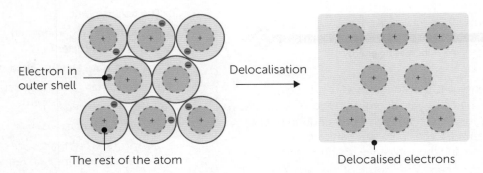

Electron in outer shell

Delocalisation

The rest of the atom

Delocalised electrons

---

1. Give the location of the electrons that become delocalised in metals. [1]

2. Compare metallic bonding with ionic bonding. [3]

   1. *The outer shells.[1]*

   2. *Both involve electrostatic forces of attraction[1] and both act in all directions[1]. However, ionic bonding involves oppositely charged ions but metallic bonding involves atoms and delocalised electrons[1].*

THE THREE STATES OF MATTER

The three **states of matter** are **solid**, **liquid** and **gas**.

The particle model

The **particle model** represents the arrangement and movement of particles in the three different states. The particles can be atoms, ions or molecules, but all of them are shown as small, solid spheres.

	Solid	Liquid	Gas
Arrangement of particles	Regular	Random	Random
Relative distance between particles	Close	Close	Far apart
Motion of particles	Vibrate about fixed positions	Move around each other	Move quickly in all directions
Particle diagram			

Energy and forces

For a given sample of a substance:

Solid   Liquid   Gas

Increasing energy

When energy is transferred to a substance by heating, particles gain energy and the temperature of the substance increases. When enough energy is transferred, forces between particles are overcome and the substance changes state:

- Melting occurs when some of the forces are overcome.
- Boiling occurs when all of the forces are overcome.

The stronger the forces are between particles, the higher the **melting point** and **boiling point** of the substance. This is because more energy is needed to overcome strong forces than weak forces.

1. Name the state changes that happen at:
   (a) The melting point. [1]
   (b) The boiling point. [1]
2. Compare the arrangement and movement of particles in:
   (a) The solid state and the liquid state. [2]
   (b) The liquid state and the gas state. [2]

1. (a) Melting and freezing.[1]
   (b) Boiling and condensing.[1]

2. (a) Particles in both states are close together[1] but they are regularly arranged in the solid state and randomly arranged in the liquid state[1].

   (b) Particles in both states are randomly arranged[1] but they are close together in the liquid state and far apart in the gas state[1].

USING THE PARTICLE MODEL

The particle model explains the temperatures at which state changes happen.

Predicting states

You can predict the state of a substance at a given temperature if you know its melting and boiling points:

- Solid below the melting point
- Liquid between the melting and boiling points
- Gas above the boiling point

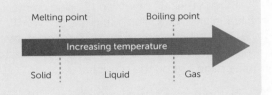

State change and temperature

The temperatures at which state changes happen depend on the type of structure and bonding present in a substance. Weak forces between small molecules are overcome during melting and boiling.

Type of substance	Forces overcome	Strength of forces	Melting and boiling points
Small molecule	Intermolecular	Weak	Low
Large molecule	Intermolecular	Medium to strong	Medium to high
Giant ionic	Ionic	Strong	High
Giant covalent	Covalent	Strong	High
Metal	Metallic	Strong	High

Bulk properties

A **bulk property** of a substance is one of its observable or measurable features. Melting point and boiling point are examples of bulk properties. Properties like these can be explained by considering the particles in a substance, but the particles themselves do not have bulk properties.

Limitations of the particle model   Higher Tier only

The simple particle model describes all particles as solid, inelastic spheres. However, particles can change shape slightly when they are close together or rebound off each other. Simple ions and individual atoms are spherical, but most ions and molecules have complex three-dimensional shapes.

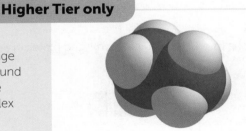

1. The melting point of ethene is −169 °C. The melting point of poly(ethene) is around 120 °C. Explain how the molecular structure of ethene and poly(ethene) influences their melting point. [4]
2. **Higher Tier only:** Describe **one** of the limitations of the simple particle model. [2]

1. *Ethene exists as small molecules[1] with few, weak forces between them[1]. However, poly(ethene) exists as large molecules[1] with many more weak forces between them[1].*
2. *It assumes that no forces exist between particles in the gas state[1] but very weak forces still exist[1].*

PROPERTIES OF IONIC COMPOUNDS

Melting point and boiling point

Ionic compounds have some typical bulk properties with high melting points and boiling points. For example, aluminium oxide (alumina) melts at 2072 °C and boils at 2977 °C. These temperatures are so high that alumina bricks are used to line the steel walls of blast furnaces (see **page 178** on the extraction of iron).

Ionic compounds have a giant ionic lattice structure. There are electrostatic forces of attraction between oppositely charged **ions**. This ionic bonding is strong, and it acts in all directions.

Large amounts of energy must be transferred to ionic compounds in order to overcome the strong electrostatic forces of attraction between the ions. This is why ionic compounds have high melting and boiling points.

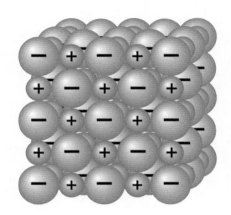

Conducting electricity

A substance can conduct **electricity** if:
- it contains charged particles, and
- these particles are free to move from place to place.

Ionic compounds do contain charged particles – their ions. However, these particles are not always able to move from place to place. They are only free to move from place when the substance is in the liquid state, but not when it is in the solid state.

Ions move to oppositely charged plates in liquids

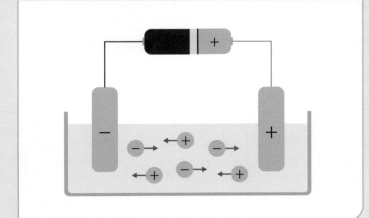

1. Explain why ionic compounds are able to conduct electricity in aqueous solution. [1]

2. Copper(II) oxide is insoluble in water. Give **one** way to allow copper(II) oxide to conduct electricity. [1]

1. Their ions are separate and free to move[1] when dissolved in water, so charge can flow[1].

2. Melt it.[1]

Electricity is conducted by ionic compounds during electrolysis. In this process, ions move to oppositely charged electrodes. Once there, they lose or gain electrons and become discharged as atoms or molecules. You can revise electrolysis on **pages 187-191**.

PROPERTIES OF SMALL MOLECULES

Small molecules

The size of molecules varies between substances. This affects the physical properties of substances. Substances that consist of small molecules have relatively low melting and boiling points. These substances may be elements or compounds, and their molecules contain few atoms.

The atoms in a small molecule are held together by strong covalent bonds. These bonds are not broken during melting or boiling. Instead, the much weaker intermolecular forces between molecules are overcome. In general, as the size of the molecule increases, intermolecular forces increase and therefore, melting and boiling points increase.

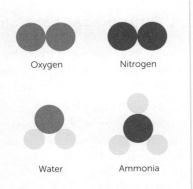

Substance	$H-\overset{\displaystyle H}{\underset{\displaystyle H}{C}}-H$	$H-\overset{\displaystyle H}{\underset{\displaystyle H}{C}}-\overset{\displaystyle H}{\underset{\displaystyle H}{C}}-H$	$H-\overset{\displaystyle H}{\underset{\displaystyle H}{C}}-\overset{\displaystyle H}{\underset{\displaystyle H}{C}}-\overset{\displaystyle H}{\underset{\displaystyle H}{C}}-H$	$H-\overset{\displaystyle H}{\underset{\displaystyle H}{C}}-\overset{\displaystyle H}{\underset{\displaystyle H}{C}}-\overset{\displaystyle H}{\underset{\displaystyle H}{C}}-\overset{\displaystyle H}{\underset{\displaystyle H}{C}}-H$
Boiling point in °C	−161	−89	−42	−1

POLYMERS

Polymers consist of large molecules, formed from many small molecules called **monomers**. The atoms in each individual polymer molecule are joined together by strong covalent bonds. Polymer molecules are attracted to each other by intermolecular forces. These forces are relatively strong because polymer molecules are so large.

You should be able to recognise polymers from diagrams like these.

Repeating unit

Tangled polymer chains

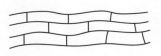

Chains with cross-links

Explain why poly(propene) is in the solid state at room temperature. [4]

Poly(propene) consists of large molecules[1]. Intermolecular forces between polymer molecules are overcome during melting[1]. These are relatively strong because the molecules are so large[1] so a lot of energy is needed to overcome them[1].

GIANT COVALENT STRUCTURES

Some elements and compounds have **giant covalent structures**.

Structure and bonding

Giant covalent structures have a lattice structure in which:
- all the atoms are linked to other atoms by covalent bonds.
- the regular lattice structure is repeated very many times.

Covalent bonds are strong and there are many of them in a giant covalent structure. A lot of energy must be transferred to melt or boil a substance with this structure. As a result of this, substances with giant covalent structures have high melting points and boiling points. For example, **sand** is mostly silica, a compound with a giant covalent structure. This melts at 1713 °C and boils at 2950 °C.

Recognising giant covalent structures

Diamond and graphite are two forms of carbon. They both have giant covalent structures, although these are different from each other (see **page 162**).

You should be able to recognise giant covalent structures from diagrams. There are very many atoms in a giant covalent structure, so you will see just enough atoms for you to understand how they are arranged. Remember that the structure you see will be repeated many times, with covalent bonds leading from atoms at the edges of the diagrams to atoms that are not shown.

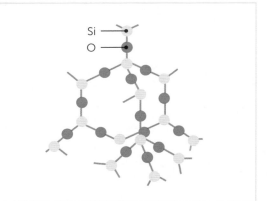

The structure of silica, SiO_2

The diagram shows the structure of a form of boron nitride, BN.

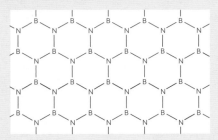

(a) Give the type of structure of boron nitride. Explain your answer. [3]

(b) Give **one** physical property that you would expect boron nitride to have. [1]

(a) Giant covalent structure[1] because it contains many atoms[1] joined by covalent bonds[1].

(b) One from: high melting point[1], high boiling point[1].

PROPERTIES OF METALS AND ALLOYS

Bending and shaping metals

Metals are good conductors of electricity and **thermal energy**. They are:
- **Malleable** – they can be bent or hammered into shape without breaking
- **Ductile** – they can be pulled to make wires without snapping.

Pure metals have these properties because their layers of atoms can slide over each other when a force is applied.

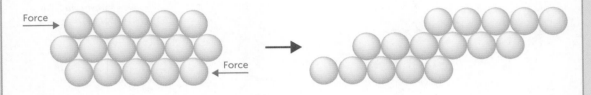

Force

Force

Hardness

A pure metal element may be too soft for a lot of uses. An **alloy** is a **mixture** of a metal element with at least one other element, usually another metal. Some alloys are harder and therefore more useful than the pure metal.

For example, steels are mixtures of iron with carbon and some other elements. Steels are much harder than iron. This is because the carbon atoms distort the regular lattice structure, making it more difficult for the layers to slide over one another.

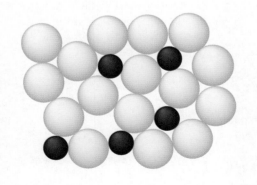

Conducting electricity

A substance can conduct **electricity** if it contains charged particles, and if these particles are free to move from place to place.

Metals contain **delocalised electrons**. These charged particles can move freely though the giant structure of metal atoms. This is why metals are good conductors of electricity.

1. Explain why metals are good conductors of thermal energy. [3]

2. Explain, in terms of structure and bonding, why metals have high melting points. [3]

   1. *When a metal is heated, energy is transferred to delocalised electrons[1]. These are free to move through the structure of the metal[1] transferring energy to atoms and other electrons[1].*

   2. *Metals in the solid state have giant structures of atoms[1] held together by metallic bonding[1]. A lot of energy must be transferred to overcome this strong bonding[1].*

DIAMOND AND GRAPHITE

Diamond and **graphite** are different forms of carbon with different giant covalent structures.

Structures

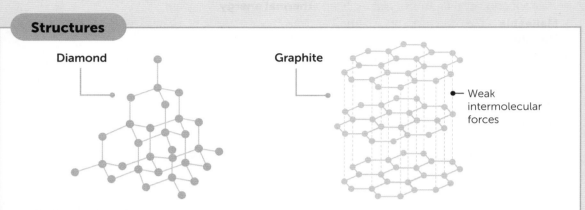

Diamond

Graphite

Weak intermolecular forces

Property	Diamond	Graphite
Contains covalent bonds	✓	✓
Atoms bonded to each C atom	4	3
Contains layers	✗	✓ Hexagonal rings of atoms
Weak forces between layers	✗	✓
Contains delocalised electrons	✗	✓

Diamond

Diamond has a very high melting point because its structure contains very many strong covalent bonds. A lot of energy is needed to break these bonds. Diamond does not contain delocalised electrons or any other charged particles that are free to move, so it does not conduct electricity.

Graphite

There are only weak intermolecular forces, rather than strong covalent bonds, between the layers of atoms in graphite. These allow the layers to slide over each other easily, so graphite can be slippery.

1. Explain why diamond is very hard. [3]
2. (a) Explain, in terms of particles, why graphite is similar to metals. [2]
   (b) Explain why graphite is a good conductor of electricity. [2]

    1. *Diamond has a giant covalent structure[1]. Its many strong covalent bonds[1] resist forces that could distort the structure[1].*

    2. (a) *Both consist of atoms[1] and contain delocalised electrons[1].*

       (b) *One electron from each carbon atom becomes delocalised.[1] These delocalised electrons are charged and are free to move.[1]*

GRAPHENE AND FULLERENES

Graphene and **fullerenes** are different forms of carbon.

Graphene

Graphene is a single layer of graphite. It does not have a layered structure, so it is transparent and flexible.

Graphene does have delocalised electrons so it conducts electricity, just like graphite does. This property is useful in electronics.

The carbon atoms in graphene are joined by strong covalent bonds, so graphene itself is strong. This property makes it useful as a replacement for carbon fibres in composite materials.

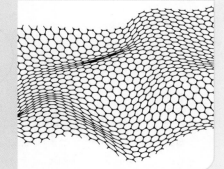

Fullerenes

Fullerenes are hollow molecules of carbon. **Buckminsterfullerene**, C_{60}, was the first fullerene to be discovered. Its molecules consist of 60 carbon atoms arranged in rings of five or six atoms. Fullerenes may also contain rings of seven carbon atoms.

Carbon **nanotubes** are cylindrical fullerenes. They have very high length to diameter ratios – this means that they are very long compared to their width.

A section of a cylindrical fullerene	A Buckminsterfullerene molecule
	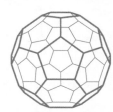

1. (a) Explain why fullerenes can conduct electricity. [2]

   (b) Give **one** potential use of fullerenes that depends on their ability to conduct electricity. [1]

2. Explain why a mixture of carbon nanotubes and plastic are used in wind turbine blades rather than plastic alone. [3]

3. Suggest **one** reason why Buckminsterfullerene may be used in lubricants. [1]

   1. (a) They contain delocalised electrons[1] which can move through their structures[1].

   (b) Electronics.[1]

   2. Carbon nanotubes are very strong[1] and stiff[1]. They can be used to reinforce the plastic[1] as a composite material[1] instead of carbon fibres. Composite materials are strong and flexible[1].

   3. Its molecules are spherical and can roll around each other.[1]

EXAMINATION PRACTICE

01   One of the stages in making steel involves adding magnesium powder. This reacts with sulfur impurities to form an ionic compound called magnesium sulfide.
      The diagram shows the outer electrons in magnesium and sulfur atoms.

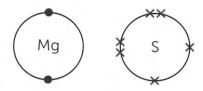

01.1   Predict the formula of magnesium sulfide.   [1]
01.2   Describe, in terms of electron transfer, what happens when magnesium reacts with sulfur. [3]
01.3   Describe the bonding in magnesium sulfide.   [2]
01.4   Predict **two** physical properties of magnesium sulfide.   [2]

02   Methane, CH_4, is the main component of natural gas.
      02.1   Complete the dot and cross diagram to show the bonding in a methane molecule.   [2]

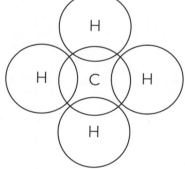

      02.2   Explain why methane is in the gas state at room temperature.   [3]

03   Diamond, graphite and graphene are three forms of carbon.
      03.1   Compare the structure and bonding in diamond and graphite.   [4]
      03.2   Explain why graphite and graphene can conduct electricity.   [2]

04   This question is about aluminium and its alloys.
      04.1   Explain why most metals have high melting points.   [2]
      04.2   Explain why pure aluminium can be bent and shaped without breaking.   [2]
      04.3   Aluminium-lithium alloys consist of aluminium and around 2.5% lithium.
             Explain why these alloys are stronger and harder than pure aluminium.   [2]
      04.4   Aluminium is used to make high-voltage overhead power lines. It is also used to make
             cooling units in computers.
             Explain why aluminium is a good conductor of thermal energy.   [2]

05   The diagram shows the crystal structure of wüstite, a mineral form of an oxide of iron.

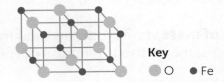

Key

◐ O   ● Fe

05.1   Describe what is meant by an empirical formula.   [2]

05.2   Determine the empirical formula of wüstite.   [1]

06   The diagram shows the structure of a compound.

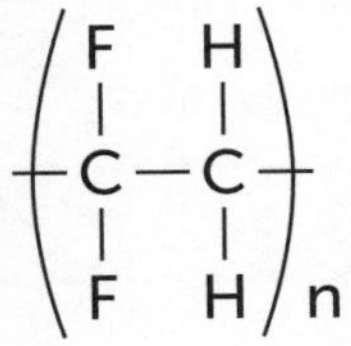

$$\left(\begin{array}{cc} F & H \\ | & | \\ -C-C- \\ | & | \\ F & H \end{array}\right)_n$$

06.1   Name the type of substance shown in the diagram.   [1]

06.2   Give the meaning of **one** of the straight lines in the diagram.   [1]

06.3   Explain why compounds like this one are solid at room temperature.   [2]

07   Dodecanoic acid is in the solid state at room temperature. A student heated a sample of this substance and recorded its temperature at regular intervals. The graph shows the results.

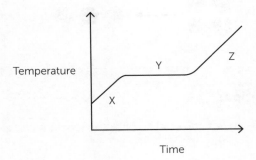

Explain the results in terms of the arrangement, energy and movement of dodecanoic acid particles.   [6]

RELATIVE FORMULA MASS

The **law of conservation of mass** states that the total mass of the products formed in a reaction is the same as the total mass of the reactants used.

Calculating relative formula mass, M_r

Covalent compounds and ionic compounds have a **relative formula mass**, and so do elements that exist as molecules. You can calculate a relative formula mass by following these steps:

- look at the **relative atomic masses** of all the elements in the formula
- add together the relative atomic masses for all the atoms in the formula.

Just like relative atomic masses, relative formula masses are numbers without units.

Law of conservation of mass

Mass is conserved during a chemical reaction because no atoms are lost or made in the reaction. This is why you need to balance symbol equations. In a balanced equation the numbers of atoms of each element must be the same on both sides of the arrow.

For example, methane burns completely in air to produce carbon dioxide and water:

$$CH_4 + 2O_2 \rightarrow CO_2 + 2H_2O$$

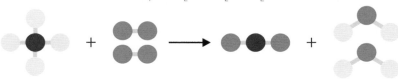

M_r and balanced equations

When a chemical equation is balanced, the total M_r of all the reactant particles equals the total M_r of all the product particles. In the example above:

Relative formula masses: $CH_4 = 16$, $O_2 = 32$, $CO_2 = 44$, $H_2O = 18$

Total M_r of reactants $= 16 + (2 \times 32) = 80$ Total M_r of products $= 44 + (2 \times 18) = 80$

---

1. Calculate the relative formula mass (M_r) of ammonium carbonate $(NH_4)_2CO_3$.
   Relative atomic masses (A_r): hydrogen = 1, carbon = 12, nitrogen = 14, oxygen = 16. [2]

2. In a chemical reaction, 6.0 g of carbon burns in excess oxygen to produce 22 g of carbon dioxide.
   Determine the mass of oxygen used in the reaction. [1]

3. Balance this equation: $Al(s) + O_2(g) \rightarrow Al_2O_3(s)$ [2]

   1. 96[1] Evidence of working[1], e.g. $(2 \times 14) + (2 \times 4 \times 1) + 12 + (3 \times 16)$
   2. $22 - 6.0 = 16$ g[1]
   3. $4Al(s) + 3O_2(g)$[1] $\rightarrow 2Al_2O_3(s)$.[1]

PERCENTAGE COMPOSITION BY MASS

The percentages by mass of each element in a compound add up to 100%. To calculate a **percentage composition** by mass, you need to know the relative atomic mass of the element you are interested in and the relative formula mass of the compound.

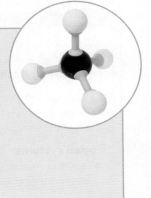

Calculating percentage composition

Methane, CH_4, is a compound of carbon and hydrogen. Its molecules each contain one carbon atom and four hydrogen atoms. You might think that the percentage of carbon is 20% and the percentage of hydrogen is 80%. However, you must take into account the relative atomic masses of each element.

$$\text{Percentage by mass of hydrogen} = \frac{(\text{number of atoms of H}) \times (A_r \text{ of H})}{M_r \text{ of } CH_4} \times 100$$

$$= \frac{4 \times 1}{12 + (4 \times 1)} \times 100 = \frac{4}{16} \times 100 = 25\%$$

A more difficult example

Ammonium nitrate, NH_4NO_3, is widely used in blast explosives and as a high-nitrogen fertiliser in agriculture. Calculate its percentage by mass of nitrogen.

Relative atomic mass (A_r) of nitrogen = 14, relative formula mass (M_r) of ammonium nitrate = 80.

$$\text{Percentage by mass of nitrogen} = \frac{2 \times 14}{80} \times 100 = 35\%.$$

1. The percentage by mass of carbon in ethane C_2H_6 is 80%.
   Determine the percentage by mass of hydrogen in ethane. [1]

2. Calculate the percentage by mass of oxygen in sulfur dioxide, SO_2.
   Relative atomic mass (A_r) of oxygen = 16, relative formula mass (M_r) of SO_2 = 64. [2]

3. Calculate the percentage by mass of hydrogen in copper(II) hydroxide, $Cu(OH)_2$.
   Relative atomic mass (A_r) of hydrogen = 1.00, relative formula mass (M_r) of $Cu(OH)_2$ = 97.5. [2]

1. 100 − 80 = 20%. [1]

2. 50% [1] Evidence of working [1], e.g. % of O = $\frac{(2 \times 16)}{64} \times 100$

3. 2.05% [1] Evidence of working [1], e.g. % of H = $\frac{(2 \times 1.00)}{97.5} \times 100$

Give your answers to a suitable number of significant figures. This is usually the lowest number of significant figures in the quantities used.

MASS CHANGES WITH GASES

Reactions involving gases may appear to involve a change in mass. This is usually explained when a reactant or product is a gas which has not been accounted for.

Closed systems

Substances cannot enter or leave a **closed system**. Common closed systems in chemistry include test tubes or flasks with bungs on and **precipitation** reactions.

In a precipitation reaction, an insoluble solid forms when two solutions are mixed. Nothing enters or leaves the reaction mixture, so its total mass stays the same.

In **open systems** (also called **non-enclosed systems**) reactants are free to enter, and products are free to leave. This gives a false impression that the law of conservation of mass has been broken.

Apparent increases in mass

Mass appears to increase when a metal reacts with oxygen from the air to form a metal oxide. This is because oxygen atoms combine with metal atoms in the reaction. If the mass of oxygen gained is included, the law of conservation of mass is not broken.

Magnesium burns in air: $2Mg(s) + O_2(g) \rightarrow 2MgO(s)$

> You need to be able to explain mass changes in non-enclosed systems, in terms of particles, when given the balanced equation for a reaction. You can revise the particle model on **page 156**.

Apparent decreases in mass

Mass appears to decrease in **thermal decomposition** reactions. In these reactions, a metal carbonate breaks down to form a metal oxide and carbon dioxide, which escapes into the air. If the mass of carbon dioxide lost is included, the law of conservation of mass is not broken.

1. Calcium oxide reacts with carbon dioxide in the air: $CaO(s) + CO_2(g) \rightarrow CaCO_3(s)$
   Give **one** reason why the mass of the solid increases. [1]
2. Sodium reacts with water in an open beaker: $2Na(s) + 2H_2O(l) \rightarrow 2NaOH(aq) + H_2(g)$
   Explain why the mass of the reaction mixture goes down. [2]

   1. *Carbon dioxide molecules combine with the calcium oxide.*[1]

   2. *Hydrogen gas is produced in the reaction*[1] *and leaves the reaction mixture*[1].

CHEMICAL MEASUREMENTS

There is always some **uncertainty** about the result of a measurement.

Resolution

The **resolution** of a measuring instrument is the smallest change that it can show. It is equal to the smallest scale division on the instrument. For example:
- 0.1 g in a digital balance reading to 1 decimal place
- 1 mm in a ruler divided into tenths of a centimetre.

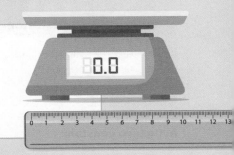

Uncertainty in readings

Uncertainty is given as ± (plus or minus) a value. The uncertainty in readings depends on the type of instrument and its resolution:
- For a digital instrument, the uncertainty is equal to plus or minus half the resolution
- For an instrument with a scale, the uncertainty is equal to the resolution (because you will be taking two readings, the start point and the end point).

1. A digital balance reads 20.4 g. Determine where the true mass lies. [1]
2. A ruler shows that a pencil is 16.8 cm long. Determine where its true length lies. [1]

   1. Between 20.35 g and 20.45 g[1]

   2. Between 16.7 cm and 16.9 cm[1].

Range and uncertainty in repeats

Uncertainty leads to random errors in readings. You cannot correct for these errors but you can reduce their effects by repeating experiments. **Anomalous** readings can be ignored when calculating a mean value, and this increases the **reliability** of your results.

The **range** of a set of values is the difference between the maximum value and minimum value. The uncertainty in a mean value obtained from a set of repeats is equal to half the range.

3. A student carried out an investigation to determine the volume of carbon dioxide produced in a reaction. The student obtained four repeats. The table shows the results.

Experiment	1	2	3	4
Volume of gas in cm^3	64.1	68.3	66.9	65.5

(a) Calculate the mean volume of carbon dioxide obtained. [1]

(b) Determine the range in volumes. [1]

(c) Give the mean volume, including the uncertainty in this value. [1]

3. (a) "Mean volume = $\frac{(64.1 + 68.3 + 66.9 + 65.5)}{4}$ = $\frac{264.8}{4}$ = 66.2 $cm^{3[1]}$

   (b) Minimum value = 64.1 cm^3. Maximum value = 68.3 cm^3.
    Range = (68.3 − 64.1) = 4.2 $cm^{3[1]}$

   (c) 66.2 ± 2.1 $cm^{3[1]}$

MOLES

The **mole** is the unit for an amount of substance. The unit symbol is mol.

Amount of substance

In everyday use, the word 'amount' can apply to quantities such as mass or volume, but it has a very particular meaning in chemistry. The **amount** of a substance refers to the number of particles it contains.

One mole, 1 mol, of any particle contains 6.02×10^{23} of these particles. The particles can be atoms, molecules, ions or electrons. The number of stated particles in 1 mol is called the **Avogadro constant**:

$$\text{Avogadro constant} = 6.02 \times 10^{23} \text{ per mole}$$

Amounts in molecules and compounds

You must be careful to identify the particles involved when you use the mole. For example, a carbon dioxide molecule consists of 1 carbon atom and 2 oxygen atoms. This means that 1 mol of CO_2 molecules contains:

- 1 mol of carbon atoms
- 2 mol of oxygen atoms
- (1 + 2) = 3 mol of atoms.

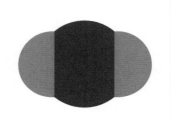

Moles and masses

The mass of 1 mol of a substance is equal to its relative formula mass (M_r) in grams. The M_r of carbon dioxide is 44, so 1 mol of carbon dioxide molecules has a mass of 44 g.

You can use this equation to calculate the mass of a substance in a given amount of it:

$$\text{mass (g)} = M_r \times \text{amount (mol)}$$

You can also use this equation to calculate the amount of a substance in a given mass.

1. The relative formula mass (M_r) of calcium hydroxide $Ca(OH)_2$ is 74.
   (a) Calculate the mass of 0.5 mol of calcium hydroxide. [1]
   (b) Determine the amount in moles of OH^- ions in 0.75 mol of calcium hydroxide. [1]
2. Calculate the amount in moles of water molecules in 45 g of water.
   Relative formula mass (M_r) of water = 18 [2]
3. Calculate the number of ammonia molecules in 2 mol of ammonia. [1]

   1. (a) Mass = 74 × 0.5 = 37 g[1]
      (b) Amount = 2 × 0.75 = 1.5 mol[1]
   2. 45 g = 18 × amount (mol), rearranging: amount = $\frac{45}{18}$ [1] = 2.5 mol[1]
   3. Number = amount (mol) × Avogadro constant = 2 × 6.02 × 10^{23} = 1.204 × 10^{24}[1]

AMOUNTS OF SUBSTANCES IN EQUATIONS

Balanced equations can be used to calculate masses of reactants and products.

Interpreting chemical equations

You can interpret balanced symbol equations in terms of moles rather than atoms and molecules. For example, nitrogen reacts with hydrogen to produce ammonia:

$$N_2 + 3H_2 \rightarrow 2NH_3$$

This shows that 1 mol of nitrogen reacts with 3 mol of hydrogen to produce 2 mol of ammonia.

Calculating masses in reactions

You can calculate the mass of a given substance involved in a reaction if you know:

- the balanced symbol equation
- the mass of one of the other substances involved, and
- the relative masses of these two substances.

In the reaction between nitrogen and hydrogen, you could calculate the mass of the nitrogen or hydrogen needed to make a given mass of ammonia, or the mass of ammonia that could be made from a given mass of nitrogen or hydrogen.

An example

Calculate the mass of hydrogen needed to make 13.6 g of ammonia.

Relative formula masses (M_r): H_2 = 2, NH_3 = 17.0

1. Work out the amount in moles of ammonia by rearranging the formula on **page 170**:

$$\text{amount (mol)} = \frac{\text{mass (g)}}{M_r} = \frac{13.6}{17}$$
$$= 0.8 \text{ mol}$$

2. From the balanced equation, 2 mol of NH_3 is made from 3 mol of H_2, so:

$$0.8 \text{ mol of } NH_3 \text{ is made from } \frac{0.8}{2} \times 3 = 1.2 \text{ mol of } H_2$$

3. Work out the mass of hydrogen: mass (g) = M_r × amount (mol) = 2 × 1.2 = 2.4 g

---

Aluminium powder reacts with oxygen to produce aluminium oxide: $4Al + 3O_2 \rightarrow 2Al_2O_3$

Calculate the mass of aluminium oxide that can be made from 8.1 g of aluminium.

Relative masses: Al = 27, Al_2O_3 = 102

[3]

Moles of Al = $\frac{8.1}{27}$ = 0.3 mol[1]

From the balanced equation, mole ratio Al : Al_2O_3 = 4 : 2 so $\frac{0.3}{4}$ × 2 = 0.15 mol of Al_2O_3 [1]

Mass of Al_2O_3 = 102 × 0.15 = 15.3 g[1] Correct answer scores 3 marks[3]

BALANCING EQUATIONS USING MOLES

You can find balancing numbers using masses of reactants and products.

Balancing numbers

Balancing numbers are used in symbol equations to ensure that the numbers of atoms of each element in the reactants and products are the same. You can calculate balancing numbers if you know:

- the masses of the reactants and products
- the relative masses of these substances.

Titanium and its alloys are used to produce strong, corrosion-resistant and lightweight parts such as replacement knee joints.

An example

One of the steps in the extraction of the metal titanium involves heating sodium with titanium(IV) chloride. In one of these reactions, 9.2 g of Na reacts with 19 g of $TiCl_4$ to produce 23.4 g of NaCl and 4.8 g of Ti. Determine the balanced symbol equation for this reaction.

Relative masses: Na = 23, $TiCl_4$ = 190, NaCl = 58.5, Ti = 48

You need to follow three steps to answer this question. Remember that:

$$\text{amount (mol)} = \frac{\text{mass (g)}}{\text{relative mass}}$$

Step	Action					
1	Calculate the amount of each substance	Na $\frac{9.2}{23}$ = 0.4 mol	$TiCl_4$ $\frac{19}{190}$ = 0.1 mol	→	Na $\frac{23.4}{58.5}$ = 0.4 mol	$TiCl_4$ $\frac{4.8}{48}$ = 0.1 mol
2	Find the smallest whole number ratio of moles	4	1		4	1
3	Write the equation			$4Na + TiCl_4 \rightarrow 4NaCl + Ti$		

Phosphorus reacts with chlorine to produce phosphorus trichloride. In one of these reactions, 6.2 g of P_4 reacts with 21.3 g of Cl_2 to produce 27.5 g of PCl_3. Determine the balanced symbol equation for this reaction. Relative formula masses (M_r): P_4 = 124, Cl_2 = 71, PCl_3 = 137.5   [3]

Amount of $P_4 = \frac{6.2}{124} = 0.05$ mol   Amount of $Cl_2 = \frac{21.3}{71}$
= 0.3 mol   Amount of $PCl_3 = \frac{27.5}{137.5} = 0.2$ mol[1]

Simplest whole number ratio (dividing all by 0.05) is
1 : 6 : 4[1]

$P_4 + 6Cl_2 \rightarrow 4PCl_3$ [1]

LIMITING REACTANTS

In a reaction between two reactants, one of them limits the amount of product.

Limiting and excess

Unless you can mix two reactants together in exactly the correct amounts according to the balanced equation:

- one reactant is described as being in **excess**
- the other reactant is the **limiting reactant**.

When magnesium reacts with dilute hydrochloric acid, magnesium is the limiting reactant if it all gets used up and some acid is left behind at the end.

The amount or mass of product is directly proportional to the amount or mass of the limiting reactant.

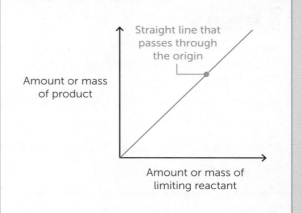

Amount or mass of product

Straight line that passes through the origin

Amount or mass of limiting reactant

Determining the limiting reactant

You can determine which reactant is limiting in a reaction if you know:

- the balanced symbol equation
- the masses of the two reactants, and
- the relative formula mass of these two substances.

For example, 675 g of aluminium powder is added to 1.6 kg of iron(III) oxide powder and heated. The equation for the reaction is: $2Al + Fe_2O_3 \rightarrow Al_2O_3 + 2Fe$

Explain why iron(III) oxide is the limiting reactant. Relative masses: Al = 27, Fe_2O_3 = 160

1. Work out the amount in moles of each reactant:

amount of Al = $\frac{675}{27}$ = 25 mol   amount of Fe_2O_3 = $\frac{1.6 \times 1000}{160}$ = 10 mol

2. Using your answer to step 1 and the balanced equation, check whether the amount of aluminium is enough to react with all the iron(III) oxide.

The mole ratio of Al : Fe_2O_3 is 2 : 1, so (2 × 10) = 20 mol of aluminium is needed. This is less than the amount used, so aluminium is in excess and iron(III) oxide must be the limiting reactant.

0.69 g of sodium is heated with 0.71 g of chlorine. The equation for the reaction is:

$$2Na + Cl_2 \rightarrow 2NaCl$$

Explain why sodium is in excess. Relative atomic masses (A_r): Na = 23, Cl = 35.5   [4]

M_r of Cl_2 = 2 × 35.5 = 71 [1]

Amount of Na = $\frac{0.69}{23}$ = 0.03 mol [1]   Amount of Cl_2 = $\frac{0.71}{71}$ = 0.01 mol [1]

Sodium is in excess because (2 × 0.01) = 0.02 mol of Na is needed and 0.03 mol is added. [1]

CONCENTRATION OF SOLUTIONS

The **concentration** of a solution is a measure of how much **solute** a given volume contains.

Volume

The volume of a substance can be measured in m^3 but this unit is too large for laboratory chemistry. Instead, use cubic decimetres, (dm^3) or cubic centimetres (cm^3).

Measuring cylinders and other laboratory apparatus are often graduated in ml:

$$1 \text{ ml} = 1 \text{ cm}^3$$

$1 \text{ dm}^3 = 1000 \text{ cm}^3$. Divide by 1000 to convert from cm^3 (or ml) to dm^3.

For example, $80 \text{ ml} = 80 \text{ cm}^3 = 80 / 1000 = 0.08 \text{ dm}^3$.

Concentration in terms of mass

You can calculate the mass of solute dissolved in a given volume of a **solution** if you know the concentration in g/dm^3 (grams per dm^3):

mass of solute (g) = concentration of solution (g/dm^3) × volume of solution (dm^3)

Higher Tier only

The concentration of solution is greater when:
- a greater mass of solute is dissolved in a given volume of solution
- a given mass of solute is contained in a smaller volume of solution.

Increasing concentration of solution

---

1. A solution of copper(II) sulfate has a concentration of 25 g/dm^3. Calculate the mass of copper(II) sulfate in 100 cm^3 of this solution. [2]

2. **Higher Tier only:** A student wants to make a 10 g/dm^3 solution of sodium chloride using 5 g of NaCl or 0.2 dm^3 of water. Determine **two** ways to make this solution. [4]

1. Volume = $\frac{100}{1000}$ = 0.1[1] Mass = 25 × 0.1 = 2.5 g.[1] Correct answer scores both marks.[2]

2. Dissolve 2 g of NaCl in water to make 0.2 dm^3 of solution[1], $\frac{2\,g}{0.2\,dm^3}$ = 10 g/dm^3[1]

Dissolve 5 g of NaCl in water to make 0.5 dm^3 of solution[1], $\frac{5\,g}{0.5\,dm^3}$ = 10 g/dm^3[1]

EXAMINATION PRACTICE

01 Calculate the relative formula mass (M_r) of the following compounds.

    01.1 Magnesium nitride, Mg_3N_2 [1]

    01.2 Iron(II) nitrate, $Fe(NO_3)_2$ [1]

          Relative atomic masses (A_r): N = 14, O = 16, Mg = 24, Fe = 56

02 Carbonic acid, H_2CO_3, is found naturally in rainwater.

    Calculate the percentage by mass of oxygen in carbonic acid.

    Relative atomic mass (A_r) of oxygen = 16, relative formula mass (M_r) of H_2CO_3 = 62   [2]

03 Copper(II) carbonate decomposes when it is heated:

$$CuCO_3(s) \rightarrow CuO(s) + CO_2(g)$$

    A boiling tube containing a sample of copper(II) carbonate was heated. Give a reason that explains why the mass of the boiling tube and its contents changed from 5.5 g to 4.9 g.   [1]

04 A solution of sodium chloride had a concentration of 8 g/dm^3.

    Calculate the mass of sodium chloride in 50 cm^3 of this solution.   [2]

Higher Tier only

05 Calculate the mass of 0.25 mol of carbon dioxide.

    Relative formula mass (M_r) = 44   [1]

06 Chlorine is manufactured by passing an electric current through concentrated sodium chloride solution:

$$2NaCl + 2H_2O \rightarrow 2NaOH + H_2 + Cl_2$$

    Calculate the theoretical mass of chlorine that can be made from 7.25 g of sodium chloride.

    Relative atomic masses (A_r): Na = 23, Cl = 35.5   [4]

07 Magnesium powder reacts with iron(III) oxide powder:

$$3Mg + Fe_2O_3 \rightarrow 3MgO + 2Fe$$

    960 g of magnesium was added to 2.0 kg of iron(III) oxide.

    Explain why iron(III) oxide was the limiting reactant.

    Relative atomic masses (A_r): O = 16, Mg = 24, Fe = 56   [4]

08 Red lead oxide, Pb_3O_4, is used in some rustproof paints. It can be prepared by heating lead(II) oxide, PbO, with oxygen. In one of these reactions, 6.69 g of PbO reacts with an excess of oxygen to produce 6.85 g of Pb_3O_4.

    08.1 Calculate the mass of oxygen used in this reaction.   [1]

    08.2 Determine the balanced symbol equation for this reaction.

          Relative formula masses (M_r): O_2 = 32, PbO = 223, Pb_3O_4 = 685   [3]

OXIDATION AND REDUCTION

Gain or loss of oxygen

Oxidation and reduction reactions often involve oxygen:
- **Oxidation** is gain of oxygen
- **Reduction** is loss of oxygen.

You can work out whether a substance is oxidised or reduced in a reaction by looking at the balanced equation for the reaction. For example, copper(II) oxide reacts with carbon:
- Carbon gains oxygen and is oxidised
- Copper oxide loses oxygen and is reduced

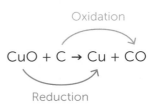

Oxidation

$$CuO + C \rightarrow Cu + CO$$

Reduction

Loss or gain of electrons   Higher Tier only

Oxidation and reduction is also described in terms of electrons:
- Oxidation is loss of electrons
- Reduction is gain of electrons

These definitions are useful for identifying oxidation and reduction in reactions that do not involve oxygen.

For example, sodium reacts with chlorine: $2Na + Cl_2 \rightarrow 2NaCl$

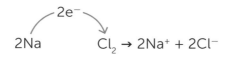

$2e^-$

$$2Na \qquad Cl_2 \rightarrow 2Na^+ + 2Cl^-$$

Rusting is an oxidation reaction

Two half equations describe what happens in this reaction:
- Sodium loses electrons and is oxidised to sodium ions: $2Na \rightarrow 2Na^+ + 2e^-$
- Chlorine gains electrons and is reduced to chloride ions: $Cl_2 + 2e^- \rightarrow 2Cl^-$

---

1. Explain why the reactions of metals with oxygen are described as oxidation reactions.   [2]
2. When powdered and heated, magnesium reacts with zinc oxide: $Mg + ZnO \rightarrow MgO + Zn$

   Identify oxidation and reduction in this reaction. Give reasons for your answer.   [2]
3. **Higher Tier only:** Bromine displaces iodine from sodium iodide. The reaction can be described by two half equations:

$$Br_2 + 2e^- \rightarrow 2Br^- \text{ and } 2I^- \rightarrow I_2 + 2e^-$$

   Explain which substance is oxidised and which substance is reduced.   [2]

   1. *Oxidation is gain of oxygen[1] and metals gain oxygen when they become metal oxides[1].*
   2. *Magnesium gains oxygen and is oxidised to magnesium oxide[1], zinc oxide loses oxygen and is reduced to zinc[1].*
   3. *Bromine is reduced because it gains electrons to form bromide ions.[1] Iodide ions are oxidised because they lose electrons to form iodine.[1]*

THE REACTIVITY SERIES

The **reactivity series** lists metals in order of their reactivity.

Losing electrons

Metals lose **electrons** to form positively charged **ions** when they react with other substances. In general, the more easily a metal forms positively charged ions, the more **reactive** it is.

The reactivity series is based on observing reactions with water, dilute acids and other substances. The more reactive the metal, the more vigorous the reaction.

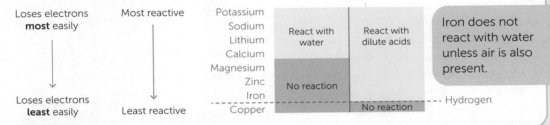

Loses electrons **most** easily — Most reactive

Loses electrons **least** easily — Least reactive

Potassium, Sodium, Lithium, Calcium, Magnesium, Zinc, Iron, Copper, Hydrogen

React with water | React with dilute acids | No reaction

Iron does not react with water unless air is also present.

Reactions with water and dilute acids

Hydrogen is given off if a metal reacts with water or with a dilute acid. The more reactive the metal, the faster the gas is given off. You can revise the reactions of potassium, sodium and lithium with water on **page 145**, and the reactions of metals with dilute acids on **page 179**. Hydrogen is a non-metal element but is often included in the reactivity series. This is because metals that are more reactive than hydrogen can react with dilute acids.

Displacement reactions

A more reactive metal can **displace** a less reactive metal from its compounds. For example, a red-brown coating of copper forms when a piece of zinc is dipped into copper(II) sulfate solution. Zinc ions replace copper ions in the solution because zinc is more reactive than copper:

$$Zn(s) + CuSO_4(aq) \rightarrow ZnSO_4(aq) + Cu(s)$$

1. Describe what happens when an iron nail is placed in water. [2]
2. Compare the reactions of zinc and iron with dilute hydrochloric acid. [4]
3. Explain why magnesium reacts with iron(III) oxide, but iron does not react with magnesium oxide. [2]

   1. *It rusts if air or oxygen is present[1] forming orange-brown hydrated iron(III) oxide[1].*

   2. *Both reactions produce hydrogen[1] but zinc gives steady bubbling and iron gives slow bubbling[1]. Both reactions produce metal chlorides[1] but the reaction with zinc produces zinc chloride and the reaction with iron produces iron(II) chloride[1].*

   3. *Magnesium is more reactive than iron.[1] This means that magnesium can displace iron from iron(III) oxide, but iron is not reactive enough to displace magnesium from magnesium oxide.[1]*

EXTRACTION OF METALS

Gold and other unreactive metals may be found in the Earth's crust as **native metals**. This is the pure form of the metals rather than one of their compounds. However, most metals are found combined with other elements such as oxygen. These metals must be extracted using chemical reactions. **Ores** contain enough of a metal or its compound for extraction to be profitable.

Extraction and the reactivity series

The method used to extract a metal is related to its position on the reactivity series:

- Metals that are more reactive than carbon must be extracted by electrolysis of one of their compounds.
- Metals that are less reactive than carbon can be extracted by heating with carbon.

The four metals shown in **black** here are commonly used metals, but you do not need to recall what happens when they are added to water or dilute acids.

Chemical reactions may be needed to purify silver, gold and platinum, but not to extract them from an ore.

Remember that carbon is a non-metal, not a metal.

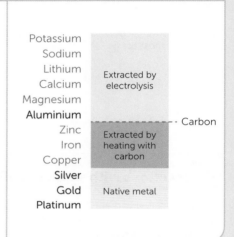

Reduction of oxides

Ores often contain metal oxides, or compounds that can be converted into metal oxides. If a metal is less reactive than carbon, its oxide can be reduced by heating with carbon. Zinc is a metal that can be extracted this way:

- Zinc oxide is reduced to zinc
- Carbon is oxidised to carbon monoxide

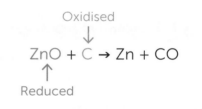

1. Give a reason why aluminium cannot be extracted from its oxide by heating with carbon. [1]

2. Haematite is an iron ore that contains iron(III) oxide, Fe_2O_3.

   Suggest a method to extract iron from haematite. Explain your answer. [3]

3. **Higher Tier only:** Rubidium is produced by reacting sodium with molten rubidium chloride:

   $$Na + RbCl \rightarrow NaCl + Rb$$

   Explain why rubidium is reduced in this reaction. Include a half equation in your answer. [3]

   1. *Aluminium is more reactive than carbon.[1]*

   2. *Heat the haematite with carbon.[1] This will reduce iron(III) oxide to iron[1] because iron is less reactive than carbon[1].*

   3. *Reduction is gain of electrons[1] and rubidium ions gain electrons to form rubidium atoms in this reaction[1]. $Rb^+ + e^- \rightarrow Rb$.[1]*

REACTIONS OF ACIDS WITH METALS

Hydrochloric acid and sulfuric acid

Metals react with **acids** to produce salts and hydrogen. The reactions with magnesium are the most vigorous and the reactions with iron are the least vigorous. The more reactive the metal, the greater the rate of bubbling.

Metal	Salts produced with:	
	Hydrochloric acid	Sulfuric acid
Magnesium	magnesium chloride (colourless solution)	magnesium sulfate (colourless solution)
Zinc	zinc chloride (colourless solution)	zinc sulfate (colourless solution)
Iron	iron(II) chloride (green solution)	iron(II) sulfate (green solution)

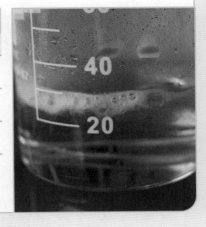

Solutions containing Fe^{2+} ions slowly turn orange-brown as these ions oxidise to Fe^{3+} ions in air.

Redox reactions   Higher Tier only

Acids in aqueous solution are a source of hydrogen ions:

$$HCl(aq) \rightarrow H^+(aq) + Cl^-(aq) \qquad H_2SO_4(aq) \rightarrow 2H^+(aq) + SO_4^{2-}(aq)$$

The reactions of metals with dilute acids involve metal atoms, and hydrogen ions from the acid. They are examples of redox reactions – reduction and oxidation happen at the same time:

- Hydrogen ions are reduced to hydrogen: $2H^+(aq) + 2e^- \rightarrow H_2(g)$
- Metal atoms are oxidised to metal ions: $M(s) \rightarrow M^{2+}(aq) + 2e^-$ (M stands for the metal)

You can combine these two half equations to obtain an ionic equation for the reaction:

$$M(s) + 2H^+(aq) \rightarrow M^{2+}(aq) + H_2(g)$$

Overall, electrons are transferred from metal atoms to hydrogen ions in the reaction.

1. (a) Describe what you would see when magnesium ribbon reacts with excess dilute hydrochloric acid. [3]
   (b) Complete the balanced equation for this reaction. Include state symbols. [2]

   $Mg(\underline{\quad}) + \underline{\quad} HCl(\underline{\quad}) \rightarrow MgCl_2(\underline{\quad}) + H_2(\underline{\quad})$

2. **Higher Tier only:** Explain why the reaction of zinc with dilute sulfuric acid is a redox reaction. [4]

   1. (a) *Rapid bubbling.*[1] *The magnesium gradually disappears*[1] *to form a colourless solution*[1].
      (b) *$Mg(s) + \mathbf{2}HCl(aq) \rightarrow MgCl_2(aq) + H_2(g)$ Correctly balanced*[1], *correct state symbols*[1].
   2. *Hydrogen ions from the acid gain electrons*[1] *and are reduced to hydrogen*[1]. *Zinc atoms lose electrons*[1] *and are oxidised to zinc ions*[1].

FORMULAE OF IONIC COMPOUNDS

You can work out the **formula** of an ionic compound if you know the formulae of the **ions** it contains.

Formulae of common ions

The table shows the formulae of some common ions.

Positive ions		Negative ions	
Name	**Formula**	**Name**	**Formula**
Ammonium	NH_4^+	Bromide	Br^-
Hydrogen	H^+	Chloride	Cl^-
Lithium	Li^+	Hydrogencarbonate	HCO_3^-
Potassium	K^+	Hydroxide	OH^-
Silver	Ag^+	Iodide	I^-
Sodium	Na^+	Nitrate	NO_3^-
Barium	Ba^{2+}	Carbonate	CO_3^{2-}
Calcium	Ca^{2+}	Oxide	O^{2-}
Copper(II)	Cu^{2+}	Sulfate	SO_4^{2-}
Magnesium	Mg^{2+}	Sulfide	S^{2-}
Zinc	Zn^{2+}	Phosphate	PO_4^{3-}
Iron(II)	Fe^{2+}		
Lead(II)	Pb^{2+}		
Iron(III)	Fe^{3+}		
Aluminium	Al^{3+}		

HCO_3^- is the hydrogencarbonate ion, but the names of compounds containing this ion end in 'hydrogen carbonate'.

You will be given the formulae of the ions needed to deduce the formula of an ionic compound.

Deducing formulae

There are equal numbers of positive and negative charges in an ionic compound. You need to take this into account when working out a formula. The charges are written in superscript.

- Sodium chloride contains Na^+ ions and Cl^- ions – its formula is NaCl.
- Ammonium sulfate contains NH_4^+ ions and SO_4^{2-} ions – its formula is $(NH_4)_2SO_4$.

A compound ion contains more than one element. Its formula goes inside brackets if you need two or more of the ionic compound in the formula. So $Mg(OH)_2$ is correct but $MgOH_2$ is not.

1. Write the formulae of the following ionic compounds: [7]
   (a) zinc bromide
   (b) iron(III) oxide
   (c) aluminium hydroxide
   (d) copper(II) carbonate
   (e) ammonium chloride
   (f) potassium phosphate
   (g) calcium nitrate

2. Name the following ionic compounds:
   (a) $PbCl_2$ (b) Ag_2O (c) CaS
   (d) $NaHCO_3$ (e) $Ba_3(PO_4)_2$ [5]

1. (a) $ZnBr_2$[1]   (b) Fe_2O_3[1]   (c) $Al(OH)_3$[1]
   (d) $CuCO_3$[1]   (e) NH_4Cl[1]   (f) K_3PO_4[1]
   (g) $Ca(NO_3)_2$[1]
2. (a) lead(II) chloride[1]   (b) silver oxide[1]
   (c) calcium sulfide[1]
   (d) sodium hydrogen carbonate[1]
   (e) barium phosphate[1]

NEUTRALISATION OF ACIDS

Bases, alkalis and metal carbonates

Acids are **neutralised** by bases, alkalis and carbonates. A **base** reacts with an acid to produce a salt and water only. Bases include metal hydroxides and metal oxides. **Alkalis** are soluble bases, usually metal hydroxides. **Carbonates** react with acids to produce a salt, water and carbon dioxide.

Predicting products

Neutralisation reactions always produce a salt and water. Reactions with a carbonate produce carbon dioxide as well. Work out which salt is formed using the rules below. You can write the formula of the salt if you know the formulae of the ions it contains.

Salts

A **salt** is a substance formed when hydrogen ions in an acid are replaced by other ions. The particular salt formed in a neutralisation reaction depends upon the two reactants:

- the positive metal ion or ammonium ion in the base, alkali or carbonate
- the acid used:
    - hydrochloric acid contains Cl^- ions and forms chlorides
    - nitric acid contains NO_3^- and forms nitrates
    - sulfuric acid contains SO_4^{2-} ions forms sulfates

Ammonium nitrate is used in fertilisers to improve the yields of crops.

The table gives some examples.

	Hydrochloric acid	Nitric acid	Sulfuric acid
Copper(II) oxide	copper(II) chloride $CuCl_2$	copper(II) nitrate $Cu(NO_3)_2$	copper(II) sulfate $CuSO_4$
Sodium hydroxide	sodium chloride $NaCl$	sodium nitrate $NaNO_3$	sodium sulfate Na_2SO_4
Ammonium carbonate	ammonium chloride NH_4Cl	ammonium nitrate NH_4NO_3	ammonium sulfate $(NH_4)_2SO_4$

1. Write word equations for the reactions between the following pairs of reactants.

   (a) Zinc oxide and nitric acid   [2]   (b) Potassium hydroxide and sulfuric acid   [2]

2. Write the formula for the salt formed when iron(II) carbonate reacts with nitric acid.   [1]

   Formulae of ions: Fe^{2+} and NO_3^-

3. Complete the equation for the reaction between calcium carbonate and hydrochloric acid. [2]

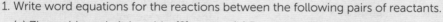

   _____ $CaCO_3$ + _____ HCl → _____ + _____ + _____

---

1. (a) zinc oxide + nitric acid → zinc nitrate[1] + water[1]

   (b) potassium hydroxide + sulfuric acid → potassium sulfate[1] + water[1]

2. $Fe(NO_3)_2$[1]

3. $CaCO_3 + 2HCl → CaCl_2 + H_2O + CO_2$ (correct formula of reactants[1] correct formulae of products[1] correctly balanced[1]).

SALTS FROM INSOLUBLE REACTANTS

Soluble salts can be made using an acid and a suitable solid, **insoluble** reactant. The insoluble reactant can be a reactive metal, a metal oxide or metal hydroxide, or a metal carbonate. Once the reaction has finished, the excess solid is filtered off leaving a salt solution. This solution can be crystallised to produce a solid salt.

Choosing the solid reactant

The choice of reactant depends on the reactivity of the metal. You need to consider whether the metal reacts with dilute acids and, if it does, whether it is too reactive to use safely.

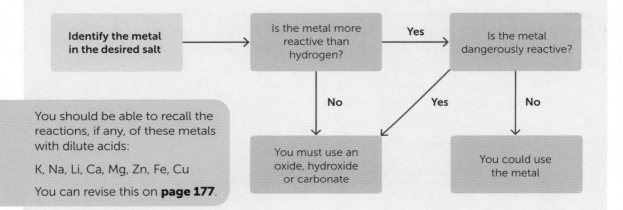

You should be able to recall the reactions, if any, of these metals with dilute acids:

K, Na, Li, Ca, Mg, Zn, Fe, Cu

You can revise this on **page 177**.

Outline of the method

Required practical 8 involves preparing a pure, dry sample of a soluble salt from an insoluble metal oxide or metal carbonate. You can revise this practical in detail on **page 183**.

In outline:

1. Add the insoluble solid to the dilute acid until no more reacts and some is left over.

2. Use filtration to remove the excess insoluble solid. Collect the filtrate (the salt solution).

3. Use crystallisation to produce the solid salt from the salt solution.

You can revise these separation methods on **pages 131 and 132**.

1. Give a reason why copper cannot be used to make copper(II) sulfate. [1]

2. Give **two** reasons why lithium, sodium and potassium salts cannot be made from insoluble reactants. [2]

3. Name **two** insoluble substances that can be used to make zinc chloride. [2]

   1. Copper does not react with dilute acids.[1]

   2. Lithium, sodium and potassium are too reactive to use with acids safely.[1] Their oxides, hydroxides and carbonates are all soluble in water.[1]

   3. Two from: zinc[1], zinc oxide[1], zinc hydroxide[1], zinc carbonate[1].

REQUIRED PRACTICAL 8

Preparing a pure, dry soluble salt

This activity helps you develop your ability to heat substances safely, to handle them safely and carefully, and to separate and purify them.

Heating water and solutions

The **Bunsen burner** is commonly used to heat substances. Adjust the rate of heating by carefully turning the gas tap once the Bunsen burner is lit and turning the collar to adjust the size of the air hole.

The orange safety flame, obtained by closing the air hole, should not be used for heating.

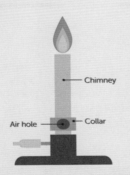

Chimney

Air hole — Collar

1. Describe a safe way to warm 25 cm³ of dilute sulfuric acid. [4]

   Place the acid in a 100 cm³ beaker[1]. Set up a tripod and gauze on a heat-resistant mat, and place the beaker on the gauze[1]. Heat the acid gently using a Bunsen burner[1], then turn off the Bunsen burner before the acid begins to boil[1].

Making a salt solution

Copper(II) oxide is an insoluble black powder. It reacts with dilute sulfuric acid to produce copper(II) sulfate and water. You should add a small mass of copper(II) oxide to warm acid using a **spatula**, then stir with a glass rod. Repeat until excess copper(II) oxide has been added.

2. (a) Describe how you will know that excess copper(II) oxide has been added. [2]

   (b) Explain why it is necessary to add excess copper(II) oxide. [2]

   (a) The copper(II) oxide will no longer disappear[1] but will form a cloudy mixture instead[1].

   (b) This ensures that all the sulfuric acid will have reacted[1] which makes it safer to carry out the next stages in the experiment[1].

Preparing crystals from a salt solution

Use **filtration** (page 131) to remove excess copper(II) oxide from the copper(II) sulfate solution. Then use **crystallisation** to evaporate water from the filtrate (page 132). Large, regularly shaped crystals are obtained if the water is evaporated slowly, and the solution is not heated to dryness over the boiling water bath.

3. Suggest **three** precautions needed for safe working in this activity. Give reasons for your answers. [3]

   Tie hair and loose clothing to avoid it catching fire in the Bunsen burner flame.[1] Let the beaker and its contents cool before filtering to avoid getting burned.[1] Wear eye protection and avoid skin contact because sulfuric acid and copper(II) sulfate are irritants.[1]

THE pH SCALE

The **pH scale** goes from 0 to 14. It is a measure of the acidity or alkalinity of a solution.

Acids

Acids produce hydrogen ions, $H^+(aq)$, when they dissolve in water. The solutions formed are described as being **acidic**, and have pH values **less than 7**.

Alkalis

Alkalis dissolve in water to form solutions described as being **alkaline**. These solutions contain hydroxide ions, $OH^-(aq)$. They have pH values **greater than 7**.

Solutions with a pH value of 7 have **equal** concentrations of H^+ ions and OH^- ions. They are described as being **neutral**.

```
0   1   2   3   4   5   6   7   8   9   10  11  12  13  14
                            Neutral
←────────────────────────          ────────────────────────→
    Increasingly acid                    Increasingly alkaline
```

Using indicators

An **indicator** is a substance that changes colour depending on the pH of the solution it is in. Universal indicator is a mixture of indicators intended to give a spectrum of colours. It is a wide range indicator because it can be used to estimate a range of pH values, not just one.

```
0   1   2   3   4   5   6   7   8   9   10  11  12  13  14
```

Neutralisation

Alkalis are soluble bases, usually metal hydroxides. When an alkali neutralises an acid:

- H^+ ions in the acidic solution react with OH^- in the alkaline solution
- water is produced in the reaction.

You can represent this neutralisation reaction with an ionic equation:

$$H^+(aq) + OH^-(aq) \rightarrow H_2O(l)$$

1. Describe how to use universal indicator paper to measure the approximate pH of a solution. [3]
2. The acidity or alkalinity of a solution can be measured using universal indicator solution or a pH probe. Compare these two methods. [4]

1. *Use a clean glass rod to add a spot of the solution to the paper[1], leave for 30 seconds for the colour to develop[1] then match it to a colour on a pH colour chart[1].*
2. *Both methods produce a pH value[1] but the value obtained by a pH probe is more accurate[1]. Universal indicator can only produce pH values with whole numbers but a pH probe can give pH values precise to 1 or 2 decimal places.[1] A pH probe must be calibrated to give accurate readings but universal indicator solution does not.[1]*

INVESTIGATING NEUTRALISATION

The **pH** of a reaction mixture changes as an acid is added to an alkali. Solutions of strong acids such as hydrochloric acid tend to have very low pH values. Solutions of strong alkalis such as sodium hydroxide tend to have very high pH values. A neutral solution forms if they are mixed together in just the right proportions.

An example method

You can investigate how pH changes as dilute hydrochloric acid is added gradually to sodium hydroxide solution.

1. Add 20 cm³ of sodium hydroxide solution to a beaker.

2. Stir with a glass rod, then measure the pH.

3. Add 1 cm³ of dilute hydrochloric acid using a dropping pipette.

4. Repeat steps 2 and 3 until you have added a total of 40 cm³ of hydrochloric acid.

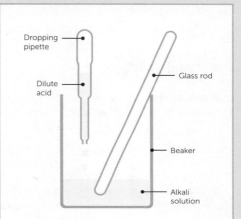

Example results

The graph shows results from an experiment like this one.

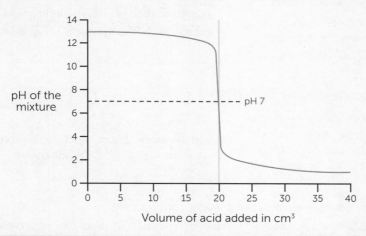

1. Determine the volume of acid needed to exactly neutralise the alkali.   [1]

2. Describe what is happening: (a) In the range 0–20 cm³ [2]   (b) In the range 20–40 cm³   [2]

> *1. 20 cm³[1]*
>
> *2. (a) The pH falls gradually up to about 18 cm³[1] then it falls rapidly to pH 7[1].*
>
>   *(b) The pH falls rapidly up to about 22 cm³[1] then it falls gradually to pH 1.5[1].*

STRONG AND WEAK ACIDS

Acids can be described as strong or weak, depending on how much they ionise in aqueous solution.

Strong acids

Strong acids fully ionise in aqueous solution – each acid molecule breaks down in the water to form ions. Common strong acids include:

- Hydrochloric acid: $HCl(aq) \rightarrow H^+(aq) + Cl^-(aq)$
- Sulfuric acid: $H_2SO_4(aq) \rightarrow 2H^+(aq) + SO_4^{2-}(aq)$

Weak acids

Weak acids only partially ionise in aqueous solution – only a few of their acid molecules break down in the water to form ions. Common weak acids include:

- Ethanoic acid: $CH_3COOH(aq) \rightleftharpoons H^+(aq) + CH_3COO^-(aq)$
- Carbonic acid: $H_2CO_3(aq) \rightleftharpoons 2H^+(aq) + CO_3^{2-}(aq)$

The equations for the strong acids have the usual arrow. The equations for the weak acids have a split arrow because these reactions are **reversible** and do not go to completion (see **page 208**).

Atmospheric carbon dioxide dissolves in rainwater to form carbonic acid. This seeps through the ground and dissolves limestone rocks, forming caves.

Concentration of H⁺ and pH

The concentration of hydrogen ions in a solution determines its pH. The greater the hydrogen ion concentration, the lower the pH.

In neutral solutions:
concentration of $H^+(aq)$ = concentration of $OH^-(aq)$

As the hydrogen ion concentration increases by a factor of 10, the pH decreases by 1. The table gives some examples of how this works for nitric acid (a strong acid) and citric acid (a weak acid).

Concentration of acid in mol/dm³	pH of nitric acid	pH of citric acid
0.01	2	2.62
0.1	1	2.08
1	0	1.57

In the exam, you only need to be able to use pH values given in whole numbers.

---

1. Explain why citric acid can be described as weak, dilute or concentrated.   [3]
2. A student discovers that 5.6 mol/dm³ ethanoic acid and 0.01 mol/dm³ hydrochloric acid have the same pH value. Explain this observation.   [2]

   1. *Citric acid is a weak acid because it is partially dissociated in aqueous solution.[1] It is dilute when a given volume contains a small amount of acid[1] and concentrated when the volume contains a larger amount of acid[1].*

   2. *Ethanoic acid is a weak acid and only partially ionises in solution[1], but hydrochloric acid is a strong acid and fully ionises in solution[1]. A high enough concentration of ethanoic acid produces the same H⁺ concentration as a lower concentration of hydrochloric acid.[1]*

ELECTROLYSIS

Electrolysis is a process that can decompose compounds to simpler substances.

Electrolytes and electrodes

An **electrolyte** is a liquid or solution that can conduct electricity. **Ionic compounds** are electrolytes when they are molten (in the liquid state) or when they are dissolved in water. This is because their ions are free to move and carry electric charge from place to place.

Electrodes are substances that conduct electricity into an electrolyte. They are usually graphite, or unreactive metals such as copper and platinum. **Inert** electrodes only provide a charged surface for reactions to happen on, and do not become chemically changed themselves.

Metals and ionic compounds conduct electricity when they are liquid, but only metals (and graphite) conduct when they are solid.

1. Give the meaning of direct current, dc. [1]
2. Explain why graphite and copper can act as electrodes. [2]
3. Explain why table sugar does not conduct electricity when molten or in solution. [2]

1. *It is electric current that flows in one direction only.*[1]

2. *They conduct electricity*[1] *because they contain delocalised electrons*[1].

3. *Table sugar consists of small molecules*[1] *and does not have charged particles that are free to move*[1].

The process of electrolysis

Ions move to oppositely charged electrodes when an electric current passes through an electrolyte. When ions reach the electrodes, they are **discharged** as elements.

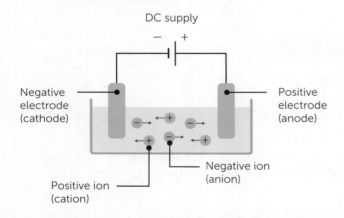

DC supply

Negative electrode (cathode)

Positive electrode (anode)

Negative ion (anion)

Positive ion (cation)

Weak acids   Higher Tier only

Reduction and oxidation processes happen at the electrodes:
- at the cathode, positively charged ions gain electrons and are **reduced**.
- at the anode, negatively charged ions lose electrons and are **oxidised**.

You can represent these reactions using **half equations**. You can revise redox reactions on **page 179** and half equations on **page 130**.

ELECTROLYSIS OF MOLTEN IONIC COMPOUNDS

Molten ionic compounds decompose into elements.

Binary ionic compounds

A **binary** ionic compound consists of a metal element and a non-metal element. This means that potassium oxide (K_2O) is a binary ionic compound, but potassium hydroxide (KOH) is not.

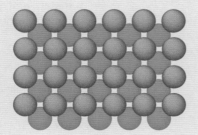

Potassium oxide structure

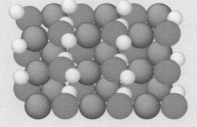

Potassium hydroxide structure

Binary ionic compounds decompose into the two elements they contain during electrolysis. The metal element forms at the negative charged electrode and the non-metal element forms at the positive electrode.

Half equations   Higher Tier only

At the cathode, potassium ions gain electrons and are reduced: $K^+ + e^- \rightarrow K$

At the anode, oxide ions lose electrons and are oxidised: $2O^{2-} \rightarrow O_2 + 4e^-$

1. A teacher passed an electric current through molten lead bromide using graphite electrodes. Orange-brown vapour was observed at the anode and silvery liquid at the cathode.

   (a) Explain the teacher's observations. [3]

   (b) Explain the function of the graphite electrodes. [2]

2. (a) Predict the product formed at each electrode during the electrolysis of molten zinc chloride. [2]

   (b) **Higher Tier:** Write half equations for the reactions that occur at the electrodes. [2]

   1. (a) *Lead ions move to the cathode[1] and are discharged as lead atoms[1]. Bromide ions move to the anode[1] and are discharged as bromine, which escapes as a vapour[1].*

      (b) *They conduct the electric current into the molten lead bromide[1] but are inert because they do not become part of either product[1].*

   2. (a) *Zinc at the negative charged electrode[1], chlorine at the positive electrode[1].*

      (b) $Zn^{2+} + 2e^- \rightarrow Zn$[1] $2Cl^- \rightarrow Cl_2 + 2e^-$[1].

EXTRACTING METALS USING ELECTROLYSIS

Aluminium is more reactive than carbon, so it is extracted using electrolysis.

Extracting metals

Metals that are less reactive than carbon can be extracted by reducing their ores with carbon.

Some metals are less reactive than carbon but cannot be extracted this way because they react with carbon. Tungsten could be extracted by reducing tungsten oxide with carbon, but tungsten carbide forms during the process.

Metals that are more reactive than carbon must be extracted using electrolysis. Large amounts of energy are used to produce the electric currents needed, so this process is usually more expensive than heating with carbon.

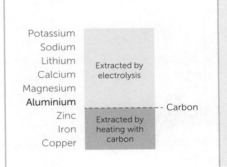

Potassium
Sodium
Lithium   Extracted by
Calcium   electrolysis
Magnesium
Aluminium - - - - - - - - - - - - - Carbon
Zinc   Extracted by
Iron   heating with
Copper   carbon

Manufacturing aluminium

Aluminium ore contains aluminium oxide, which is insoluble in water. It would be very expensive to produce molten aluminium oxide because its melting point is very high (around 2072 °C). To reduce the amount of energy needed, it is dissolved in molten **cryolite** (which melts at 950 °C).

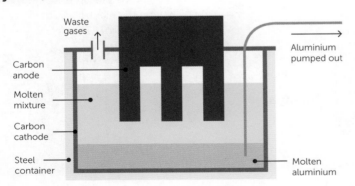

Waste gases

Carbon anode

Molten mixture

Carbon cathode

Steel container

Aluminium pumped out

Molten aluminium

During the electrolysis of the molten mixture of aluminium oxide and cryolite, aluminium ions are discharged at the cathode and oxide ions are discharged at the anode.

Higher Tier only

At the cathode: $Al^{3+} + 3e^- \rightarrow Al$ **At the anode:** $2O^{2-} \rightarrow O_2 + 4e^-$

Explain why the carbon anodes must be replaced frequently during the manufacture of aluminium.   [2]

Oxide ions are discharged as oxygen at the anodes. This reacts with the carbon in the anodes to produce carbon dioxide[1]. The anodes wear away as carbon escapes as carbon dioxide[1].

The huge electrical currents passing through the electrolysis cells in an aluminium smelter create powerful magnetic fields.

ELECTROLYSIS OF AQUEOUS SOLUTIONS

Ions from water

The electrolysis of aqueous solutions may produce gases at both electrodes. Some of the water molecules in water break down to produce hydrogen ions and hydroxide ions:

$$H_2O(l) \rightarrow H^+(aq) + OH^-(aq)$$

Depending on the other ions present, these ions may be discharged:

- H^+ at the cathode to make hydrogen
- OH^- at the anode to make oxygen.

Higher Tier only

An oxidation reaction happens at the anode:

$$4OH^- \rightarrow 2H_2O + O_2 + 4e^-$$

At the cathode

Hydrogen, H_2, is produced unless the metal is less reactive than hydrogen, when the metal is produced instead.

Potassium
Sodium
Lithium
Calcium   Hydrogen at
Magnesium   the cathode
Zinc
Iron   - - - - - - - - - - - - - - - - Hydrogen
Copper   Metal at the
Silver   cathode

At the anode

Oxygen, O_2, is produced unless halide ions are present, when a halogen is produced instead:

- chlorine if chloride ions are present
- bromine if bromide ions are present
- iodine if iodide ions are present.

1. Predict the products formed at the cathode during the electrolysis of these solutions:
   (a) silver nitrate   [1]
   (b) potassium bromide   [1]
   (c) iron(III) chloride   [1]
2. Predict the products formed at the anode during the electrolysis of these solutions:
   (a) zinc nitrate   [1]
   (b) sodium carbonate   [1]
   (c) zinc chloride   [1]
3. Predict the products formed during the electrolysis of dilute sulfuric acid, H_2SO_4(aq).   [2]

1. (a) silver[1]   (b) hydrogen[1]
   (c) hydrogen[1]
2. (a) oxygen[1]   (b) oxygen[1]
   (c) chlorine[1]
3. Hydrogen at the negative electrode[1] and oxygen at the positive electrode[1].

Examples

Aqueous solution	Positive ions	At the cathode	Negative ions	At the anode
Sodium hydroxide	H^+ Na^+	Hydrogen	OH^-	Oxygen
Sodium chloride	H^+ Na^+	Hydrogen	OH^- Cl^-	Chlorine
Copper(II) chloride	H^+ Cu^{2+}	Copper	OH^- Cl^-	Chlorine
Copper(II) sulfate	H^+ Cu^{2+}	Copper	OH^- SO_4^{2-}	Oxygen

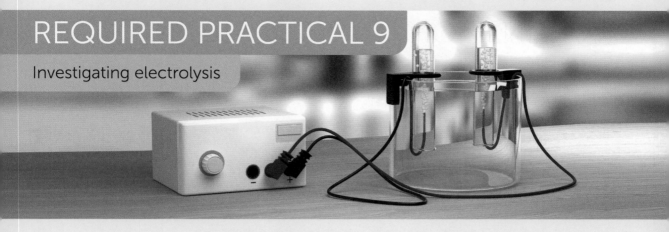

REQUIRED PRACTICAL 9

Investigating electrolysis

This required practical activity helps you develop your ability to draw, set up and use electrochemical cells, and to analyse the products formed.

Drawing electrochemical cells

A **scientific diagram** is a labelled representation of objects, not an artist's drawing. Make sure you:

- use a pencil and ruler
- draw straight lines where possible
- avoid unnecessary shading.

When you add a label to your drawing:

- draw a single straight line away from the object
- write a clear label at the end of the line.

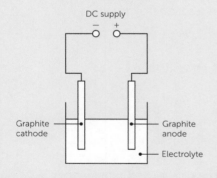

DC supply

Graphite cathode

Graphite anode

Electrolyte

Gas tests

It is difficult to obtain enough hydrogen or oxygen for gas tests with simple electrochemical cells, but you should be able to test for the presence of **chlorine**. You can revise gas tests on **page 226**.

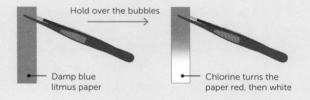

Hold over the bubbles

Damp blue litmus paper

Chlorine turns the paper red, then white

1. Predict what you would see during the electrolysis of copper(II) chloride solution. [2]
2. Explain why the graphite electrodes are described as inert in these experiments. [2]

   1. *A brown-red coating on the negative electrode.[1] Bubbles coming off the positive electrode.[1]*
   2. *They provide a surface for the electrode reactions to happen on[1] but the carbon does not take part in the electrode reactions[1].*

Electrolysis of aqueous solutions

The table shows the expected products of the electrolysis of four different **electrolytes** using graphite **electrodes**. You can revise how to predict these products on **page**s 188 and 190.

Electrolyte	Product at cathode (negative electrode)	Product at anode (positive electrode)
Copper(II) chloride solution	Copper	Chlorine
Copper(II) sulfate solution	Copper	Oxygen
Sodium chloride solution	Hydrogen	Chlorine
Sodium sulfate solution	Hydrogen	Oxygen

EXAMINATION PRACTICE

01 A student carried out an experiment to determine the relative reactivity of three metals (X, Y and Z). Mixtures of metals and metal oxides were heated, and the presence or absence of a reaction was recorded. The table shows the student's results.

	X oxide	Y oxide	Z oxide
Metal X		Reaction	No reaction
Metal Y	No reaction		No reaction
Metal Z	Reaction	Reaction	

01.1 Determine the order of reactivity, starting with the most reactive. [1]

01.2 Give a reason why the student did not heat metal X with X oxide. [1]

01.3 In a separate experiment, the student heated magnesium with copper(II) oxide. Magnesium oxide and copper were produced.
Explain which substance was oxidised and which substance was reduced. [2]

02 This question is about copper(II) nitrate.
02.1 Name the dilute acid needed to produce copper(II) nitrate. [1]
02.2 Give a reason why copper(II) nitrate cannot be made using copper. [1]
02.3 Name **two** insoluble solids that could be used to make copper(II) nitrate. [2]

03 A student added dilute hydrochloric acid to sodium hydroxide solution.
03.1 Compare the pH values of these two reactants before the experiment began. [1]
03.2 Write the balanced ionic equation for the neutralisation reaction that occurred. Include state symbols. [2]

04 Write the chemical formula of aluminium sulfide and sodium phosphate.
Formulae of ions: Al^{3+}, Na^+, S^{2-}, PO_4^{3-} [2]

05 Aluminium is extracted from aluminium oxide using electrolysis.
05.1 Give a reason why aluminium cannot be extracted by heating aluminium oxide with carbon. [1]
05.2 Explain why a molten mixture of aluminium oxide and cryolite is used in the process. [3]
05.3 Explain why the positive electrode must be replaced frequently. [3]

06 A teacher demonstrated electrolysis using molten zinc chloride and carbon electrodes.
06.1 Explain why molten zinc chloride conducts electricity. [2]
06.2 Name the products given off at each electrode. [2]
06.3 The teacher repeated the experiment using zinc chloride solution instead.
Compare the products with the products obtained from molten zinc chloride. [2]

07 Copper(II) chloride solution, $CuCl_2$(aq), is electrolysed using carbon electrodes.

Balance these half equations for the reactions that take place at each electrode.

07.1 Cu^{2+} + ... e^- → Cu [1]

07.2 ...... Cl^- → Cl_2 + ... e^- [1]

08 Magnesium reacts with dilute hydrochloric acid:

$$Mg + 2HCl \rightarrow MgCl_2 + H_2$$

08.1 Identify the substance that is oxidised in the reaction.

Explain your answer with the aid of a suitable half equation. [2]

08.2 Give a reason why this reaction is an example of a redox reaction. [1]

09 This question is about weak and strong acids.

09.1 Carbonic acid, H_2CO_3, is a weak acid. Explain what is meant by a weak acid. [2]

09.2 1.0×10^{-2} mol/dm³ sulfuric acid has a pH of 2. Sulfuric acid, H_2SO_4, is a strong acid.

Determine the pH of 1.0×10^{-5} mol/dm³ sulfuric acid. [1]

EXOTHERMIC AND ENDOTHERMIC REACTIONS

Reactions can be **exothermic** or **endothermic**, depending on whether they transfer energy to or from the surroundings.

Energy changes in reactions

The amount of energy in the universe stays the same during a chemical reaction. This means that energy is conserved – the amount of energy transferred to or from the surroundings during a reaction is equal to the amount of energy lost or gained by the products.

Exothermic reactions

Exothermic reactions include:
- Combustion
- Many oxidation reactions
- Neutralisation

Energy is transferred **to** the surroundings in an exothermic reaction. The temperature of the surroundings increases if energy is transferred by heating.

Endothermic reactions

Endothermic reactions include:
- Thermal decomposition
- Citric acid and sodium hydrogen carbonate reacting together

Energy is transferred **from** the surroundings in an endothermic reaction. The temperature of the surroundings decreases if energy is transferred by heating.

Practical uses

Some cold packs used to treat sports injuries use endothermic reactions. Exothermic reactions are used in handwarmers for cold days. A pouch contains iron powder, and salty water held in carbon. Air enters when the pouch is opened. Oxygen reacts with the iron powder, and the salty water speeds up the reaction. As the iron oxidises, energy is transferred to the surroundings by heating.

1. Suggest the type of reaction (exothermic or endothermic) that is needed in a self-heating food can. Give a reason for your answer. [1]

2. Energy transfers occur when salts dissolve in water. Plan an experiment to distinguish between a salt that produces an exothermic change when it dissolves, and a salt that produces an endothermic change when it dissolves. [6]

   1. *Exothermic, because an exothermic reaction transfers energy to the surroundings and increases their temperature.[1]*

   2. *Add the same volume of water to two polystyrene cups.[1] Measure and record the temperature in each cup.[1] Add a different salt to each cup and stir to dissolve.[1] Measure and record the new temperature in each cup.[1] The exothermic change causes the temperature to go up.[1] The endothermic change causes the temperature to go down.[1]*

Exothermic and endothermic reactions

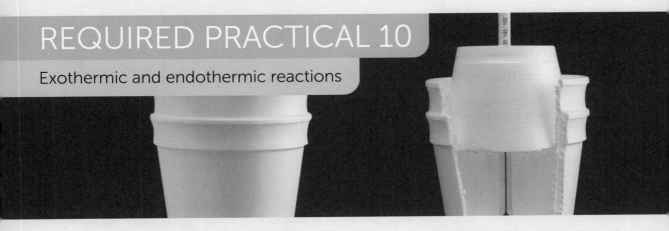

This required practical activity helps you develop your ability to carry out and monitor chemical reactions, including using substances carefully and safely.

Simple calorimeters

Reactions in solution cause changes in temperature. A **calorimeter** is used in experiments where these temperature changes are measured. Energy transfer between the surroundings and the reaction mixture is the greatest source of error in these experiments.

A simple calorimeter consists of a polystyrene cup held securely inside a beaker. The polystyrene and the air gap reduce energy transfers by **conduction**.

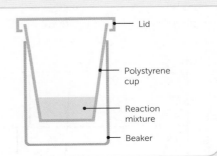

An example experiment

A typical experiment involves:

- recording the starting temperature of a solution of an acid, alkali or salt
- mixing the reactants
- recording the temperature of the reaction mixture over several minutes
- determining the maximum change in temperature.

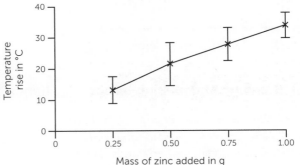

The graph shows the results of an investigation in which different masses of zinc powder were added to copper(II) sulfate solution. The reactions at each mass were repeated several times:

- each × shows a mean rise in temperature
- each I-shaped bar is an **error bar** that shows the **uncertainty** in the results.

1. Give a reason why there is a lid on the polystyrene cup in the diagram. [1]
2. Explain why a line graph was used, rather than a bar chart. [2]

   1. The lid reduces energy transfers by convection.[1]
   2. The temperature rise and mass of zinc added can both be measured.[1] *They are continuous variables*[1], *so a line graph is suitable. Bar charts are suitable if a variable is categoric.*[1]

REACTION PROFILES

Reaction profiles represent the energy changes during a chemical reaction.

Colliding particles

For a chemical reaction to happen:
- reactant particles must collide with one another, and
- they must have enough energy.

The activation energy is the minimum amount of energy needed for a reaction to happen.

Reaction profiles are also called energy level diagrams. They show the relative amounts of energy involved in reactions.

Exothermic reactions

In the reaction profile for an exothermic reaction:
- the energy level of the reactants is higher than the energy level of the products.

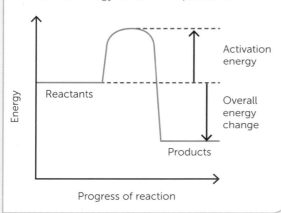

Endothermic reactions

In the reaction profile for an endothermic reaction:
- the energy level of the products is higher than the energy level of the reactants.

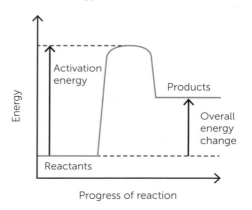

Breaking and making bonds   Higher Tier only

In a chemical reaction, energy is transferred:
- to the reactants in order to break bonds
- to the surroundings when bonds form in the products.

The overall energy change is the difference between the amounts of energy involved.

Compare the activation energy with the overall energy change in exothermic and endothermic reactions. [2]

The activation energy is always positive[1], but the overall energy change is negative in exothermic reactions and positive in endothermic reactions[1].

THE ENERGY CHANGE OF REACTIONS

Bond energies

Bond energies can be used to calculate the overall energy change of a reaction. A **bond energy** is the amount of energy needed to break one mole of a given type of covalent bond. Different bonds have different bond energies, so chemists use tables of data rather than remembering their values. The table below shows the bond energies of some common bonds.

Bond	C–C	C–H	H–H	O–H	O=O	C=C	C=O
Energy in kJ/mol	347	413	436	464	498	612	805

As a chemical reaction happens, energy is:
- supplied to break bonds in the reactants – this is an endothermic process
- released when bonds form in the products – this is an exothermic process.

overall energy change of a reaction = (total energy supplied) – (total energy released)

	Exothermic	Endothermic
Amounts of energy involved	(energy in) < (energy out)	(energy in) > (energy out)
Sign of the overall energy change	Negative	Positive

1. Hydrogen reacts with oxygen to produce water: 2(H–H) + O=O → 2(H–O–H)
   (a) Calculate the overall energy change of this reaction. [4]
   (b) Explain what your answer to part (a) shows about the reaction. [2]
2. Propane undergoes a cracking reaction to produce ethene and methane:
   (a) Calculate the overall energy change of this reaction. [4]

   (b) Explain what your answer to part (a) shows about the reaction. [2]

1. (a) Energy in to break bonds = (2 × 436) + 498 = 1370.[1]
      Energy out when bonds form = (4 × 464) = 1856.[1]
      Overall energy change = 1370 – 1856[1] = –486 kJ/mol.[1]
   (b) The reaction is exothermic[1] because the overall energy change is negative[1].
2. (a) Energy in to break bonds = (2 × 347) + (8 × 413) = 3998.[1]
      Energy out when bonds form = 612 + (4 × 413) + (4 × 413) = 3916.[1]
      Overall energy change = 3998 – 3916[1] = +82 kJ/mol.[1]
   (b) The reaction is endothermic[1] because the overall energy change is positive[1].

> The eight C–H bonds in question 2 are unchanged. You may leave out unchanged bonds from your calculations if you are confident.

EXAMINATION PRACTICE

01   Describe what is meant by an exothermic reaction.   [2]

02   A student added some water to a polystyrene cup, then recorded its temperature. The student
      dissolved some ammonium nitrate in the water, then recorded the temperature of the solution
      formed. The table shows the results.

Temperature at start in °C	Temperature at end in °C
19.4	16.7

02.1   Explain what the results show about the process of dissolving ammonium nitrate.   [2]

02.2   Suggest **one** practical use of the observed changes.   [1]

03   The diagram is a reaction profile for the reaction:

$$CaCO_3 \rightarrow CaO + CO_2.$$

It is not drawn to scale.

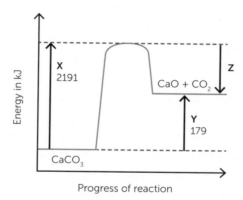

Progress of reaction

Explain what this reaction profile represents.

In your answer, you should refer to energy transfers and chemical bonds.

                                                                                                          [6]

Higher Tier only

04   Hydrogen reacts with iodine to form hydrogen iodide:

$$H-H + I-I \rightarrow 2(H-I)$$

Bond energies in kJ/mol: $I-I$ = 151, $H-I$ = 298. The overall energy change is −9 kJ/mol.

Calculate the bond energy for the H–H bond.   [4]

CHEMISTRY
TOPICS FOR PAPER 2

Information about Paper 2:

Trilogy 8464:

Written exam: 1 hour 15 minutes
Foundation and Higher Tier
70 marks
16.7% of the qualification grade
All questions are mandatory

Specification coverage

The content for this assessment will be drawn from topics on: The rate and extent of chemical change; Organic chemistry; Chemical analysis; Chemistry of the atmosphere; and Using resources.

Questions

A mix of multiple-choice, structured, closed short answer and open response questions. They may include calculations.

Questions assess knowledge, understanding and skills.

CALCULATING RATES OF REACTION

The **rate of reaction** is a measure of how quickly a reaction takes place.

Mean rate of reaction

The more quickly a reaction happens, the greater its rate of reaction. A reaction with a high rate finishes sooner than a reaction with a low rate.

You can calculate the **mean rate** by recording the loss of a reactant during the reaction:

$$\text{mean rate of reaction} = \frac{\text{quantity of reactant used}}{\text{time taken}}$$

You can also calculate the mean rate of reaction by recording the formation of a product:

$$\text{mean rate of reaction} = \frac{\text{quantity of product formed}}{\text{time taken}}$$

1. A student recorded the volume of gas produced in a reaction. The table shows the results.

Time in s	0	30	60	90	120	150	180	210	240
Volume of gas in cm³	0	24	34	59	70	76	79	80	80

(a) Plot these results on the grid. Draw a line of best fit. [3]

(b) Calculate the mean rate of reaction. Give the units. [3]

(a) *All points correct to within ± ½ a small square[2] (allow 1 mark if 6 or 7 are correct), best fit (must be smooth and ignore the anomalous result)[1].*

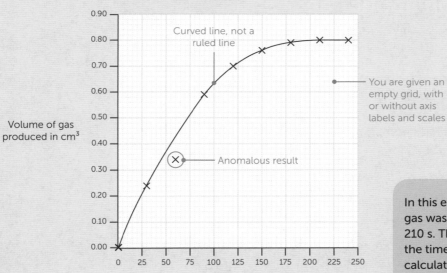

Curved line, not a ruled line

Volume of gas produced in cm³

Anomalous result

You are given an empty grid, with or without axis labels and scales

Time in s

In this example, no more gas was produced after 210 s. This means that the time taken for the calculation was 210 s, rather than 240 s. The student continued to record results after the reaction had finished.

(b) *Mean rate =* $\dfrac{80 \ cm^3}{210 \ s}$ [1] *= 0.38[1] cm³/s.[1]*

Interpreting graphs

The **gradient** of the line on a rate of reaction graph gives the instantaneous rate.

Tangents and gradients

As the rate of reaction increases, the line on a rate of reaction graph becomes steeper and the gradient increases.

You can see this by drawing tangents to the curve. The line is horizontal when the reaction has stopped, and the gradient is zero.

If a rate of reaction is measured by recording changes in mass, its units are g/s.

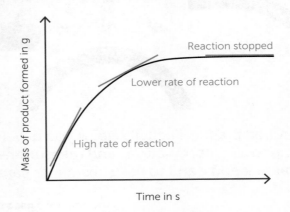

Rates from graphs   Higher Tier only

You can find the rate of reaction at a given time during a reaction:
- Draw the tangent to the curve at that time.
- Calculate the gradient of the tangent.

When you draw your tangent, make sure that your ruler edge is the same distance either side of the curve at that time.

Draw two lines to your tangent so that you make a triangle as large as you can. This improves the accuracy of your gradient.

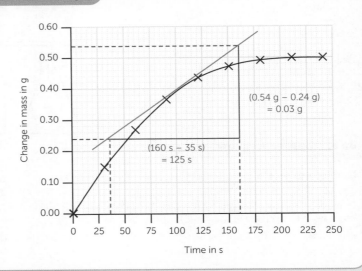

---

2. Describe what happens to the rate of reaction as the reaction proceeds. [2]

   Higher Tier only: In the second graph, calculate the rate of reaction at 100 s. Give the units. [3]

3. In a reaction, 0.10 mol of magnesium are used up in 50 s. Calculate the mean rate of reaction. Give the units. [2]

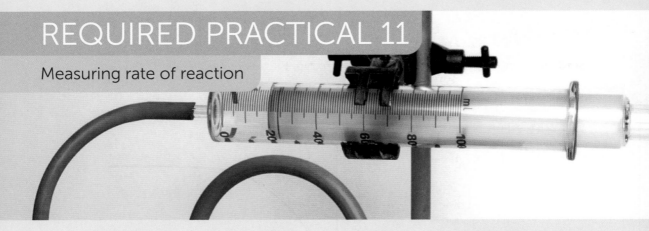

This required practical activity helps you develop your ability to accurately make and record observations and measurements, to carry out and monitor chemical reactions, and to use substances carefully and safely.

Measuring volumes of gases

You can use a **gas syringe** to measure the volume of any gas more accurately. They are made of glass and are expensive, so you may be given a measuring cylinder instead.

Fill the measuring cylinder with water, then turn it upside down under water in a trough.

As bubbles of gas go into the measuring cylinder, the water it contains is pushed downwards before you read the new volume.

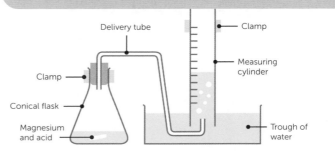

An example experiment

Magnesium reacts with dilute hydrochloric acid to produce magnesium chloride and hydrogen. You can use the rate of production of hydrogen to compare the rate of reaction at different concentrations of acid. The faster the gas syringe or measuring cylinder fills with gas, the greater the rate of reaction. You should control variables such as:

* the mass of magnesium – the greater the mass, the more gas is produced
* the size of the piece of magnesium – powders react more quickly than ribbon.

A table is the best way to record the results. For example:

Time in s	0	10	20	30	40	50	60	70	80	90
0.5 mol/dm³ acid										
1.0 mol/dm³ acid										

1. Explain why the temperature of the acid is a variable that should be controlled. [2]

   1. The rate of reaction depends upon the temperature of the reaction mixture.[1] If this was not kept the same each time, the rate of reaction would not only depend upon the concentration of the hydrochloric acid.[1]

Turbidity

The cloudier a liquid is, the greater its **turbidity**. Some reactions produce a **precipitate** that makes the reaction mixture more turbid. The greater the rate of reaction, the less time it takes for the reaction mixture to become so cloudy that you cannot see through it.

> You must carry out two different investigations for this practical activity. One like the experiment described on the opposite page, and another where you measure a change in colour or turbidity as the reaction carries on.

The disappearing cross investigation

> You need to develop a **hypothesis** of your investigations – a proposal intended to explain observations or facts in advance of any experiments.

Sodium thiosulfate solution reacts with dilute hydrochloric acid to produce four products, including sulfur. This is an insoluble yellow substance that makes the reaction mixture cloudy.

You can get a measure of the rate of reaction if you:

- place a conical flask containing the reaction mixture on top of piece of paper with a cross drawn on it, then
- time how long it takes before you can no longer see the cross through the reaction mixture.

The total volume of the reaction mixture must be controlled, otherwise its depth will also affect the time taken. You can achieve this, and vary the concentration of sodium thiosulfate, by adding measured volumes of water.

The table shows some examples. A fixed volume of acid is needed to begin the reaction.

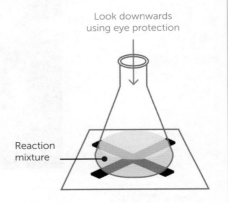

Look downwards using eye protection

Reaction mixture

Volume of sodium thiosulfate solution in cm³	10	20	30	40	50
Volume of water added in cm³	40	30	20	10	0
Concentration of sodium thiosulfate solution in g/dm³	8	16	24	32	40

2. Name **one** piece of apparatus that can be used to measure a change in colour. [1]

3. Refer to the disappearing cross investigation in this question.

   (a) Explain **one** precaution needed to work safely. [3]

   (b) Describe the expected results. [1]

   (c) Suggest **one** way in which the investigation could be extended. [1]

2. Colorimeter.[1]

3. (a) One of the products is sulfur dioxide gas[1], which is toxic and can cause breathing problems[1]. Avoid breathing in fumes when looking down at the reaction mixture.[1]

*   (b) As the concentration of sodium thiosulfate increases, the reaction time decreases.[1]*

*   (c) Heat the sodium thiosulfate solution before adding the acid[1] so that the effect of temperature on the rate of reaction can be investigated[1].*

FACTORS AFFECTING THE RATE OF REACTION

Five common factors that affect the rates of chemical reaction include the **concentrations** of reactants, the **presence of catalysts**, the **surface area** of solid reactants, the **pressure** or gases and the **temperature**.

Surface area

Explosions involve powdered substances that react together at very high rates.

As the surface area of a solid reactant increases, the rate of reaction increases. This is why powders react much more quickly than sheets and lumps.

Pressure

The **pressure** of a gas is a measure of how much force it exerts on its container walls and on objects inside it. If a chemical reaction involves a reactant in the gas state, the rate of reaction increases as the pressure increases.

1. Describe how the rate of a reaction depends on the concentration of a reactant in solution. [1]
2. Describe what a catalyst is. [2]

1. As the concentration of the dissolved reactant increases, the rate of reaction increases.[1]

2. A catalyst is a substance that increases the rate of a reaction[1], but by the end of the reaction is unchanged chemically[1] or in mass[1].

Temperature

Cooking involves chemical reactions in food. It takes longer to cook eggs in hot water than it does to cook them in boiling water.

The **temperature** of a substance is a measure of the average **kinetic energy** of its particles. For a substance in a given state (solid, liquid or gas), heating causes its particles to move more quickly, and its temperature increases. As the temperature of a reaction mixture increases, the rate of reaction increases.

You need to know about the effects of concentration, surface area, pressure, temperature, and catalysts.

COLLISION THEORY

Collision theory is a scientific model that explains how different factors affect the rate of a reaction.

Colliding particles

A chemical reaction can only happen when:
- reactant particles collide with one another, **and**
- the colliding particles have enough energy.

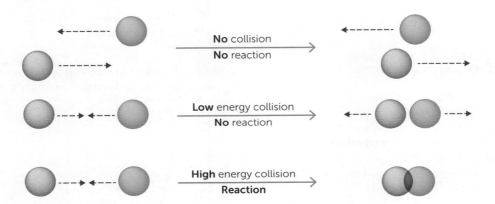

The **activation energy** is the minimum amount of energy that particles need for a reaction to happen. A collision that results in a reaction is a **successful collision**. The greater the **frequency** of successful collisions, the greater the rate of reaction. The table summarises how this works for increases in the concentration or pressure of a reacting substance.

Factor that increases	Frequency of collisions	Energy of particles	Frequency of successful collisions	Rate of reaction
Concentration	Increases	Does not change	Increases	Increases
Pressure	Increases	Does not change	Increases	Increases

1. For the same mass of a solid reactant, the rate of a reaction is greater if the reactant is a powder than if it is a lump. Explain this observation in terms of collision theory. [4]

2. A student explains the effect of the pressure of a reacting gas on reaction rate:
   "As the pressure increases, the number of collisions increases, so the rate of reaction increases."
   Explain why the student's explanation is **not** correct. [2]

   1. *As the particle size decreases, the surface area to volume ratio increases.[1] More of the reacting solid is exposed to particles of the other reactant.[1] The energy of the particles does not change[1] but the frequency of successful collisions increases.[1]*

   2. *The student says 'number' of collisions instead of 'frequency' / 'rate' of collisions.[1] The number of collisions can become large over a long time, even in a slow reaction with a low rate of collisions.[1]*

ACTIVATION ENERGY AND CATALYSTS

Effects of concentration

You might see two different effects of concentration on the rate of reaction:

- **Line 1** if the rate depends on the concentration of one of the reactants
- **Line 2** if the rate depends on the concentrations of two or more reactants

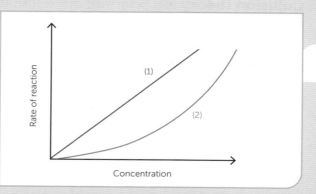

Effect of temperature

As the temperature of a reactant increases, the average energy of its particles increases. The percentage of particles that have the **activation energy** or more is much greater at high temperatures than it is at low temperatures.

Unlike increases in concentration, pressure or surface area, an increase in temperature causes:

- an increase in the frequency of collisions, **and**
- an increase in the energy of reactant particles.

These two factors combine to produce very large increases in the frequency of successful collisions, and in the rates of reactions. You are likely to observe reactions starting, or happening much faster, when the reaction mixture is warmed up.

This graph is here just to help you understand what is happening. You do not need to remember it.

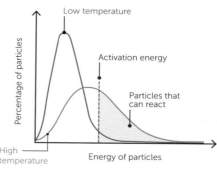

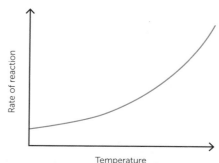

---

1. Explain what line 1 on the first graph shows. [3]

2. Predict how the rate of a reaction will depend on the surface area of a solid reactant. Give a reason for your answer. [2]

1. *The graph shows that the rate of reaction is directly proportional to the concentration of a reactant.*[1] *This is because the line is straight, has a positive gradient*[1] *and passes through the origin*[1].

2. *The rate of reaction will be directly proportional to surface area.*[1] *This is because the number of reactant particles exposed at the surface increases as the surface area increases.*[1]

Catalysts

A catalyst changes the rate of a chemical reaction without being used up in the reaction. Different reactions are catalysed by different catalysts.

You can tell if a substance in a reaction mixture acts as a catalyst because:

- it increases the rate of reaction
- it is **not** shown in the chemical equation for the reaction.

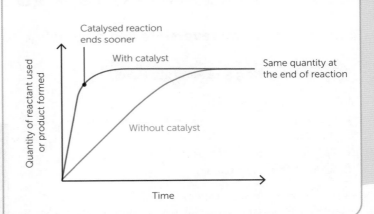

Activation energy

A catalyst provides a pathway that has a **lower activation energy** than the uncatalysed reaction. This can be shown on a **reaction profile** (you can revise these diagrams on **page 196**).

At a given temperature, a greater percentage of reactant particles will have the activation energy or more. The frequency of collisions will be the same as in the uncatalysed reaction, but a greater percentage of these collisions will be successful.

In the middle diagram on the opposite page, the effect is as if the activation energy line is moved to the left, letting more particles react.

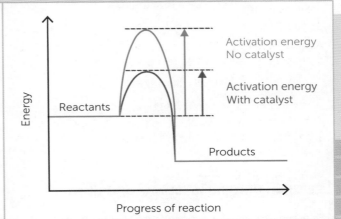

Cars are fitted with catalytic converters which contain precious metals. Hot exhaust gases heat these up and increase the rate of reaction to change harmful gases into CO_2 and water vapour.

1. A mixture of hydrogen peroxide solution and manganese(IV) oxide powder forms water and oxygen:

$$2H_2O_2(aq) \rightarrow 2H_2O(l) + O_2(g)$$

   Identify the catalyst in this reaction. Give a reason for your answer. [2]

2. (a) Describe what is meant by an enzyme. [2]

   (b) Give **one** example of the use of an enzyme in chemistry. [2]

   1. Manganese(IV) oxide is the catalyst[1] because it does not appear in the balanced equation as a reactant or as a product[1].

   2. (a) A protein that acts as a catalyst[1] in living things and other biological systems[1].

   (b) Fermentation to make ethanol[1] using enzymes in yeast[1].

REVERSIBLE REACTIONS

Some reactions are **reversible** and may not go to completion.

Representing a reversible reaction

In a reversible reaction, there is:
- a forward reaction – the reactants react together to produce products
- a reverse reaction – the products react together to produce the original reactants.

A split arrow symbol is used in equations for reversible reactions. In general:

Reverse reaction:
$C + D \rightarrow A + B$

$A + B \rightleftharpoons C + D$

Forward reaction:
$A + B \rightarrow C + D$

Ammonia reacts with hydrogen chloride to form ammonium chloride. The reaction is reversible:

$$NH_3(g) + HCl(g) \rightleftharpoons NH_4Cl(s)$$

White clouds of ammonium chloride form in the reaction. The photo shows how the reaction also happens with ammonia solution and hydrochloric acid.

A reaction between ammonia solution and hydrochloric acid

The reverse reaction is sometimes called the backward reaction.

Changing the reaction conditions

You can change the direction of a reversible reaction if you change the reaction conditions:
- change the temperature of the reaction mixture
- change the concentration of a reacting substance
- change the pressure of a reacting gas that is involved.

The reaction between anhydrous copper(II) sulfate and water is reversible:

anhydrous copper(II) sulfate (white) + water $\rightleftharpoons$ hydrated copper(II) sulfate (blue)

1. When heated in a test tube, solid ammonium chloride decomposes to form ammonia and hydrogen chloride.
   Explain what you expect to see where the test tube is cold. [2]

2. The reaction between anhydrous copper(II) sulfate and water is exothermic.
   Describe the reverse reaction. [2]

   1. *Solid ammonium chloride will form on the cold sides of the test tube[1] because heating makes the reaction go in one direction, and cooling makes it go in the opposite direction[1].*

   2. *It is endothermic[1] and transfers the same amount of energy as the forward reaction[1].*

EQUILIBRIUM

Reversible reactions can reach **equilibrium** in closed systems.

Using up reactants

In any chemical reaction, the amounts of reactants gradually decreases as the reaction carries on. This means that the **rate of reaction** also decreases.

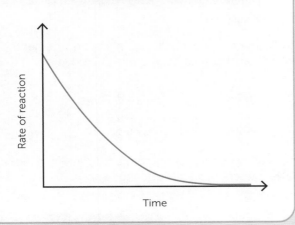

1. Magnesium reacts with dilute hydrochloric acid to produce magnesium chloride solution and hydrogen.
   Give **two** reasons why this reaction cannot reach equilibrium in an open beaker. [2]

2. A student says that the concentrations of all the reacting substances in a reaction at equilibrium are equal. Explain why the student is incorrect. [2]

   1. The reaction is not reversible[1] and one of the reacting substances, hydrogen, can escape[1].

   2. The concentrations of each substance stay the same[1] but different from each other[1].

Reaching equilibrium

In a reversible reaction, the amounts of products gradually increases as the reaction carries on. This means that the rate of the reverse reaction also increases. In a closed system such as a test tube or conical flask with a stopper on, or a reacting solution in a beaker, the rates of the forward reaction and reverse reaction eventually become exactly the same.

The reaction is described as having reached equilibrium.

In a reversible reaction at equilibrium:

- the forward and reverse reactions still keep going, and at the same rate
- the relative amounts of all the reacting substances do not change.

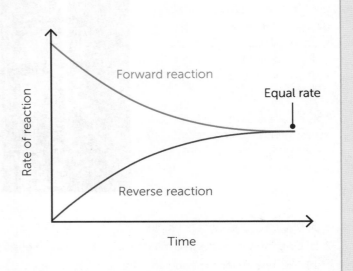

You might imagine equilibrium as like walking the wrong way on an escalator or travelator. You appear to stay in the same position if the rate of movement in each direction is the same.

LE CHATELIER'S PRINCIPLE

You can use **Le Chatelier's principle** to predict how changing the conditions can affect a reaction at equilibrium.

Equilibrium position

The equilibrium position gives you an idea of the relative amounts of the reacting substances in a reaction at equilibrium. In the Haber process, nitrogen reacts with hydrogen to make ammonia:

$$N_2(g) + 3H_2(g) \rightleftharpoons 2NH_3(g)$$

The equilibrium position can lie to the left or to the right of the equation, depending on whether the relative amounts of reactants are greater than, or less than, the relative amounts of products.

$$N_2(g) + 3H_2(g) \rightleftharpoons {\small 2NH_3(g)}$$

Equilibrium position lies to the left
High relative amounts of reactants

$${\small N_2(g) + 3H_2(g)} \rightleftharpoons 2NH_3(g)$$

Equilibrium position lies to the right
High relative amounts of products

A reaction at equilibrium responds to a change in reaction conditions, causing a movement in the equilibrium position. It moves to the left or to the right to counteract the change imposed.

Catalysts

In a reversible reaction, a **catalyst** increases the rates of the forward and reverse reactions by the same percentage, so:

- the equilibrium position stays the same, **but**
- equilibrium is reached more quickly.

This means that catalysts are still useful in industrial processes that involve reversible reactions, such as the Haber process.

An industrial ammonia plant

Predictions

Le Chatelier's principle does not explain why a change in a reaction condition leads to a change in the equilibrium position, but it does let you predict the outcome:

- you do not have to predict the actual amounts of each reacting substance, **but**
- you do have to predict whether the relative amounts of products will increase or decrease.

---

1. Describe Le Chatelier's principle. [2]
2. (a) Give **three** changes in conditions that can cause a change in equilibrium position. [3]
   (b) Give **one** change in conditions that cannot change the equilibrium position. [1]

   1. It is the idea that when a change in conditions is imposed on a system at equilibrium[1], the system responds in a way that counteracts this change[1].
   2. (a) Concentration[1], temperature[1], pressure[1].
      (b) Adding a catalyst.[1]

CONCENTRATION CHANGES AND EQUILIBRIUM

You can also use **Le Chatelier's principle** to predict how changing the concentration of a reacting substance can affect the equilibrium position.

Rate of reaction

The rate of reaction increases as you increase the concentration of a reacting substance in solution.

For a reaction at equilibrium, if you add more of one of the reactants:

- the rate of the forward reaction increases
- more reactants are used up and more product forms
- the rates of the forward and reverse reactions gradually become equal again
- a new equilibrium position is reached, further to the right than before.

If you add more of one of the products instead, a new equilibrium position is reached again, but this time it is further to the left than before.

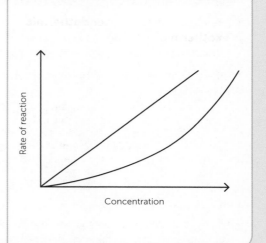

Making predictions

If you increase the concentration of a reacting substance, the equilibrium position moves in a direction away from this substance.

Think about the general equation for a reversible reaction:

$$A + B \rightleftharpoons C + D$$

If you increase the concentration of A or B, the position of equilibrium will move to the right. The effect of this is to use up some A and B, and to produce more C and D. In this way, the system responds to counteract the addition of extra reactants.

Remember that the reaction does not 'know' how to respond to changes in conditions.

Ethanol reacts with ethanoic acid to produce ethyl ethanoate and water. The reaction is reversible and takes place in aqueous solution. Sulfuric acid is a catalyst for this reaction.

Predict the effect on the equilibrium concentration of ethyl ethanoate of:

(a) increasing the concentration of ethanol. [1]

(b) decreasing the concentration of ethanoic acid. [1]

(c) decreasing the concentration of sulfuric acid. [1]

(a) It will increase.[1]

(b) It will decrease.[1]

(c) There will be no change.[1]

TEMPERATURE CHANGES AND EQUILIBRIUM

An increase in temperature shifts the equilibrium position in the direction of the endothermic change.

Exothermic and endothermic changes

If a reversible reaction is **endothermic** in one direction, it is **exothermic** in the opposite direction. The amount of energy transferred is the same in each direction.

Methanol can be manufactured from carbon monoxide and hydrogen:

exothermic
−91 kJ/mol
↓
$$CO(g) + 2H_2(g) \rightleftharpoons CH_3OH(g)$$
↑
+91 kJ/mol
endothermic

If the temperature of this reaction mixture is increased:
- the equilibrium position shifts to the left, because this is in the direction of the endothermic change
- the relative amount of methanol at equilibrium is decreased.

If you wanted to increase the relative amount of methanol at equilibrium, you would need to decrease the temperature of the reaction mixture. The equilibrium position would then shift to the right, in the direction of the exothermic change.

Steam reforming is used to manufacture hydrogen. It involves the reaction between methane and steam:

$$CH_4 + H_2O \rightleftharpoons CO + 3H_2$$

Energy change for forward reaction = −206 kJ/mol

Explain how the temperature of the reaction mixture may be changed to increase the equilibrium yield of hydrogen. [3]

Reduce the temperature[1] because this favours the exothermic reaction[1], and the forward reaction that produces hydrogen is exothermic[1].

Interpreting graphs

You can tell from a suitable graph whether a manufacturing process involves an exothermic reaction or an endothermic reaction. Ammonia is manufactured from nitrogen and hydrogen:

$$N_2 + 3H_2 \rightleftharpoons 2NH_3$$

The forward reaction, the one that produces ammonia, is exothermic. As the temperature is increased, the equilibrium moves to the left and the equilibrium yield of ammonia is decreased. A graph of equilibrium yield of product against temperature will have a negative gradient.

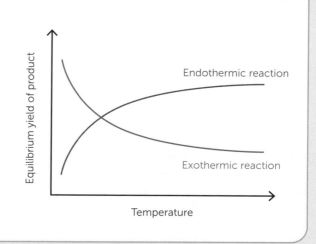

PRESSURE CHANGES AND EQUILIBRIUM

You must look at the balanced equation to predict the effect of changing the pressure of a reversible reaction.

Making predictions

If you increase the pressure of a reacting gas, the equilibrium position moves towards the side of the balanced equation that has the smallest amount of gas molecules.

Changing the pressure has no effect on the equilibrium position if there are no reacting gases.

For example, sulfur dioxide reacts with oxygen to form sulfur trioxide:

2 gas molecules
$$2SO_2(g) + O_2(g) \rightleftharpoons 2SO_3(g)$$
2 + 1 = 3 gas molecules

There are fewer molecules of gas on the right-hand side. If the pressure is increased:

- the equilibrium position shifts to the right
- the relative amount of sulfur trioxide at equilibrium increases.

If you reduce the pressure instead, the equilibrium position shifts to the left, and the relative amounts of sulfur dioxide and oxygen at equilibrium increase.

Summary

Change in conditions	Effect on equilibrium position
Increase in concentration	Shifts away from the substance that is increased in concentration
Increase in temperature	Shifts in the direction of the endothermic change
Increase in pressure	Shifts towards the side of the equation with fewest gas molecules
Catalyst added	No change

The time taken to reach equilibrium is decreased in each situation shown in the table.

1. Calcium carbonate decomposes when heated:

$$CaCO_3(s) \rightleftharpoons CaO(s) + CO_2(g)$$

Explain why the yield of calcium oxide is reduced if the pressure is increased. [2]

2. Hydrogen reacts with iodine:

$$H_2(g) + I_2(g) \rightleftharpoons 2HI(g)$$

Explain the effect of decreasing the pressure on the equilibrium position. [2]

1. *The equilibrium position moves to the left[1] because there are molecules of gas on the product side of the equation but no molecules of gas on the reactant side[1].*

2. *No change[1] because the number of gas molecules is the same on each side of the equation[1].*

EXAMINATION PRACTICE

01 A lump of calcium carbonate reacts with dilute hydrochloric acid. Give **one** change that could
    be made to the lump to increase the rate of reaction.   [1]

02 A student reacted 0.08 g of calcium with excess dilute hydrochloric acid.
    The graph shows how the volume of hydrogen given off changed during the reaction. Line C
    represents the student's results.

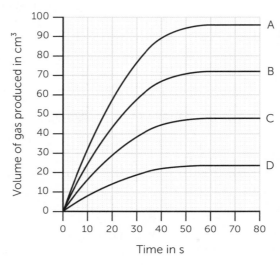

02.1 Describe how the student could measure the volume of hydrogen given off.   [1]
02.2 Calculate the mean rate of reaction for the results shown by line C. Give the units. [3]
02.3 The student repeated the experiment using 0.12 g of calcium. Identify the line on the
      graph that represents the results from this experiment. Give a reason for your answer. [3]
02.4 **Higher Tier only:** Determine the rate of reaction at 40 s shown in line A.   [3]

03 Give the meaning of the term activation energy.   [1]

04 Sodium thiosulfate solution reacts with dilute hydrochloric acid:

$$Na_2S_2O_3(aq) + 2HCl(aq) \rightarrow 2NaCl(aq) + H_2O(l) + SO_2(g) + S(s)$$

04.1 Explain why the rate of this reaction decreases when water is added to the
      reaction mixture.   [2]
04.2 Explain why the rate of this reaction increases when the temperature is increased.   [3]

05 One of the stages in the manufacture of sulfuric acid involves producing sulfur trioxide from a
    mixture of sulfur dioxide, oxygen and vanadium(V) oxide:

$$2SO_2 + O_2 \rightleftharpoons 2SO_3$$

05.1 Identify the catalyst in this reaction. Give a reason for your answer.   [2]
05.2 Explain how a catalyst works.   [2]

06   Cobalt chloride is a substance used to test for the presence of water:

anhydrous cobalt chloride + water $\rightleftharpoons$ hydrated cobalt chloride
        (Blue)   (Pink)

06.1   Give the meaning of the $\rightleftharpoons$ symbol.   [1]

06.2   Predict what you would see if anhydrous cobalt chloride became damp.   [1]

06.3   Suggest a way to produce anhydrous cobalt chloride from hydrated cobalt chloride.   [1]

06.4   The forward reaction is exothermic. Describe **two** features of the reverse reaction that
       can be deduced from this statement.   [2]

07   A reversible reaction may reach equilibrium.

07.1   Give a reason why a reversible reaction in a stoppered boiling tube may reach
       equilibrium, but the same reaction cannot reach equilibrium if the stopper is removed.   [2]

07.2   The rate of the forward reaction in a certain reversible reaction at equilibrium is 0.50 g/s.
       Explain what the rate of the reverse reaction will be.   [2]

08   Magnesium ribbon reacts with hydrochloric acid: $Mg(s) + 2HCl(aq) \rightarrow MgCl_2(aq) + H_2(g)$

Plan an investigation to show how the concentration of hydrochloric acid affects the rate
of reaction.   [6]

Higher Tier only

09   Converters change harmful waste gases from car engines into less harmful ones:

$$2NO_2(g) + 4CO(g) \rightleftharpoons N_2(g) + 4CO_2(g)$$

Give a reason why these converters need catalysts.   [1]

10   Ethene reacts with chlorine to produce dichloroethane:

$$C_2H_4(g) + Cl_2(g) \rightleftharpoons C_2H_4Cl_2(g)$$

Energy change for forward reaction = −218 kJ/mol

10.1   Explain the effect of increasing the pressure on the equilibrium yield of dichloroethane.   [2]

10.2   Explain the effect of increasing the temperature on the equilibrium yield
       of dichloroethane.   [2]

11   Iron(III) ions and thiocyanate ions react together in aqueous solution:

The mixture quickly reaches equilibrium.

Explain what you expect to happen when iron(III) chloride solution is added.   [3]

iron(III) ion + thiocyanate ion $\rightleftharpoons$ thiocyanatoiron(III) ion
  $Fe^{3+}(aq)$ $SCN^-(aq)$ $FeSCN_2+(aq)$
   (Brown)   (Colourless)   (Deep red)

CRUDE OIL, HYDROCARBONS AND ALKANES

Crude oil consists of many compounds, most of which are **hydrocarbons**.

Crude oil

Crude oil and **natural gas** formed over millions of years from once-living plants and animals (mainly plankton). They are **finite resources** because they take so long to form or may not be being formed anymore. They are being used faster than they are being replaced.

Millions of years

Plankton and algae die and are buried in mud

Ancient remains are exposed to high pressures and temperatures

Crude oil and natural gas in rocks

Hydrocarbons and alkanes

1. Butane is an alkane. Its molecules contain four carbon atoms.

1. (a) Determine the molecular formula of butane. [1]

(b) Draw the displayed structural formula of butane. [1]

1. (a) Number of hydrogen atoms = (2 × 4) + 2 = 10, so the formula is C_4H_{10} [1]

(b) [1]

```
    H   H   H   H
    |   |   |   |
H — C — C — C — C — H
    |   |   |   |
    H   H   H   H
```

Hydrocarbons are compounds of hydrogen and carbon only. There are several types of hydrocarbon, but the ones in crude oil are mostly **alkanes**.

The alkanes are a **homologous** series, a 'family' of compounds with similar chemical properties and the same general formula:

$$C_nH_{2n+2}$$

(where n stands for the number of carbon atoms in the molecule)

The table shows the first three alkanes.

Name of alkane	Methane	Ethane	Propane
Molecular formula	CH_4	C_2H_6	C_3H_8
Displayed structural formula	H—C—H with H above and H below	H—C—C—H with H H above and H H below	H—C—C—C—H with H H H above and H H H below

FRACTIONAL DISTILLATION AND PETROCHEMICALS

The different hydrocarbons in crude oil are separated by **fractional distillation**.

Fractional distillation

Fractional distillation relies on the different **boiling points** of the different components in a mixture. See fractional distillation of mixtures on **page 135**.

It involves:

- heating the mixture to **evaporate** its components
- cooling the vaporised components so they **condense** at different temperatures.

The fractional distillation of crude oil happens in a metal tower called a fractionating column.

The column gets cooler towards the top, so it has a **temperature gradient**. Different **fractions** have different ranges of boiling points, so they condense at different heights in the column.

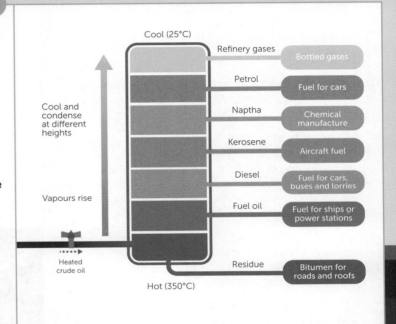

Cool (25°C)

Cool and condense at different heights

Vapours rise

Heated crude oil

Hot (350°C)

Refinery gases — Bottled gases

Petrol — Fuel for cars

Naptha — Chemical manufacture

Kerosene — Aircraft fuel

Diesel — Fuel for cars, buses and lorries

Fuel oil — Fuel for ships or power stations

Residue — Bitumen for roads and roofs

A substance occupies less volume when it is in the liquid state than when it is in the gas state. This makes LPG more suitable for transporting petroleum gases.

Fuels and petrochemicals

Crude oil provides many of the **fuels** that we use in everyday life. It also provides **feedstock** (starting materials) for the petrochemical industry. **Petrochemicals** are useful substances and materials produced from crude oil. They include detergents, lubricants, polymers (plastics) and solvents.

2. Place the following fractions in decreasing order of boiling point: diesel oil, heavy fuel oil, kerosene, LPG (liquefied petroleum gases), petrol. [1]

3. Give a reason why a huge number of different natural and synthetic carbon compounds are possible. [1]

2. Correct order: heavy fuel oil, diesel oil, kerosene, petrol, LPG.[1]

3. Carbon atoms can form homologous series or families of similar compounds.[1]

PROPERTIES OF HYDROCARBONS

Some **physical properties** of hydrocarbons depend on the size of their molecules.

Boiling point, viscosity and flammability

As the size of a hydrocarbon's molecules increases:
- the boiling point and viscosity increase
- the flammability decreases.

The **viscosity** of a substance is a measure of how easily it flows. A very viscous liquid has a high viscosity. It is 'thick' rather than runny, and does not flow easily.

These properties influence the use of hydrocarbons as fuels. For example, propane is in the gas state at room temperature, so it also has a very low viscosity. Propane ignites very easily. It is used in bottled gases for camping and for homes that are not connected to the mains gas supply.

Heavy fuel oil contains very large hydrocarbon molecules. It is a very viscous liquid that ignites with difficulty. This makes it suitable as a fuel for large ships, rather as a fuel for cars.

Getting more difficult to boil

Getting 'thicker' and less runny

Getting more difficult to ignite

Complete combustion

Combustion or burning is an **oxidation** reaction. It transfers energy to the surroundings, mainly by heating. Combustion in excess oxygen or air is called complete combustion. In general, for the complete combustion of a hydrocarbon:

hydrocarbon + oxygen → carbon dioxide + water

The hydrogen and carbon in the molecules gain oxygen and are oxidised.

Gases have very low viscosities, so they flow easily to fill their containers.

1. Write a balanced equation for the complete combustion of propane, C_3H_8. [2]
2. The hydrocarbon molecules in diesel oil are larger than the hydrocarbon molecules in petrol.
   Compare the boiling point, viscosity and flammability of these two fuels. [2]

   1. $C_3H_8 + 5O_2 \rightarrow 3CO_2 + 4H_2O$ 1 mark for correct formulae[1], 1 mark for correct balancing[1].
   2. Two from: Petrol has a lower boiling point[1], lower viscosity[1] and higher flammability[1] than diesel oil.

CRACKING

Cracking is a chemical process that breaks down hydrocarbons with large molecules to produce more useful substances with smaller molecules.

Breaking bonds

During cracking, **covalent bonds** are broken and new bonds form. A mixture of hydrocarbons with smaller molecules than the reactants is produced. For example:

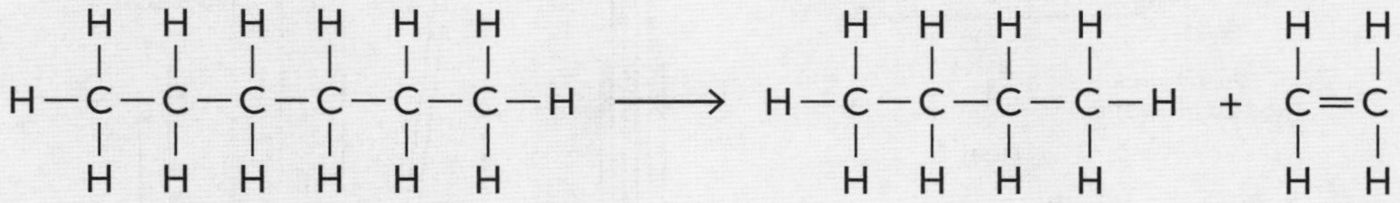

One of these products, ethene, is a type of hydrocarbon called an **alkene**. Unlike *alkane* molecules, alk**ene** molecules contain a carbon-carbon double bond, C=C.

Types of cracking

Steam cracking and catalytic cracking are two types of cracking. The table summarises the general conditions that these types of cracking need.

	Steam cracking	Catalytic cracking
Temperature in °C	750–900	450–550
Pressure in atmospheres	2–4	1
Catalyst	✗	✓

The catalysts in catalytic cracking are 'zeolites', substances that contain aluminium oxide.

1. Write a balanced equation for the cracking of octane, C_8H_{18}, to produce butane, C_4H_{10}, and one other product. [1]

2. Give **two** examples of how modern, everyday life depends on hydrocarbons. [2]

1. $C_8H_{18} \rightarrow C_4H_{10} + C_4H_8$ [1]

2. *Hydrocarbons are used as fuels for cars[1], and for making polymers[1].*

Usefulness of cracking

The petrochemical industry carries out cracking for two reasons:
- It helps to balance the supply of different hydrocarbons with the demand for them
- It produces alkenes, which are a feedstock for making **polymers**.

Hydrocarbons with large molecules are less useful as fuels than hydrocarbons with small molecules. Fractional distillation usually produces too much of the larger hydrocarbons and too little of the smaller hydrocarbons. Cracking helps to meet the high demand for fuels with smaller molecules.

The **supply** of something is how much can be made. The **demand** for something is a measure of how much people need or want.

ALKENES

Alkenes are hydrocarbons that contain carbon-carbon double bonds.

Making alkenes in the lab

Cracking can be carried out in the laboratory using liquid paraffin as the feedstock.

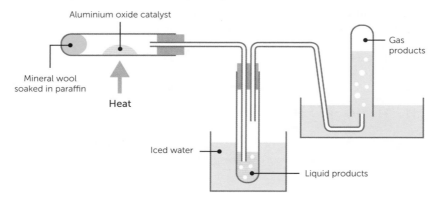

The catalyst must be heated very strongly before the paraffin is evaporated and passed over it. The iced water bath cools and condenses hydrocarbons with larger molecules. Hydrocarbons with smaller molecules carry on and are collected as gases.

Testing for alkenes

Alkene molecules contain C=C bonds. Alkanes do not. These C=C double bonds make alkenes more reactive than alkanes. This is the basis of a laboratory test for alkenes that uses **bromine water**.

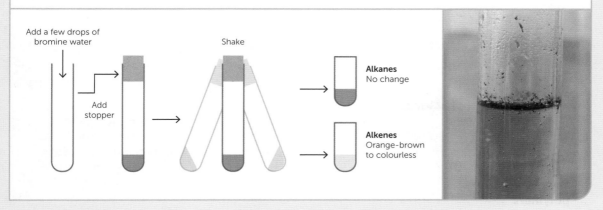

Liquid paraffin was cracked in the laboratory. Explain why the products of cracking decolourised bromine water but the liquid paraffin did not. [2]

Alkenes can decolourise bromine water, but it stays orange-brown when mixed with alkanes.[1]
The liquid paraffin did not contain alkenes but the products of cracking did.[1]

EXAMINATION PRACTICE

01 Decane, $C_{10}H_{22}$, is found in crude oil.

    01.1 Explain why decane is a hydrocarbon. [1]

    01.2 Explain, as fully as you can, why decane is an alkane. [2]

02 Draw the fully displayed formula of propane, C_3H_8. [2]

03 Crude oil is separated into useful fractions by fractional distillation.

    03.1 Name **two** crude oil fractions used as fuels. [2]

    03.2 Explain how fractional distillation of crude oil works. [4]

04 Hydrocarbons have different physical properties.

    04.1 Describe what is meant by the **viscosity** of a substance. [1]

    04.2 Describe how the boiling point and flammability of hydrocarbons change as their molecules become larger. [2]

05 Write a balanced equation for the complete combustion of pentane, C_5H_{12}. [2]

06 Hexane and hexene are liquid hydrocarbons. Hexane is an alkane and hexene is an alkene. Describe a chemical test to distinguish between these two liquids. [2]

07 Oil refineries crack hydrocarbon fractions from crude oil.

    07.1 Give **one** reason why cracking is carried out. [1]

    07.2 Compare the general reaction conditions in steam cracking and catalytic cracking. [2]

    07.3 Balance this equation for the cracking of hexadecane: [2]

$$\_\_\_\_\_C_8H_{18} \rightarrow C_7H_{16} + C_3H_8 + \_\_\_\_\_C_2H_4$$

PURE SUBSTANCES

The chemical meaning of the word **pure** is different from its everyday meaning.

Elements, compounds and mixtures

An **element** consists of atoms with the same **atomic number**. A **compound** consists of two or more different elements, chemically combined in fixed proportions as a result of a chemical reaction.

A **mixture** contains two or more elements or compounds:
- the individual substances are not chemically combined
- the individual components are not in fixed proportions.

Elements

Carbon   Hydrogen   Oxygen

Compounds

Carbon dioxide   Water

Pure vs impure substances

To a chemist, a pure substance consists of one element or one compound only, with no other element or compound mixed with it. In everyday life, a pure substance is usually a substance with nothing else added to it.

Orange juice is often described as 'pure' because the carton only contains orange juice. The juice itself is impure because it is a mixture of water, sugar, acids, 'bits' and many other substances.

Melting points and boiling points

A pure substance has a specific **melting point** and **boiling point**. An impure substance:
- melts and boils at different temperatures than a pure substance
- may melt or boil over a range of temperatures instead of at a single temperature.

You can see the difference when the temperature of a substance is recorded as it is heated until it melts or boils.

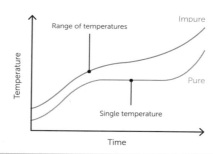

The table shows the melting points and boiling points of two samples of water.

Sample	Melting point in °C	Boiling point in °C
Distilled water	0.0	100.0
Seawater	−1.9	100.5

Explain why the seawater cannot be pure water.   [2]

Distilled water is pure water.[1] The melting and boiling points of seawater are different from those of pure water[1], so it cannot be pure water.

FORMULATIONS

A **formulation** is a mixture in which each component is there for a good reason.

Formulations vs simple mixtures

The table shows similarities and differences between formulations and simple mixtures.

	Formulation	Simple mixture
Number of components	Two or more, usually many	Two or more, often a few
Proportion of each component	Fixed	Not fixed

Examples of formulations include:

Cleaning products

Fertilisers

Foods

Fuels

Medicines

Paints

Components of formulation

Each individual substance in a formulation has a particular purpose. They are carefully chosen and mixed in controlled quantities to produce a formulation with the desired properties.

The table shows the composition of two alloys used in jewellery.

Alloy	% Gold	% Copper	% Silver
Yellow-coloured gold	75.00	12.50	12.50
Rose-coloured gold	75.00	22.25	2.75

Explain why these alloys are formulations. [2]

They contain different metal elements in carefully measured percentages[1] which gives them different, useful properties[1].

EXPLAINING CHROMATOGRAPHY

Chromatography separates a mixture of coloured solutes in a solution.

Phases

Chromatography relies on two 'phases':
- A **stationary phase** that does not move
- A **mobile phase** that moves past the stationary phase

Required practical activity 6 involves paper chromatography. The stationary phase is in the paper, and the mobile phase is a solvent such as water or propanone.

R$_f$ values

A substance in a solution forms chemical bonds with both phases. The relative strengths of these bonds determine how far the substance travels up the paper with the solvent.

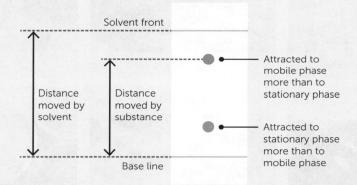

The **R$_f$** value of a substance is a measure of the relative distance it travels up the paper:

$$R_f = \frac{\text{Distance moved by substance}}{\text{Distance moved by solvent}}$$

Remember to measure from the centre of a spot to the base line.

Gas-liquid chromatography is an advanced type of chromatography in which the mobile phase is a gas, and the stationary phase is a liquid on a solid support.

1. Explain whether the diagram shows a chromatogram of a pure substance or a mixture. [2]
2. In the diagram, the centre of the blue spot is 72 mm from the base line. The solvent travelled 94 mm. Calculate the R$_f$ value of the blue spot.
   Give your answer to an appropriate number of significant figures. [3]

   1. *(It is a mixture because) there are two spots[1] and a pure substance would produce one spot[1].*
   2. *R$_f$ = $\dfrac{72 \text{ mm}}{94 \text{ mm}}$ [1] = 0.76595[1] = 0.77[1] to 2 significant figures.*

REQUIRED PRACTICAL 12

Paper chromatography

This required practical activity helps you develop your ability to separate mixtures, and to accurately make and record measurements.

Separating mixtures of dyes

Food colourings are soluble in water, so water can be used as the **mobile phase** when analysing them using paper chromatography. You can revise the method and how it works on **page 133** and opposite.

Recording the results

A table is a good way to record several measurements or observations. A food colouring may contain more than one substance, so you could make a separate table for each food colouring.

The important thing is that the results are recorded clearly. In this example, the table includes a column for the R_f values after they have been calculated as part of the **analysis**.

Food colouring: Green
Distance travelled by solvent = 80 mm

Colour of spot	Distance travelled by spot in mm	R_f value of spot
Green	40	0.50
Blue	60	0.75

You can calculate the R_f value of each spot using the equation opposite.

Food colourings make food more interesting

1. Describe **two** features of identical compounds on a chromatogram. [2]
2. A student uses paper chromatography to analyse a sample of ink. Suggest **two** precautions needed when drawing the base line.
   Give a reason for each suggestion. [4]

   1. *They will have the same R_f value[1] and the same colour[1].*
   2. *Draw the line with pencil rather than pen[1] because pencil will not dissolve in the solvent[1]. Make sure the line will be above the surface of the solvent[1] otherwise the ink will leave the paper[1].*

Evaluating the investigation

In your **evaluation**, you could discuss the method and how well it worked, including any difficulties you might have had when analysing the results. You could also suggest improvements to the method, and ideas for extending the investigation.

IDENTIFYING COMMON GASES

Some gases can be identified using simple laboratory tests.

When you describe a test for a gas:
- Say what you need to do
- Say what you observe if the test is positive.

Hydrogen

Hydrogen burns rapidly in air.

To test for hydrogen:
- Hold a lighted wooden splint near the open end of a test tube
- If hydrogen is present, it burns with a 'pop' sound.

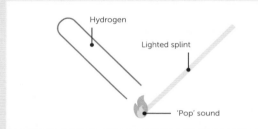

Oxygen

The greater the relative amount of oxygen present, the more rapidly a substance will burn.

To test for oxygen:
- Hold a glowing wooden splint inside a test tube
- If oxygen is present, the splint relights.

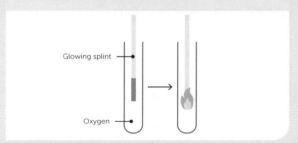

Carbon dioxide

To test for carbon dioxide:
- Shake the gas with limewater
- If carbon dioxide is present, the limewater turns milky.

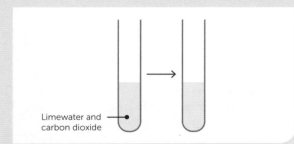

Chlorine

Chlorine has a distinctive, choking smell. It turns some coloured substances colourless.

To test for chlorine:
- Put damp litmus paper into the gas
- If chlorine is present, the litmus is bleached and the paper turns white.

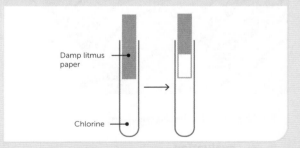

A student bubbled carbon dioxide through limewater. The limewater turned cloudy white. Explain why this reaction happened. [3]

Limewater is aqueous calcium hydroxide solution.[1] It reacted with the carbon dioxide to produce an insoluble white precipitate[1] of calcium carbonate[1].

EXAMINATION PRACTICE

01   Vigorous bubbling is seen when potassium iodide is added to hydrogen peroxide solution.
Describe a test to confirm that the gas produced is oxygen. [2]

02   Bubbling is seen when calcium carbonate reacts with dilute hydrochloric acid.
Describe a test to confirm that the bubbles contain carbon dioxide. [2]

03   A student carried out electrolysis using sodium chloride solution. Bubbles of gas were seen at each electrode. The student predicted the production of hydrogen at the cathode and chlorine at the anode. Describe how the student could test these predictions. [4]

04   Pure iron is soft but alloys of iron with carbon and other elements are much harder.
04.1   Describe how the chemical definition of 'pure' differs from its everyday definition. [2]
04.2   Give a reason why alloys are formulations rather than simple mixtures. [1]

05   The diagram shows the results of a paper chromatography experiment.

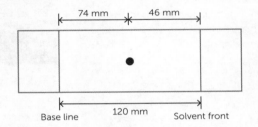

Calculate the R_f value for the substance on the chromatogram. Use the equation:

$$R_f = \frac{\text{Distance moved by substance}}{\text{Distance moved by solvent}}$$

Give your answer to an appropriate number of significant figures. [3]

06   Dodecanoic acid is a white solid at room temperature. Its melting point is 43.8 °C.
Explain how a student could determine whether a sample of dodecanoic acid is pure. [2]

07   Different fertilisers are described by three numbers. For example, 5-3-3 fertilisers contain 5% nitrogen compounds, 3% phosphorus compounds, 3% potassium compounds, with the remaining ingredient being carefully chosen to give the fertiliser its desired properties.
Explain how this information tells you that fertilisers are formulations. [3]

08   Explain how paper chromatography separates mixtures. [4]

09   A student uses paper chromatography to separate the coloured substances in ballpoint pen ink. These substances are soluble in propanone but not in water.
Describe how the student could use paper chromatography to determine whether black ink contains more than one coloured substance. In your answer, include essential apparatus and methods, and how the student could use their results. [6]

THE ATMOSPHERE

The Earth's **atmosphere** has changed over billions of years but has remained stable over the last 200 million years.

- About 4/5 nitrogen, N_2
- About 1/5 oxygen, O_2
- Smaller proportions of other gases

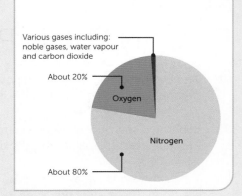

Various gases including: noble gases, water vapour and carbon dioxide

About 20%

Oxygen

Nitrogen

About 80%

The early atmosphere

The Earth formed just under 4.6 billion years ago. Volcanic activity released gases such as water vapour and carbon dioxide. As the Earth cooled water vapour **condensed**, fell as rain, and formed oceans. Carbon dioxide then dissolved in the oceans.

Nitrogen was also released from volcanoes. This is a relatively **unreactive** gas, so it gradually built up in the atmosphere. Small proportions of ammonia and methane may also have been in the atmosphere.

A volcanic eruption

Comparisons

The modern atmospheres of Venus and Mars are mainly carbon dioxide, with little or no oxygen. The Earth's early atmosphere may have been like these atmospheres.

The atmosphere of Venus has strongest greenhouse effect in the Solar System. Surface temperatures reach over 460 °C. You can revise greenhouse gases and the greenhouse effect on **pages 231** and **232**.

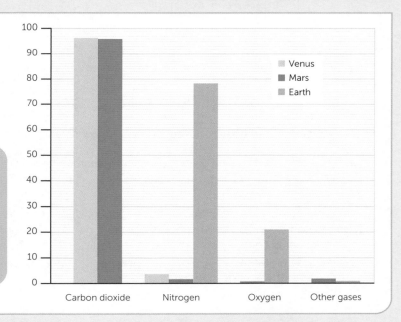

1. Suggest a reason why evidence for the composition of the early atmosphere is limited. [1]
2. Compare the Earth's atmosphere today with the Earth's early atmosphere. [3]

1. It was billions of years of ago and no-one was there to observe the atmosphere then.[1]

2. Today there is more nitrogen[1], more oxygen[1], less carbon dioxide[1].

HOW OXYGEN INCREASED

Photosynthesis

Photosynthesis is an important process in the development of the atmosphere. Plants and **algae** make their own food by photosynthesis. Overall:

$$\text{carbon dioxide + water} \xrightarrow{\text{light}} \text{glucose + oxygen}$$

$$6CO_2 + 6H_2O \xrightarrow{\text{light}} C_6H_{12}O_6 + 6O_2$$

These living things produce the oxygen that is in the modern atmosphere.

Billions of years ago

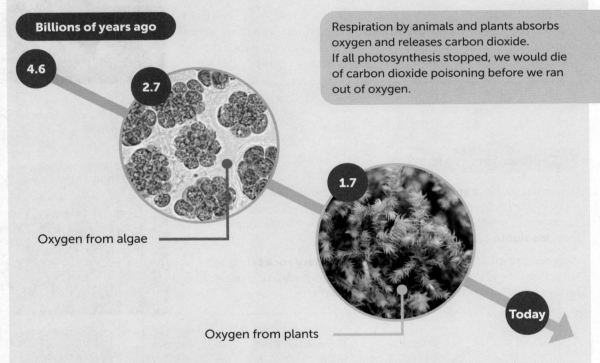

Respiration by animals and plants absorbs oxygen and releases carbon dioxide.
If all photosynthesis stopped, we would die of carbon dioxide poisoning before we ran out of oxygen.

4.6

2.7

1.7

Oxygen from algae

Today

Oxygen from plants

Animals could start to evolve when there was enough oxygen in the atmosphere, around 600 million years ago.

The photo shows a type of rock called banded iron. It is about 2.7 billion years old. The red layers consist of iron oxide.

(a) Suggest how banded iron provides evidence for the existence of oxygen in the atmosphere billions of years ago. [2]

(b) Describe how the oxygen was produced. [2]

(a) Iron reacted with oxygen to form the red layers.[1] The rock is about 2.7 billion years old, so oxygen must have been present billions of years ago.[1]

(b) Oxygen was produced by photosynthesis[1] in algae and plants[1].

HOW CARBON DIOXIDE DECREASED

Several processes caused the proportion of carbon dioxide to decrease.

Fossil fuels

Fossil fuels formed over millions of years from the ancient remains of living things:

- Coal formed from trees and other plants
- Crude oil and natural gas formed from plankton and algae

You can revise crude oil and natural gas on **page 216**.

The plants and algae contained carbon compounds formed from the glucose produced by **photosynthesis**. The plankton contained carbon compounds because they ate algae. This means that fossil fuels contain carbon that was absorbed from the atmosphere millions of years ago.

Coal

There are four different types of coal, depending on how long it has been forming.

Sedimentary rocks

Carbon dioxide is a **soluble** gas. It dissolved in the Earth's early oceans, reducing its percentage in the atmosphere. Dissolved carbon dioxide releases carbonate ions, which can react with dissolved metal ions to form **insoluble** carbonates.

Over millions of years, these became **sedimentary rocks** such as limestone. You can sometimes see fossils of sea creatures that became trapped in the sediments as the limestone formed.

Fossils in limestone

Insoluble carbonates sink ┄➤ Layers build up as sediments ┄➤ Water squeezed out from the sediments ┄➤ Sediment particles get stuck together ┄➤ Sedimentary rocks form

---

1. Explain why the appearance of plants and algae decreased the percentage of carbon dioxide in the atmosphere. [2]
2. Explain how deposits of coal formed. [3]
3. **Higher Tier only:** Write an ionic equation for the production of calcium carbonate. [2]

   1. *They carried out photosynthesis[1] which needs carbon dioxide absorbed from the air[1].*
   2. *Trees died and were buried under sediment.[1] Their remains were exposed to high pressures and temperatures[1] and were converted to coal over millions of years[1].*
   3. *$Ca^{2+}(aq) + CO_3^{2-}(aq) \rightarrow CaCO_3(s)$ Correct formulae and balancing[1], state symbols[1].*

GREENHOUSE GASES

Greenhouse gases include carbon dioxide, methane and water vapour.

The greenhouse effect

Greenhouse gases keep the Earth warmer than it would be without them. They keep its temperature high enough for life to exist.

1

Energy is transmitted from the Sun to the Earth by **radiation**.

2

Some radiation is re-radiated into space by the Earth's atmosphere, clouds, and surface.

3

The remaining radiation is absorbed by the Earth's surface and warms it up.

4

The warm surface emits radiation, which warms the atmosphere.

5

Greenhouse gases in the atmosphere absorb radiation.

Nitrogen and oxygen comprise about 99% of the atmosphere, but they are not greenhouse gases.

Take care not to mention the ozone layer when discussing greenhouse gases and the greenhouse effect.

In terms of wavelength, describe the radiation that passes through the atmosphere, and the radiation emitted by the Earth's surface and atmosphere. [2]

The radiation that passes through the atmosphere is short wavelength radiation.[1] The radiation that is emitted by the Earth's surface and atmosphere is long wavelength radiation.[1]

6

Absorbed radiation is emitted in all directions, keeping the atmosphere and the Earth's surface warm.

HUMAN ACTIVITIES AND GREENHOUSE GASES

Human activities release additional quantities of greenhouse gases into the atmosphere.

Carbon dioxide

Carbon dioxide accounts for most of the greenhouse gas emissions because of human activities, such as:

- burning fossil fuels, for example for transport and generating electricity
- cement manufacture.

Methane

Methane is a much more powerful greenhouse gas than carbon dioxide. Human activities that release methane include:

- cattle farming
- coal mining, and oil and gas production.

Methane is also released from decaying animal and plant waste on farms and in landfill sites.

Linking greenhouse emissions to global warming

As carbon dioxide levels in the atmosphere have increased, the average global temperature has increased.

The Intergovernmental Panel on Climate Change (IPCC) makes scientific conclusions based on large amounts of scientific research. It believes that it is highly likely that human activities are the main cause of global warming.

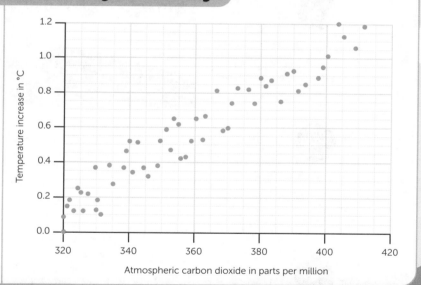

1. Describe the relationship shown in the graph. [2]
2. The IPCC uses 'peer reviewed' scientific evidence. Describe what this means. [2]

   1. There is a positive correlation[1] between the increase in temperature and the atmospheric carbon dioxide concentration[1].
   2. Peer reviewed evidence is evaluated by other scientists.[1] This means that findings reported by a scientist or research group are checked before being accepted or rejected.[1]

GLOBAL CLIMATE CHANGE

Increasing global temperatures are a major cause of **climate change**.

Impacts of climate change

The **weather** describes atmospheric conditions over a short period, such as a few hours or a day. Climate is weather in an area over a long period, such as 30 years. Many scientists believe that **global warming** will lead to **global climate change** – changes in the long-term weather conditions across the whole world.

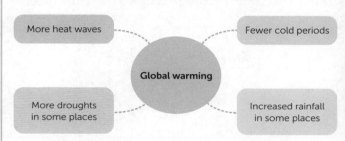

- More heat waves
- Fewer cold periods
- **Global warming**
- More droughts in some places
- Increased rainfall in some places

The Thames Barrier protects London from flooding during very high tides

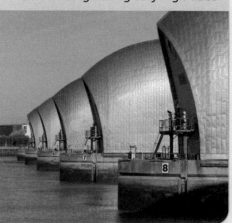

1. Explain **two** problems that climate change may cause farmers. [3]
2. Climate change is leading to increased sea levels.
   (a) Give **two** reasons why this is happening. [2]
   (b) Describe **one** problem caused by increased sea levels. [2]

   1. Changes in temperature and rainfall[1] may mean that some crops can no longer grow[1]. Farmers may need to plant different crops to cope with these changes.[1]

   2. (a) As the oceans warm up, they expand.[1] Water from melting ice in the Antarctic and in glaciers flows into the oceans.[1]

   (b) Land near the coast will become flooded.[1] This will reduce the amount of land available for dwellings and farms.[1]

Climate change is predicted to affect different parts of the world in different ways.

Uncertainties

It is difficult for scientists to develop accurate climate **models**.
- Changes in conditions in the atmosphere are very complex.
- Scientists cannot collect evidence from every place in the world.
- The effects of global warming may be more severe in some parts of the world.

Climate models become more accurate as scientists include more factors and data.

3. Suggest **two** reasons why some people do not believe that climate change is happening. [2]

   3. They may only think about parts of the evidence[1], which may be biased[1].

CARBON FOOTPRINT

A person's **carbon footprint** is a measure of their greenhouse gas emissions.

Carbon dioxide equivalent

Carbon dioxide and methane are **greenhouse gases** that contain carbon. They are released from human activities such as the production and combustion of **hydrocarbon** fuels. Sulfur dioxide and oxides of nitrogen do not contain carbon, but they are greenhouse gases. They are also released because of hydrocarbon fuels. You can revise these **pollutants** on **page 235**.

When a carbon footprint is calculated, it takes into account the greenhouse effect of these gases relative to carbon dioxide. It gives the total amount of carbon dioxide and other greenhouse gases given off over the lifetime of products, services and events. The chart shows examples.

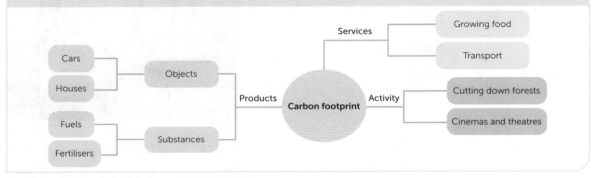

Reducing the carbon footprint

A carbon footprint is reduced if emissions of carbon dioxide and methane are reduced. This can be done by individuals such as you, by manufacturers, and by the actions of governments.

For example, you could:
- walk or cycle instead of taking journeys in cars
- make changes to your diet, such as reducing the quantity of meat you eat while maintaining a healthy diet.

Manufacturers can adopt new methods and products that use less energy and raw materials. **Life cycle assessments** help them to make decisions. (See **page 242**.)

TURN OFF WHEN NOT IN USE

---

1. Suggest **two** ways in which governments can encourage reductions in carbon footprint. [2]
2. Suggest **two** reasons why the actions of individuals to reduce their carbon footprint may be limited. Give suitable examples in your answer. [4]

1. They can pass laws to ban or control the use of products, services or events that release a lot of greenhouse gases.[1] They can reduce taxes on things with low carbon footprints.[1]

2. People may be reluctant to make changes[1] or they may forget to make the changes needed to reduce their carbon footprint[1]. For example, they may still want to fly to foreign countries for their holidays[1] or may forget to turn off the lights when they leave a room[1].

ATMOSPHERIC POLLUTANTS FROM FUELS

Combustion of fuels

Atmospheric **pollutants** are harmful substances released into the air. The **combustion** of **hydrocarbon** fuels releases several substances.

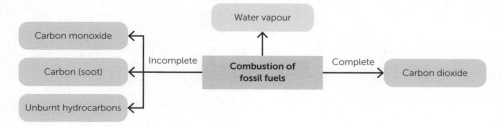

These substances can act as pollutants.

Pollutant	Properties	Problem
Carbon dioxide	**Greenhouse gas**	Increased levels lead to **global warming**
Carbon monoxide	Odourless, colourless, **toxic**	Causes breathing difficulties and even death
Carbon / unburnt hydrocarbons	Form soot and other **particulates**	Causes breathing problems and **global dimming**

Some fuels contain some sulfur. This reacts with oxygen when the fuel is used, forming sulfur dioxide. This gas causes breathing difficulties. It also dissolves in clouds to form an acidic solution that falls as **acid rain**. This harms trees and crops, and living things in rivers and lakes.

1. Oxides of nitrogen, NO$_x$, are a cause of acid rain.

   (a) Give **one** other problem caused by oxides of nitrogen. [1]

   (b) Explain how oxides of nitrogen form when hydrocarbon fuels are used. [2]

2. Suggest why carbon monoxide is difficult to detect. [1]

1. (a) *Breathing difficulties in people.*[1]

   (b) *Oxygen and nitrogen in the air*[1] *react together at high temperatures in engines*[1].

2. *You cannot smell carbon monoxide or see it.*[1]

EXAMINATION PRACTICE

01   The pie chart shows the proportions of the two main gases in the atmosphere of Venus today.

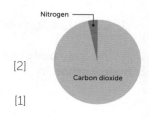

02.1   Give **two** ways in which the atmosphere of Earth and Venus today are similar.   [2]

02.2   Give **one** way in which the atmosphere of Earth and Venus today are different.   [1]

02   Give the approximate proportions of nitrogen and oxygen in today's atmosphere.   [2]

03   One theory about the formation of the Earth's early atmosphere involved volcanoes.
Describe how volcanic activity led to oceans forming.   [2]

04   Write a word equation for photosynthesis.   [2]

05   Describe how the presence of plants and algae changed the proportions of carbon dioxide and oxygen in the Earth's early atmosphere.   [3]

06   Carbon dioxide is a greenhouse gas.

06.1   Name **one** other greenhouse gas.   [1]

06.2   Give **one** human activity that increases the amount of carbon dioxide in the Earth's atmosphere on a global scale.   [1]

06.3   Describe how greenhouse gases keep temperatures high enough for life on Earth.   [3]

07   Describe briefly **two** possible effects of global climate change.   [2]

08   Some people are very concerned about 'carbon footprint'.

08.1   Describe what is meant by a 'carbon footprint'.   [2]

08.2   Suggest **one** way in which an individual person may reduce their carbon footprint.   [1]

09   Hydrogen and methane, CH_4, can replace petrol in some vehicle engines. Name **one** product of the combustion of methane that is not a product of combustion of hydrogen.   [1]

10   Propane, C_3H_8, is a hydrocarbon fuel used for heating and cooking. Predict **three** products of incomplete combustion of propane.   [3]

11   Give **two** reasons why carbon monoxide is not easily detected.   [2]

12   Name the type of pollutant that is a cause of global dimming.   [1]

13   Pollutants caused by the use of hydrocarbon fuel include sulfur dioxide and oxides of nitrogen. These pollutants can cause health problems for humans, and can cause damage to the environment.
Describe how sulfur dioxide and oxides of nitrogen are formed, and the harm they cause to our health and the environment.   [6]

SUSTAINABLE DEVELOPMENT

Sustainable development means meeting our needs today without making it difficult or impossible for people in the future to meet their needs.

The Earth's resources

The Earth's crust, oceans and atmosphere provide us with the **resources** we need for food, shelter, warmth and transport. Farming can provide us with more of some of these resources. Natural resources can be processed to provide energy and new materials.

Farming plants for biodiesel and bioethanol fuels competes with farming for food.

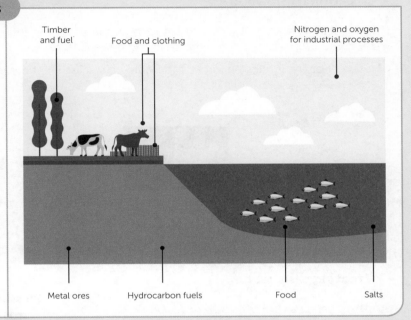

Timber and fuel

Food and clothing

Nitrogen and oxygen for industrial processes

Metal ores   Hydrocarbon fuels   Food   Salts

Finite and renewable resources

Metal ores, and fossil fuels such as crude oil and natural gas, are **finite resources.** They are no longer being made or are being made very slowly. This means that they will run out one day if we keep on using them.

Renewable resources are being made quickly enough to be replaced as we use them. They include biodiesel, trees and other types of **biomass**, and energy resources such as wind power and solar power.

1. Natural materials can be replaced by artificial materials made by chemical processes. Give **one** example of this type of replacement for:

   (a) material for clothing [2]

   (b) construction materials for buildings. [2]

2. Give **one** example of a natural food that can be replaced by manufactured food. [2]

   1. (a) Cotton and wool[1] can be replaced by artificial polymers such as nylon[1].

   (b) Wood from trees[1] can be replaced by bricks and steel[1].

   2. Meat from animals[1] can be replaced by mycoprotein from fungi grown in fermenters[1].

POTABLE WATER

Potable [POE-tuh-bull] water is water that is safe to drink. Unlike **pure** water, potable water contains dissolved substances, including mineral **ions** and gases.

Treating fresh water

Most potable water in the UK comes from ground water, and water in rivers and lakes. This fresh water comes from rainfall. It must be treated to make it potable.

Chlorine or ozone is bubbled through the water to sterilise it. Ultraviolet light may also be used.

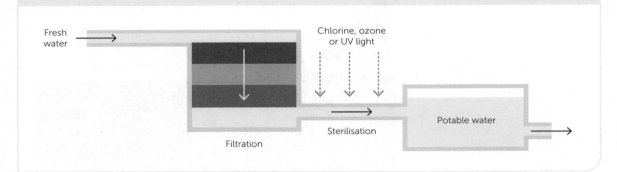

Treating salty water

Salty water and sea water contain dissolved salts that make them unsafe to drink. These **salts** are removed by desalination.

Desalination may be used if ground water supplies are limited. It can be carried out using:

- simple **distillation**
- or **reverse osmosis**.

In biology, osmosis involves water moving through a **partially permeable membrane**. In reverse osmosis, water is forced through a partially permeable membrane. The membrane lets water molecules through but not larger molecules or ions, bacteria, viruses or insoluble particles.

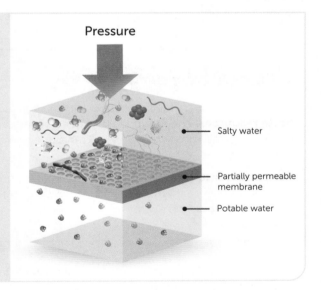

1. Fresh water is treated to make it potable by filtration and sterilisation.
   Give reasons for these **two** processes. [2]
2. Suggest **one** reason that explains why desalination is usually only carried out if fresh water supplies are limited. [1]

   1. Filtration removes insoluble particles[1] and sterilisation kills harmful microbes[1].

   2. Distillation and reverse osmosis need a lot of energy[1].

REQUIRED PRACTICAL 13

Analysing and purifying water samples

This required practical activity helps you develop your ability to measure pH, to heat substances safely, and to separate and purify them.

Analysing water samples

You will have some samples of impure water to test. Estimate their **pH** using **universal indicator paper** and a pH colour chart. You can then determine the mass of dissolved solids in each sample:

Measure and record the mass of an empty evaporating basin → Add a known volume of a water sample → Place the evaporating basin and its sample on a tripod and gauze

Allow to cool, then measure and record the mass of the basin ← Stop heating just before all the water has evaporated ← Heat from below using a Bunsen burner

Purifying a water sample

You should use **simple distillation** to obtain pure water from one of the samples. This diagram shows one way to carry out this separation method. **Page 134** shows a more complex way to do it.

You can check the purity of your **distillate** by measuring its **boiling point**. Heat a sample of a distillate in a boiling tube until it boils. Pure water boils at 100 °C.

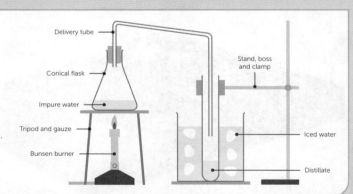

Delivery tube

Conical flask

Impure water

Tripod and gauze

Bunsen burner

Stand, boss and clamp

Iced water

Distillate

1. Suggest why a sample of water may **not** have a pH of 7. [2]
2. Describe how to use the results to calculate the mass of dissolved solids in a sample. [1]
3. Suggest **one** reason why a thermometer may not show 100 °C in boiling distilled water. [1]

   1. It contains dissolved substances[1] which dissolve to form an acidic or alkaline solution[1].
   2. Calculate: (mass of basin and contents at end) – (mass of empty basin at start).[1]
   3. The thermometer may not be calibrated properly.[1]

WASTE WATER TREATMENT

Waste water must be treated before it can be released back into rivers.

Waste water

Large volumes of waste water come from sources such as farms, homes and factories.

Depending on the source, this water may contain harmful substances, solids containing **organic compounds**, and harmful **microbes**. These will damage the environment and may cause disease unless the waste water is treated to remove them.

Treatment

Waste water goes through several stages before it is released into the environment.

Solid materials such as toilet paper and grit are removed. Sewage **sludge** and liquid **effluent** are produced by sedimentation, then digested by bacteria. Further biological treatment removes excess nitrates and phosphates that can cause excessive, harmful growth of algae in rivers.

1. Describe what is meant by:
   (a) anaerobic digestion [2]
   (b) aerobic biological
       treatment. [2]
2. Suggest why it is easier to obtain potable water from fresh water rather than from waste water. [1]

   1. (a) Breaking down substances[1] in the absence of air or oxygen[1].

   (b) Treatment that uses living things such as bacteria[1] in the presence of air or oxygen[1].

   2. Many more stages are needed to treat waste water to make it safe to drink.[1]

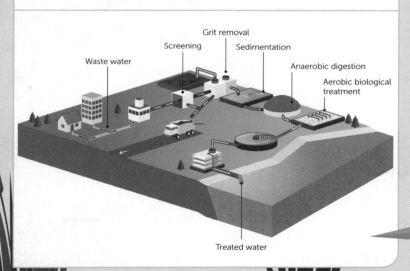

Waste water
Screening
Grit removal
Sedimentation
Anaerobic digestion
Aerobic biological treatment
Treated water

Methane produced by anaerobic digestion can be used as a fuel. You can revise hydrocarbon fuels on **pages 216** and **217**.

ALTERNATIVE METHODS OF EXTRACTING METALS

Alternative extraction methods avoid mining, transporting and disposing of vast amounts of rock.

Phytomining

Phytomining involves plants. It is useful for obtaining copper from low-grade ores.

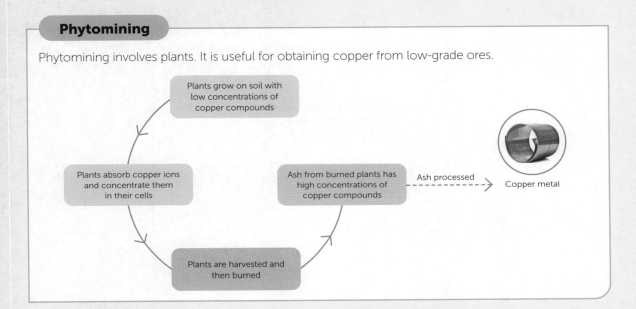

- Plants grow on soil with low concentrations of copper compounds
- Plants absorb copper ions and concentrate them in their cells
- Plants are harvested and then burned
- Ash from burned plants has high concentrations of copper compounds
- Ash processed → Copper metal

Bioleaching

Bioleaching involves bacteria. Some bacteria can absorb copper compounds and obtain energy from them through chemical reactions. These bacteria produce a solution called a **leachate**. It has copper compounds in high enough concentrations to make processing worthwhile.

Processing copper solutions

Solutions of copper compounds can be processed to extract copper in two ways:
- **Electrolysis** (See electrolysis on **pages 187-190**).
- **Displacement** reactions using scrap iron (See the reactivity series on **page 177**).

Bioleaching avoids environmental damage due to mining, but it produces acidic wastes that can also harm the environment.

---

1. Explain why low-grade ores, rather than high-grade ores, are being increasingly used. [3]
2. Suggest **one** disadvantage of using alternative methods of extraction. [1]
3. **Higher Tier only:** Write an ionic equation for the displacement of copper by iron. [3]

　1. *High-grade ores contain relatively high concentrations of metals[1] but are limited resources and are becoming scarce[1]. Low-grade ores with low concentrations are what is left.[1]*
　2. *They are slower compared to traditional methods such as mining ores.[1]*
　3. *$Cu^{2+}(aq) + Fe(s) \rightarrow Cu(s) + Fe^{2+}(aq)$ Correct formulae[1], balanced[1], state symbols[1].*

LIFE CYCLE ASSESSMENT

A **life cycle assessment (LCA)** assesses the **environmental impact** of four different stages in the lifetime of a product, and of the transport and distribution used between them.

Aims

The aims of carrying out an LCA include researching alternative methods of manufacture, maintenance, and disposal. This includes adapting the designs and sources of energy for a product.

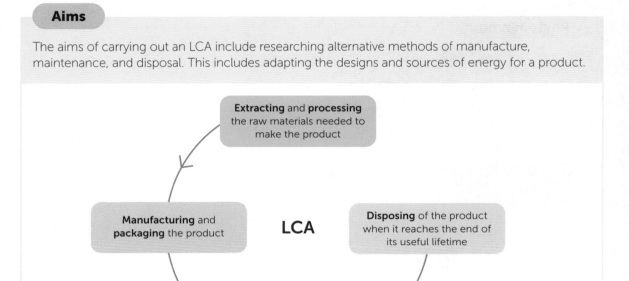

Limits of an LCA

Some aspects of an LCA can be **quantified** (given values). These can include the amount of energy used, the volume of water used, the mass of raw materials and similar resources, and some waste substances produced.

Some aspects cannot easily be quantified, such as the effects of some pollutants released because of a product. The effects of one pollutant may need to be weighed against the effects of another pollutant. Some judgement is needed, so LCAs are not entirely objective.

 You must be able to carry out and compare simple LCAs for plastic and paper shopping bags.

1. Suggest why LCAs are sometimes described as assessing the 'cradle to grave' environmental impacts of a product. [2]

2. Describe how life cycle assessments might be misused. [2]

1. A product is assessed from when the raw materials needed to make it are extracted and processed (its 'cradle')[1] to when the product reaches the end of its life (its 'grave')[1].

2. Manufacturers might use part of an LCA or choose a more favourable LCA[1] in order to support claims about their product or processes in advertising[1].

REUSE AND RECYCLING

Recycling can reduce the use of limited resources and reduce waste.

Raw materials and energy

The Earth's crust, oceans and atmosphere are the sources of the raw materials needed to make materials and products. Most raw materials and energy resources are **limited resources**, so it is important to conserve them. We can all help by:

- using less of these resources
- **re-using** products where possible, and
- recycling products and their parts when they have reached the end of their useful lives.

Doing this reduces harm to the environment caused by mining and quarrying. It also reduces waste.

Waste being dumped at a landfill site

Recycling

Some products can be re-used. For example, glass bottles may be returned for washing and refilling, but most products cannot be re-used. They should be recycled instead.

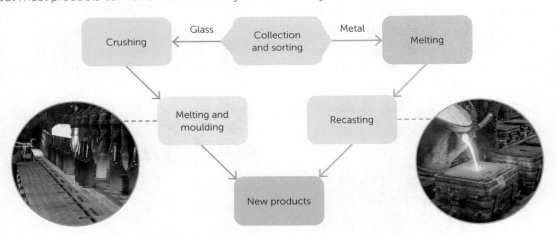

Crushing ← Glass — Collection and sorting — Metal → Melting

Melting and moulding

Recasting

New products

Used products usually need sorting in order to separate different materials from one another. Some contain many different materials which may be more difficult to separate.

1. Iron is extracted from iron ore in a blast furnace. Some scrap steel may be added to the liquid iron. Explain why this is done. [2]

2. Suggest an explanation why broken bathroom tiles are more difficult to recycle than plastic shampoo bottles. [2]

   1. *It recycles used steel[1] which reduces the amount of iron ore that must be used[1].*

   2. *Tiles are made from clay ceramics which cannot be melted and reformed[1] but most plastics can be melted and reformed into new products[1].*

EXAMINATION PRACTICE

01 Fresh water is treated to make it potable or safe to drink.

    01.1 Give **one** reason why filtration is used to treat fresh water to make it potable. [1]

    01.2 Explain why chlorine may be added during water treatment. [2]

    01.3 Give **one** way in which sea water may be treated to make it potable. [1]

    01.4 Give a reason why sea water is not used to make potable water in the UK. [1]

02 The diagram shows three of the processes used to treat waste water.

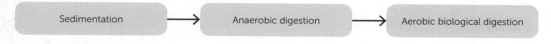

    02.1 Describe what happens in the stage before sedimentation. [2]

    02.2 Describe the function of microbes in the treatment of waste water. [2]

03 Aluminium is obtained from aluminium ore or by recycling used aluminium items.

    03.1 Explain why aluminium ore is described as a limited resource. [2]

    03.2 Suggest **two** reasons, other than conserving supplies of aluminium ore, why we should recycle aluminium instead of making aluminium from bauxite. [2]

04 This question is about sustainable development.

    04.1 Describe what sustainable development means. [2]

    04.2 Shirts can be made from polyester or cotton fibres. Crude oil is the raw material for polyester. Suggest why using polyester alone is less sustainable than using a mixture of both types of fibre. [2]

05 Shopping bags can be made from paper or plastic. The table shows information from life cycle assessments of these bags.

	Paper bags	Plastic bags
Mass per bag in g	50	30
Source of raw material	Trees	Crude oil
Energy used to make one bag in kJ	135	120
Biodegradable	Yes	No
Recyclable	Yes	Yes
Reusable	No	Yes

Evaluate the use of paper to make bags compared to the use of plastic. Use information in the table and your own knowledge and understanding. [6]

Higher Tier only:

06 Gold can be extracted from gold ores taken from mines, and by phytomining.

    06.1 Describe how phytomining is carried out. [3]

    06.2 Give **two** advantages of phytomining gold compared to extracting gold from ores. [2]

PHYSICS
TOPICS FOR PAPER 1

Information about Paper 1:

Trilogy 8464:

Written exam: 1 hour 15 minutes
Foundation and Higher Tier
70 marks
16.7% of the qualification grade
All questions are mandatory

Specification coverage

The content for this assessment will be drawn from topics on: Energy; Electricity; Particle model of matter; and Atomic structure.

Questions

A mix of calculations, multiple-choice, closed short answer and open response questions assessing knowledge, understanding and skills.

Questions assess skills, knowledge and understanding of Physics.

ENERGY STORES AND SYSTEMS

Energy is stored in **systems**. A system is an object or a group of objects.

Common energy stores

Common energy stores are: **chemical**, **kinetic**, **elastic potential**, **gravitational potential** and **thermal**. Further stores are **magnetic**, **electrostatic** and **nuclear**. When a system changes, the way the energy is stored in the system changes.

Scenario	Store that decreases	Store that increases
An object thrown upwards	Kinetic	Gravitational potential
A moving object hitting a wall	Kinetic	Thermal
A bow releasing an arrow	Elastic potential	Kinetic
A vehicle slowing down when braking	Kinetic	Thermal (caused by friction)
Heating water on a camping stove	Chemical (in gas)	Thermal
Using a battery-operated fan	Chemical (in battery)	Kinetic and thermal (of the motor and surroundings)

Change in systems

Systems can be changed by:

Heating

Work done by forces

Work done when current flows

1. Complete the sentences using answers from below.

   chemical   elastic potential   gravitational potential
   kinetic   thermal

   (a) A cyclist accelerates. Energy is transferred from a chemical store to a _____ store. [1]

   (b) When a stone falls from a cliff, energy is transferred from a _____ store to a _____ store. [2]

2. A fully charged electric scooter is ridden down a street at a constant speed. Explain how the energy stores change in this system. [3]

   1. (a) kinetic[1], (b) gravitational potential[1], kinetic[1]

   2. Three points from: Chemical energy[1] stored in the battery is transferred to the kinetic energy[1] store of the scooter and then to the thermal energy[1] store of the surroundings due to air resistance and friction between the wheels and the ground[1].

KINETIC ENERGY (E_k)

Calculating kinetic energy

A moving object has a store of **kinetic energy**. The kinetic energy of a moving object can be calculated using the equation:

Kinetic energy = 0.5 × mass × (speed)^2, or $E_k = \frac{1}{2} mv^2$

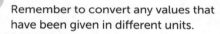

You need to be able to recall and apply this equation.

E_k = kinetic energy in joules, J m = mass in kg v = velocity (speed) in metres per second.

The equation can be rearranged to calculate mass and speed.

1. A cyclist and bike have a combined mass of 75 kg. The cyclist is moving at 10 m/s. Calculate the kinetic energy of the cyclist and bike. [2]

2. A ball has a mass of 400 g. The ball moves with a speed of 2 m/s.

   Calculate the kinetic energy of the ball. Give the unit. [4]

3. **Higher only:** The mass of a runner is 70 kg. The kinetic energy of the runner is 875 J.

   Calculate the speed of the runner in m/s. [3]

Remember to convert any values that have been given in different units.

For example:

1 MJ = 1 000 000 J

0.1 kg = 100 g

1 km = 1000 m; 1 cm = 0.01 m

1 minute = 60 seconds;
1 hour = 60 × 60 = 3600 seconds.

1. E_k = $\frac{1}{2} mv^2$

   = 0.5 × 75 × 10$^{2[1]}$

   = 3750 J$^{[1]}$

2. m = 400 g = 0.4 kg$^{[1]}$

   E_k = $\frac{1}{2} mv^2$

   = 0.5 × 0.4 × 2$^{2[1]}$

   = 0.8$^{[1]}$ J$^{[1]}$

3. E_k = $\frac{1}{2} mv^2$

   875 = 0.5 × 70 × $v^{2[1]}$

   v = $\sqrt{\dfrac{2 \times 875}{70}}^{[1]}$

   v = 5 m/s$^{[1]}$

ELASTIC POTENTIAL ENERGY (E_e)

A stretched object, such as a spring or an elastic band under tension, stores **elastic potential energy**.

Calculating elastic potential energy

The elastic potential energy of an object can be calculated using the equation:

elastic potential energy = 0.5 × spring constant × (extension)²

$$E_e = \frac{1}{2} ke^2$$

You need to be able to select this equation from the equation sheet and apply it.

E_e = elastic energy in joules, J

k = spring constant in newtons per metre, N/m

e = extension in metres, m

The equation can be rearranged to calculate the spring constant and extension.

A spring also has a store of elastic potential energy if it is squashed. The amount of energy can be calculated using the same equation, but with compression replacing extension.

The limit of proportionality

The **limit of proportionality** refers to the point at which an object will no longer return to its original shape or length when stretched or squashed. Beyond this point, you cannot calculate the elastic potential energy. See **page 304**.

1. A toy pops out on a compressed spring when a box is opened.

   (a) Describe the change in energy stores as the box opens. [2]

   (b) The spring extends by 0.1 m when the box opens. The spring constant is 4.0 N/m .
   Calculate the elastic potential energy, E_e, (J) stored in the spring when it is inside the box. [2]

2. **Higher only:** A training band stretched by 0.05 m has an elastic potential energy store of 0.02 J.
   Calculate the spring constant of the band in N/m. [3]

   1. (a) *Elastic potential energy[1] is transferred to kinetic energy[1].*

   (b) $E_e = \frac{1}{2} ke^2$

   $= 0.5 \times 4.0 \times 0.1^{2[1]} = 0.02\ J^{[1]}$

   2. $E_e = \frac{1}{2} ke^2$

   $0.02 = 0.5 \times k \times 0.05^{2[1]}$

   $k = \frac{2 \times 0.02}{0.05^2}$ [1]

   $= 16\ N/m^{[1]}$

GRAVITATIONAL POTENTIAL ENERGY (E_p)

When an object is raised it gains **gravitational potential energy (g.p.e.)**.

Calculating gravitational potential energy

The gravitational potential energy gained by an object raised above the ground can be calculated using the equation:

g.p.e. = mass × gravitational field strength × height

$$E_p = mgh$$

You need to be able to recall and apply this equation.

E_p = gravitational potential energy in joules, J

m = mass in kilograms, kg

g = gravitational field strength in newtons per kilogram, N/kg

h = height in metres, m.

In any calculation the value of the gravitational field strength (g) will be given. Remember that h is the change in **vertical** height.

The equation can be rearranged to calculate mass, gravitational field strength and height.

Use g = 9.8 N/kg for all questions.

1. Calculate the increase in gravitational potential energy when a mass of 1.2 kg is lifted to a height of 5 m. [2]

2. A skier has a mass of 80 kg. Calculate how much energy is transferred from the gravitational potential energy store as the skier travels 150 m down a mountain. [2]

3. A ball is dropped from a height of 2.0 m. Calculate the speed at which the ball hits the ground. [3]

1. g.p.e. = mgh
   = 1.2 × 9.8 × 5[1]
   = 58.8 J[1]

2. g.p.e. = mgh
   = 80 × 9.8 × 150[1]
   = 117 600 J[1]

3. decrease in g.p.e. = increase in kinetic energy
   $m \times g \times h$ = 0.5 × $m \times v^2$[1]
   0.5 × v^2 = 9.8 × 2.0[1]
   $v = \sqrt{\dfrac{19.6}{0.5}}$ = 6.3 m/s[1]

ENERGY CHANGES IN SYSTEMS

Specific heat capacity

The amount of **thermal energy** needed to increase the temperature of a substance depends on its mass and **specific heat capacity**.

The specific heat capacity of a substance is the amount of energy needed to increase the temperature of one kilogram (1 kg) of the substance by one degree Celsius (1 °C).

When a substance is heated, there is an increase in its thermal energy store. When a substance cools down, there is a decrease in its thermal energy store. The change in the thermal energy store when the temperature is increased is the same as the change in the thermal energy store when the temperature decreases by the same amount.

The change in thermal energy store as the temperature of a system changes, can be calculated using the equation:

change in thermal energy = mass × specific heat capacity × temperature change

$$\Delta E = m\, c\, \Delta\theta$$

ΔE = change in thermal energy in joules, J

> Δ means there is a change. You need to be able to select this equation from the equation sheet and apply it.

m = mass in kilograms, kg

c = specific heat capacity in joules per kilogram per degree Celsius, J/kg °C

$\Delta\theta$ = temperature change in degrees Celsius, °C

The equation can be rearranged to calculate mass, specific heat capacity, and temperature change.

1. The specific heat capacity of copper is 385 J/kg °C

   Calculate the change in thermal energy when a 2.5 kg block of copper cools down by 20 °C.

   Give your answer in kJ.   [3]

2. In an investigation to find the specific heat capacity of water, the temperature of 0.2 kg of water increased by 30 °C. The energy transferred to the thermal energy store of the water was 25.8 kJ.

   Calculate the specific heat capacity of water.   [3]

1. $\Delta E = mc\Delta\theta$	2. $\Delta E = mc\Delta\theta$
$= 2.5 \times 385 \times 20$ [1]	$25\,800 = 0.2 \times c \times 30$ [1]
$= 19\,250$ J [1]	$c = \dfrac{25\,800}{0.2 \times 30}$ [1]
$= 19.25$ kJ [1]	$= 4300$ J/kg °C [1]

POWER

Calculating power

Power is the **rate** at which energy is transferred, or the rate at which work is done. This means power is the amount of energy transferred, or how much work is done, in any given amount of time.

$$\text{power} = \frac{\text{Energy transferred}}{\text{time}} = \frac{\text{Work done}}{\text{time}}$$

$$P = \frac{E}{t} = \frac{W}{t}$$

P = power in watts, W

E = energy transferred in joules, J

W = work done in joules, J

t = time in seconds, s

1 watt = 1 joule per second

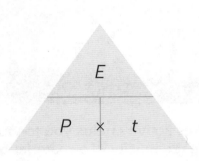

You could use W instead of E in the formula triangle.

Work done is represented by an italicised W so be careful not to muddle it with the unit watts, W.

1. Choose the equation used to calculate work done.
   Tick **one** box. [1]

   ☐ $W = Pt$ ☐ $W = \frac{P}{t}$ ☐ $W = \frac{E}{t}$

2. Sara and Rob both do the same 45-minute cycling spin class.
   (a) Sara does more work than Rob in the class.
       Describe what this means in terms of power. [2]
   (b) Sara does 594 000 J of work in the 45-minute class. Calculate her average power during the class. [3]
   (c) Explain why the power is calculated as an average. [2]

   1. $W = Pt$[1]

   2. (a) Sara does more work/transfers more energy than Rob in the same time[1] so Sara's power is higher than Rob's power in this class[1].

       (b) t = 45 × 60 = 2700s[1]
           $P = \frac{W}{t}$
           P = 594 000 / 2700[1]
           = 220 Watts (W)[1]

       (c) The amount of energy transferred by Sara each second will not be constant[1] so the power/rate of transfer also changes and an average is calculated for the time given[1].

REQUIRED PRACTICAL 14

Specific heat capacity

This activity helps you to work safely and accurately in calculating the specific heat capacity of a material or substance.

The change in the **thermal energy store** of a substance (ΔE) must be linked with a measurable change in another energy store.

For greater accuracy it is important to minimise unwanted energy transfers. It is assumed that all of the energy is transferred to the substance being tested.

This practical involves heating up a substance with an electric heater. The **mass** (m) and the **temperature change** ($\Delta\theta$) of the substance are measured and recorded. The amount of energy supplied (the electrical **work done**) by the heater (ΔE) can be calculated. This could also be measured using a device called a joulemeter.

The **specific heat capacity** (c) can then be calculated using: $c = \dfrac{\Delta E}{m\,\Delta\theta}$

Finding the specific heat capacity of a solid metal

1. Measure the mass of the metal block in kilograms.

2. Heat the block using an electric heater placed inside a hole in the block.

3. Record the temperature of the block every minute for 10 minutes.

4. Calculate the work done by the heater using $\Delta E = P \times t$ (See **pages 251 and 272**) where P is the power of the heater in watts, and t is the time in seconds the heater has been on.

5. Plot a graph of $\Delta\theta$ against ΔE and draw a line of best fit.

6. The gradient of the straight part of the line = $\Delta\theta/(\Delta E)$.

7. $c = \dfrac{1}{\text{gradient} \times m}$

Note: The calculated specific heat capacity is often higher than the actual value. This is because not all of the energy supplied by the heater will be transferred to the substance. Some will be transferred to the thermal energy store of the heater and thermometer and some will be transferred to the thermal energy store of the surroundings. See **pages 254-255**.

To reduce errors

- Insulate the metal thoroughly to reduce thermal energy transfer away from the metal.
- Add oil or water into the thermometer hole for better thermal contact.
- Use a datalogger to record the temperature and time simultaneously. This can also help to reduce errors reading the thermometer scale.
- Use a variable resistor to keep the heater current constant.

Safety factors

- Place the metal block on a heat proof mat.
- Leave the heater to cool before handling.

If the power of the heater is not known, ΔE can be found by recording the potential difference, V, across the heater in volts, and the current, I, through it, in amps.

So, $\Delta E = V \times I \times t$. See **page 273**.

1. Name the equipment used to measure:

   (a) Mass. [1]   (b) Temperature. [1]

2. Give **two** safety factors to consider when investigating the specific heat capacity of water.   [2]

3. Describe an experiment to find the specific heat capacity of olive oil in J/kg °C using an immersion heater and a joulemeter.

   Include the other apparatus used, ways to reduce errors and how to calculate the specific heat capacity of olive oil.   [6]

   1. (a) Balance.[1]   (b) Thermometer.[1]

   2. Any two from: care handling hot heater / hot water[1]; keep water away from electrical supply[1]; place the metal block on a heat proof mat[1]; leave the heater to cool before handling[1].

   3. Place a beaker on a balance and zero it.[√] Add some oil to the beaker[√] record the mass of the oil in kg[√]. Remove the beaker from the balance.[√] Connect the joulemeter and immersion heater in a series circuit with a power supply and a switch.[√] Put the immersion heater into the oil[√] ensure it is fully submerged, so all the thermal energy is transferred to the oil.[√] Add a thermometer to the oil.[√] Insulate the beaker[√] and add a lid to reduce heat loss to the surroundings[√]. Record the starting temperature of the oil[√] and check that the joulemeter is reading zero. Switch on the circuit to begin heating the oil.[√] Stir the oil to ensure the thermal energy is evenly distributed.[√] Record the temperature and reading on the joulemeter after about 20 minutes.[√]

   Calculate the specific heat capacity using:

   $c = \dfrac{\Delta E}{m \, \Delta\theta}$ [√]

   ΔE is the reading from the joulemeter in J.[√]

   $\Delta\theta$ = final temperature - starting temperature in °C.[√]

   m is the mass of oil heated in kg.[√]

   This is an extended response question that should be marked in accordance with the levels based mark scheme on page 374.

ENERGY TRANSFERS IN A SYSTEM

Energy can only be stored, transferred or dissipated (spread out to the surroundings). It cannot be created or destroyed.

Energy is transferred from one store into one or more other stores. Sometimes the energy is transferred in a **useful** way. When energy is described as **wasted** it means it has been transferred less usefully. It has been **dissipated**.

Systems

Energy within a **closed system** transfers between stores keeping the net energy change in the closed system at zero. This means that no energy is dissipated or wasted from the system and the total amount of energy in the system is constant.

Example system:

A hot drink is in an insulated cup with a lid. Energy is transferred from the thermal energy store of the drink to the cup and lid. Energy is then dissipated to the thermal energy store of the surroundings. The energy is less useful and cannot be gathered back together again.

There is no net change in energy in the system.

Usually we have to include the surroundings in a system to consider it as a closed system. Very few systems are completely isolated from their surroundings.

1. A playground swing is released with a child sitting still on it. It swings a few times and eventually stops.

   Describe the motion of the swing in terms of energy transfers. [5]

   Before the swing is released it has a store of gravitational potential energy.[1] As the swing moves, energy is transferred to the kinetic energy store which reaches a maximum at the bottom of the swing.[1] As the swing moves upwards, energy is transferred to the gravitational potential energy store, and the kinetic energy store of the swing decreases.[1] The swing goes less high each time until it stops[1] because energy is dissipated to the surroundings due to the friction of the swing/child (with the air and where the swing is attached)[1].

Reducing unwanted energy transfers

Unwanted energy transfers happen for many reasons. It is useful to look for ways to reduce these, and therefore transfer energy as usefully as possible.

Any moving parts in a machine will waste energy because of resistance, or friction. **Friction** can be **reduced** by lubricating these moving parts using oils and powders for example.

When an electric current flows through a resistor, energy in the chemical store of a battery is transferred to the thermal energy store of the surroundings by electrical work.

When a device transfers energy to a thermal store, such as water in a kettle, the energy has been transferred to a useful store. However, energy is often wasted by devices dissipating energy to the thermal store of the surroundings. Good **thermal insulation** can reduce this loss. For example, insulation around a hot water tank reduces the amount of energy transferred to the thermal energy store of the surroundings.

Materials with a higher **thermal conductivity** have a higher rate of energy transfer by **conduction** across the material.

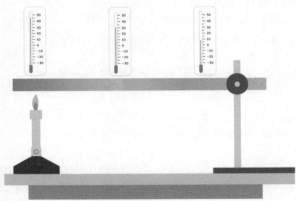

Good **conductors**, such as metals, transfer energy easily by the movement of free electrons. Materials with a low thermal conductivity are good **insulators**.

To prevent heat loss from buildings, the rate of cooling through the walls can be reduced by using insulating materials with thicker layers.

2. Complete the sentence:

   Thermal energy travels from a _____ place to a _____ place. [1]

3. New houses have walls with a low thermal conductivity. Explain how this helps to keep the inside of the building warm in winter. [2]

4. Suggest why the wires used to connect an electrical device to the mains supply should have a very low electrical resistance. [2]

   2. warm; cooler.[1]

   3. A low thermal conductivity means that the rate of energy transfer is low.[1] This means that the rate of energy transfer from the inside to the outside of the building is reduced[1] so not as much energy is needed to keep the inside warm.

   4. A lower resistance means that less energy is transferred by electrical work,[1] so less energy is transferred to the thermal energy store of the surroundings[1].

EFFICIENCY

Efficiency shows how much of the total energy transferred is transferred to useful energy stores, and how much is transferred to less useful stores, or wasted.

Calculating efficiency

More efficient appliances, homes and vehicles cost less to run as more of the energy is transferred usefully. This can reduce carbon dioxide emissions if fossil fuels are used.

Efficiency can be calculated using the equations:

> You need to be able to recall and apply these equations.

$$\text{efficiency} = \frac{\text{useful output energy transfer}}{\text{total input energy transfer}} = \frac{\text{useful power output}}{\text{total power input}}$$

Efficiency is given as a decimal value between 0 and 1.0, or as a percentage % (decimal × 100). Efficiency does not have a unit.

Power is the rate at which energy is transferred in watts (W) so the alternative equation above can be used.

useful output

efficiency × total input

Higher Tier only

The efficiency of an energy transfer is increased when the ratio of useful energy to wasted energy is increased. The amount of useful energy output transfer is increased, and the amount of energy wasted is reduced.

For example, a light bulb can be made more efficient by reducing the amount of energy transferred by heating. Reducing friction by lubricating moving parts in a machine also increases the efficiency.

1. A lamp is rated at 12 W and transfers 4.8 W as visible light.
   Calculate the efficiency of the light bulb. [2]
2. A solar panel is 23% efficient. Calculate the energy wasted when 500 J is input. [3]
3. Explain why it is not possible to have an efficiency greater than 1.0. [2]
4. **Higher Tier only:** Double glazed windows are made of two panes of glass with a layer of gas between them. Explain why double glazing is more thermally efficient than a single pane of glass. [2]

1. Efficiency $= \dfrac{\text{useful power output}}{\text{total power input}}$

   $= \dfrac{4.8}{12}$ [1]

   $= 0.4$ or 40% [1]

2. Efficiency $= \dfrac{\text{useful output energy transfer}}{\text{total input energy transfer}}$

   $0.23 = \dfrac{\text{useful output energy transfer}}{500}$ [1]

   useful output energy transfer $= 0.23 \times 500$
   $= 115$ J [1]

   energy wasted $= 500 - 115 = 385$ J [1]

3. This would mean that the amount of useful energy output is greater than the total input energy transferred [1] which is not possible as energy cannot be created [1].

4. The gas is a poor thermal conductor and is trapped between two layers of glass so the thermal conductivity of the window is lower. [1] The rate of energy transfer is lower than for the single pane, so it is more thermally efficient. [1]

NATIONAL AND GLOBAL ENERGY RESOURCES

Energy resources store a large amount of energy. They can be used to transfer energy from one store to another, usually by generating electricity, and also for heating and transport.

Non-renewable energy	Renewable energy
A non-renewable energy resource is a resource that is being used up faster than it can form. It has a finite supply. This is either because it forms very slowly or it is no longer being formed. It will eventually run out when all the reserves have been used up.	A renewable energy resource is one that can be replenished as it is used, either by human action or natural processes.

Example

Burning fossil fuels releases carbon dioxide into the atmosphere which contributes to global warming and climate change. However, some countries are heavily dependent on the distribution of fossil fuels, so they resist change to alternative resources.

Plans for tidal barrages have been rejected for the Severn Estuary because of potential environmental damage to marine habitats, but mainly because of the huge costs. Offshore floating wind farms are now proposed as an alternative to the tidal barrage to increase the output from renewables.

Nuclear fuel does not produce carbon dioxide but does create radioactive waste which raises issues of disposal and the safety of nuclear power stations.

1. Circle the non-renewable energy resources. [1]

    Biofuel   Coal   Gas   Geothermal   Wind   Oil   Sun   Nuclear fuel

2. Hydroelectric, tidal and wave generators are three renewable energy resources that use water to produce electricity. Complete the table below showing information about the resources. [3]

	Energy store	Power output	Reliability	Environmental impact
Hydroelectric power	Gravitational potential energy	High	Reliable	
Tidal power	Gravitational potential energy	High		Damage to marine and estuary habitats
Wave power		Low	Reliable	Has a low impact

1. *Coal   Gas   Nuclear Fuel   Oil*[1].

2. *Hydroelectric power – Flooding (dams) damages land / habitats*[1]; *Tidal – Reliable*[1]; *Wave – Kinetic.*[1]

IMPACTS AND USES OF ENERGY RESOURCES

Non-renewable energy sources

Energy resource	Advantages	Disadvantages
Fossil fuels Oil Gas Coal	Relatively low cost, easy to use for transport, very adaptable	Produce carbon dioxide (contributes to global warming) and sulfur dioxide (causes acid rain and can cause breathing problems; resources are finite (will run out).
Nuclear Nuclear	No greenhouse gases produced. A large amount of energy produced from a small amount of fuel	Disposal of uranium is costly and hazardous. Huge cost to build, run and decommission power stations. Risk of catastrophic accidents.

Renewable energy sources

Energy resource	Advantages	Disadvantages
Biofuel	Can use up plant and animal waste. Replacement engine fuel for transport.	Some biofuels are grown especially which uses a lot of land that could be used for food production.
Geothermal	Can be used for heating local areas. Electricity can be generated from hot rocks.	Not available everywhere and sometimes in dangerous or inaccessible areas.
Hydroelectricity	Available at any time and can be started and stopped quickly.	Areas of natural beauty can be flooded to make the dams and reservoirs required.
Sun	Solar cells on roof tops and solar farms generate electricity. Also used for heating water. Solar power stations are in development.	Not available at night. Works better in places with more intense sunlight. Visual impact on the landscape raises objections.
Tides	Electricity can be generated at predictable times.	Could damage marine and estuary habitats. Tends to be extremely expensive with few suitable places for barrages.
Water waves	Can generate electricity on a small scale.	Maintenance, impact on marine environment and inconvenience to shipping.
Wind	Many suitable offshore sites in the UK. Fairly low maintenance when established. Turbines can be built out at sea.	Unreliable unless the wind speed is within the right range. Visual pollution of turbines on land. Many turbines are needed to generate suitable amounts of electricity.

EXAMINATION PRACTICE

01 Choose the correct equation to calculate kinetic energy. Tick (✓) **one** box. [1]

☐ $E_k = \frac{1}{2} mv^2$

☐ $E_k = ke^2$

☐ $E_k = \frac{1}{2} ke$

02 The diagram shows a spring before and after it is stretched.
Calculate the value of e, the spring extension, when it is stretched. [1]

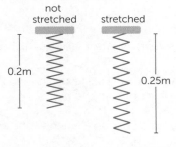

03 Calculate the kinetic energy store of a bird with a mass of 0.4 kg flying at 8 m/s. Give the unit. [3]

04 A mountain biker rides along a path and jumps into the air. The cyclist and bike have a total mass of 85 kg.

04.1 Calculate the gravitational potential energy when the cyclist has jumped 2 m above the ground. Use $g = 9.8$ N/kg. [2]

04.2 Describe the energy transfers as the bike falls and then hits the ground. [2]

04.3 **Higher Tier only:** The cyclist does another jump and has a maximum gravitational potential energy of 680 J. Calculate the speed at which the cyclist hits the ground.
Assume that there are no energy transfers due to friction or other forces. [3]

05 The table below shows diesel and electric car sales in the UK for the same month in 2018 and 2019.

Year	Diesel car sales	Electric car sales
2018	51 108	9971
2019	36 941	15 132

Identify the trends in both car sales. Suggest a reason for each trend. [4]

Diesel car sales: Electric car sales:

Trend: Trend:

Reason: Reason:

06 Give **one** advantage and **one** disadvantage of using nuclear fuel instead of coal to generate electricity. [2]

07 The thermal conductivity of brick is approximately six times higher than the thermal conductivity of air.

Explain why the walls of houses are often built using two layers of bricks with a gap in between them, rather than a single layer of bricks. [3]

08 A device with a total input power of 30 W has an efficiency of 0.45.

Calculate how much **useful energy** the device transfers in 20 s [3]

09 Explain why a more powerful heater is better at warming up a cold room than a less powerful heater. [2]

10 The specific heat capacity of water is 4200 J/kg°C

10.1 Define the specific heat capacity of water. [1]

A student does an experiment to find the specific heat capacity of water using an electric heater with a power of 30 W.

10.2 What does the student need to measure to find the energy transferred by the heater? [1]

The student completes the experiment and finds the specific heat capacity of water to be 4500 J/kg°C. All measurements and calculations were done accurately and repeated.

10.3 Suggest how the experiment could be improved to get a value nearer to 4200 J/kg°C. [2]

10.4 **Higher Tier only:** Calculate the expected rise in temperature when 2100 J is transferred to 250 g of water. Use c = 4200 J/kg°C. [4]

STANDARD CIRCUIT DIAGRAM SYMBOLS

Electric circuit diagrams are drawn using standard **circuit symbols** for each of the different **components** that make up a circuit.

Standard symbols

The symbols below are universally recognised. This saves time drawing or describing them in other ways.

Some symbols are quite similar. You will need to be able to recall and draw them accurately, preferably with a pencil and ruler.

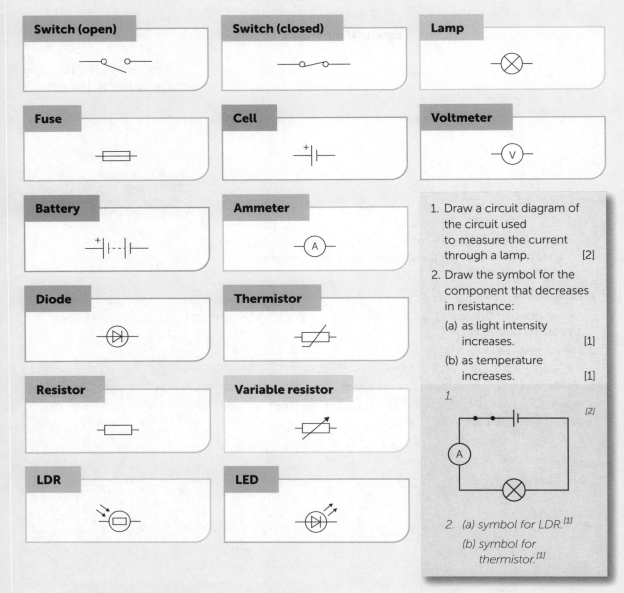

Switch (open)

Switch (closed)

Lamp

Fuse

Cell

Voltmeter

Battery

Ammeter

Diode

Thermistor

Resistor

Variable resistor

LDR

LED

1. Draw a circuit diagram of the circuit used to measure the current through a lamp. [2]

2. Draw the symbol for the component that decreases in resistance:

   (a) as light intensity increases. [1]

   (b) as temperature increases. [1]

1.

[2]

2. (a) symbol for LDR. [1]

   (b) symbol for thermistor. [1]

ELECTRICAL CHARGE AND CURRENT

Electric **current** is the **rate of flow of electrical charge**. In metal wires, this charge is carried by **electrons**.

Calculating charge

The rate of flow of charge, or size of the current, can be measured by determining how much charge passes by a point in the circuit each second. The current in a component is measured using an **ammeter** –Ⓐ– connected in **series** with the component. The current has the same value anywhere in a single closed loop.

For an electric charge to flow, the circuit must be a **closed circuit**. This means it is an unbroken loop, so any switches must be closed (on).

The circuit also needs a source of **potential difference** so must include a cell ⊣⊢, battery ⊣⊦··⊢ or power supply.

The charge flow can be calculated using the equation:

> You need to be able to recall and apply this equation.

$$charge\ flow = current \times time$$

$$Q = I \times t$$

Q = charge flow in coulombs, C

I = current in amperes (amps), A

t = time in seconds

> Remember: It is the charge that flows, not the current.

 Learn the symbols for the quantities Q and I and the units C and A. Make sure not to get these quantities and units confused.

1. Complete the sentence.
   One coulomb is the _____ that passes a point in a circuit when there is a _____ of one ampere for one second. [2]

2. The current in a lamp is 0.7 A.
   Calculate the charge, in C, that flows through the lamp in 1.5 minutes. [3]

3. A charge of 0.12 C flows through a resistor in 86 s. Calculate the current, in A, in the resistor. [3]

4. There is a current of 0.80 A in a circuit. Calculate the time taken, in s, for 32 C of charge to flow past a point in the circuit. [3]

1. *Charge[1], current.[1]*

2. $t = 1.5 \times 60 = 90\ s$[1]
   $Q = It$
   $= 0.7 \times 90$[1]
   $= 63\ C$[1]

3. $Q = It$
   $0.12 = I \times 86$[1]
   $I = \dfrac{0.12[1]}{86}$
   $= 0.0014\ A$[1]

4. $Q = It$
   $32 = 0.80 \times t$[1]
   $t = \dfrac{32}{0.80}$[1]
   $= 40\ s$[1]

CURRENT, RESISTANCE AND POTENTIAL DIFFERENCE

Resistance is a measure of how hard it is for charge to flow through a component or material.

Potential difference and resistance

Materials with extremely high resistance are **electrical insulators**. The current in a component depends on the **potential difference** across the component and the resistance of the component. The potential difference across a component can be thought of as the difference in energy carried by electrons before and after they have flowed through a component. It is a measure of the electrical work done.

For a given potential difference, the current decreases as the resistance increases.

The greater the resistance of the component, the smaller the current for a given potential difference across the component.

Calculating potential difference

Potential difference, current and resistance are linked by the equation:

potential difference =
current × resistance

$$V = I \times R$$

V = potential difference in volts, V
I = current in amperes (amps), A
R = resistance in ohms, Ω

You need to be able to recall and apply this equation.

Potential difference (pd) is also known as voltage. In an exam, the term 'potential difference' will be used but the correct use of either term will gain marks.

1. The resistance of a resistor is 250 Ω. The current in the resistor is 1.2 A. Calculate the potential difference across the resistor, in volts. [2]

2. The potential difference across a component is 4.5 V. The resistance of the component is 2.5 Ω. Calculate the current, in amps, in the component. [3]

3. The potential difference across a resistor is 30 V .The current in the resistor is 0.50 A . Calculate the resistance, in ohms, of the resistor. [3]

4. Explain why the resistance in a wire changes as the wire gets hotter. [3]

1. $V = IR$
   $= 1.2 \times 250$[1]
   $= 300$ V[1]

2. $V = IR$
   $4.5 = I \times 2.5$[1]
   $I = 4.5 / 2.5$[1]
   $= 1.8$ A[1]

3. $V = IR$
   $30 = 0.50 \times R$[1]
   $R = 30 / 0.50$[1]
   $= 60 \Omega$[1]

4. As the wire gets hotter the atomic vibrations are larger[1], so electrons collide with atoms more frequently[1] and the resistance increases[1].

REQUIRED PRACTICAL 15

Investigating resistance

This practical activity helps you develop your ability to use appropriate apparatus and circuit diagrams to measure electrical quantities.

Investigating the effect of length on the resistance of a wire at a constant temperature

- Set up a circuit to measure the potential difference and current for a wire at different lengths along the wire.

- Calculate the resistance for the different lengths of wire using $R = V / I$

- Plot a graph of resistance against length of wire and draw a line of best fit.

- Make a conclusion about the relationship between resistance and length.

The investigation should lead to the conclusion that the longer the wire the higher the resistance.

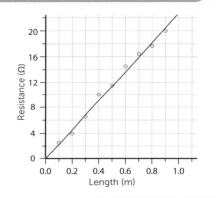

1. (a) Describe what the results in the graph show. [1]

   (b) Explain how changing the length of the wire affects the resistance. [2]

2. Give **two** ways to reduce errors caused by the heating effect of a current when investigating the resistance of a wire. [2]

3. At a wire length of 200.0 cm, the pd is measured at 0.53 V, 0.58 V and 0.54 V.

   (a) Calculate the mean potential difference (pd). [1]

   (b) The mean current is 0.22 A. Calculate the resistance of the wire at 200.0 cm. [3]

> 1. (a) The resistance increases as the length of wire increases.[1]
>
>   (b) Resistance is directly proportional to the length of the wire.[1] For a shorter wire, electrons do not have to travel so far through the wire so resistance is less.[1]
>
> 2. Use a low potential difference to keep current low and avoid heating the wire too much.[1]
>   Only turn current on briefly to take the reading / allow the wire to cool between readings.[1]
>
> 3. (a) (0.53 + 0.58 + 0.54) / 3 = 0.55 V[1]
>
>   (b) V   = IR
>   0.55 = 0.22 x R[1]
>   R   = 0.55 / 0.22[1]
>   = 2.5 Ω[1]

To increase the reliability of the results:

Repeat the values and calculate a mean resistance for each length. Use smaller gaps between lengths.

Sources of error:

- The heating effect of the wire.
- Measuring the length between the crocodile clips consistently.

Resistance of resistors arranged in series and parallel

- Use circuit diagrams to set up circuits connecting two **identical resistors**, **R₁** and **R₂**, in series and then in parallel.
- Measure the potential difference across the resistors and the current in both circuits.
- Calculate the total resistance in the circuit.
- Compare the resistance of each arrangement of resistors.

2 resistors in series

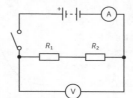

2 resistors in parallel

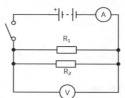

It is important to connect the components correctly. This makes sure the resistances can be compared accurately and safely, and also protects the components. There should be positive readings on both the ammeter and the voltmeter. If you see a negative value, swap the connections around. The smaller the **resolution** of the ammeter and voltmeter, the more decimal places the readings can be given to.

Circuit	Potential difference (V)	Current (A)
Series	6.4	0.16
Parallel	5.9	0.59

4. The table above shows the results for two identical resistors connected in series and in parallel.

Use the equation resistance = potential difference / current.

(a) Calculate the total resistance in the series circuit. [1]

(b) Calculate the total resistance in the parallel circuit. [1]

(c) Compare the resistance of the two arrangements. [2]

4. (a) Series: $V = IR$
$6.4 = 0.16 \times R$
$R = 6.4 / 0.16$
$= 40 \ \Omega$ [1]

(b) Parallel: $V = IR$
$5.9 = 0.59 \times R$
$R = 5.9 / 0.59$
$= 10 \ \Omega$ [1]

(c) *The total resistance in the circuit is higher when the resistors are connected in series.* [1]
The resistance in series is four times higher than the resistance when they are connected in parallel. [1]

RESISTORS

For some resistors, the value of R (resistance) remains constant as the current changes. These resistors are **ohmic conductors**.

Non-ohmic resistors do not follow this relationship and the value of R can change as the current changes. Filament lamps, diodes, thermistors and LDRs are non-ohmic resistors. Their resistance changes with the current in the component.

Ohmic and non-ohmic conductors

Ohmic conductor

Current / Potential difference

At a **constant temperature**, the current through an ohmic conductor is directly proportional to the potential difference across the resistor.

The resistance stays constant when the current changes.

Non-ohmic conductors

Filament lamp

The resistance of a filament lamp increases as the temperature of the filament increases, so the gradient of the graph decreases as the potential difference increases.

Current / Potential difference

Diode

The current through a diode flows in **one direction** only.

The diode has such a high resistance in the reverse direction that no current can go that way.

Current / Potential difference

Remember the basic circuit for measuring the resistance of a component:

- Ammeter in series with the component
- Voltmeter in parallel (across the component).

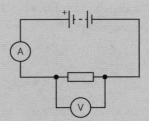

Thermistor

The resistance of a thermistor decreases as the temperature increases.

These can be used as temperature sensors in fire alarm systems. Alarms or sprinklers are switched on when a particular temperature is exceeded. When the resistance becomes low enough, a current can flow in the circuit which is used to switch on the systems. They are also used in thermostats for switching on devices at required temperatures.

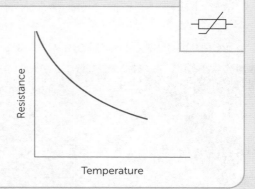

LDR (Light Dependent Resistor)

The resistance of an LDR decreases as light intensity increases. In bright light, the resistance of an LDR is low so more current can flow through it. In the dark, or at low light levels, the resistance of an LDR is high, so much less current can flow through it.

This means LDRs can be used to detect light levels. They will switch on security and street lamps automatically at low light, and switch them off when light levels rise again.

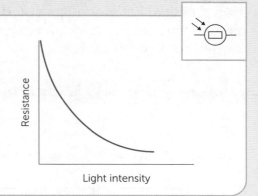

Explain the I-V graphs of component X and component Y. Include resistance and temperature in your answer.

[6]

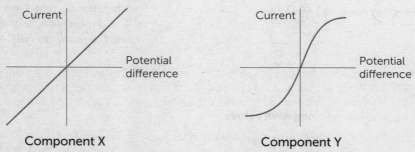

Component X is an ohmic conductor[√]. The current in it is directly proportional to the potential difference across it[√]. As $V = I \times R$[√], the resistance is constant whatever the potential difference is across it. The value of R can be found from the gradient of the graph[√] which is I / V[√] which equals 1 / R[√]. For this relationship, the temperature of the component is constant.[√]

Component Y is a non-ohmic resistor.[√] The gradient of the graph decreases as potential difference increases[√] which means that the resistance increases as the potential difference increases.[√] This is caused by an increasing temperature[√] from the heating effect of a current[√].

This is an extended response question that should be marked in accordance with the levels based mark scheme on page 374.

REQUIRED PRACTICAL 16

Investigating *I-V* characteristics of components

This practical activity helps you develop your ability to use appropriate circuits and circuit diagrams to measure current and potential difference for different components.

Measuring current and potential difference

Investigate what happens to the current in (i) a **resistor**, (ii) a **filament lamp** and (iii) a **diode** when the potential difference across it is changed.

This is the circuit used for finding the characteristics of the diode.

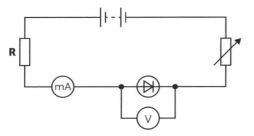

When testing the resistor and filament lamp, replace the diode and remove resistor *R*.

Use a normal ammeter instead of a milliammeter (mA) to measure current when testing these two components. This depends on the values of the current, and the resolution of the ammeters.

Set up the circuit for each component. Adjust the variable resistor and record several pairs of ammeter and voltmeter readings in a suitable table. Swap the connections on the power supply – connect the ammeter to the negative terminal and connect the component being tested to the positive terminal. This will give readings with **negative values**.

Plot a graph of current against potential difference for each component.

The I-V graphs plotted for each component (current against potential difference) should look similar to those on page 267. The shape of the graph is the *I-V* characteristic of the component.

1. The gradient of the graph for the resistor is a straight line that goes through the point (0, 0). Explain what this tells you about the resistor. [2]

2. Give reasons why the circuit to test the diode has an extra resistor in it and uses a milliammeter. [2]

1. *The resistor is an ohmic conductor[1] because the graph is directly proportional which means that the resistance is constant[1].*

2. *The extra resistor is used to protect the diode from high currents[1] and a milliammeter is used because the current values are low[1].*

SERIES AND PARALLEL CIRCUITS

There are two ways of joining electrical components. A single continuous loop is a **series** circuit. A **parallel** circuit has components that are connected across each other. Circuits can include series and parallel parts.

Comparing series and parallel circuits

Compare a cell and two lamps when connected in series and in parallel.

	Series	Parallel
Circuit diagram		
Potential difference	Total potential difference of the power supply is shared between the lamps. $V_{cell} = V_1 + V_2$	Potential difference across each lamp is the same: $V_{cell} = V_1 = V_2$
Current	Same through each lamp. $I_1 = I_2$	Total current through whole circuit is the sum of the current through each lamp. $I_{total} = I_1 + I_2$
Resistance	Total resistance is the sum of the resistance of each lamp. $R_{total} = R_1 + R_2$ The current passes through all the resistors in series so adding resistors decreases the current and increases the total resistance of the circuit	The total resistance is less than the resistance of the bulb with the smallest resistance in the circuit. The total current splits between resistors in parallel. Adding resistors means there are more paths for the current, so the total current increases and the total resistance decreases.
Lamps	Both off or both on.	Can be switched on and off separately.

1. Look at the series circuit in the text above. V_{cell} = 1.5 V, and the resistance of each lamp is 2 Ω.

   (a) Calculate the total resistance, in ohms, in the circuit. [1]

   (b) Calculate the current, in amperes, in the circuit. [3]

2. A series circuit consists of three identical lamps connected with a power supply of 12 V.
   Calculate the potential difference across each lamp. [1]

3. Look at the parallel circuit in the text above. The resistance of each lamp is 2 Ω.
   Describe the equivalent resistance in this circuit. [1]

1. (a) $R_{total} = R_1 + R_2$
   $= 2 + 2$
   $= 4$ Ω

   (b) $V = IR$
   $6.0 = I \times 4$[1]
   $I = 6.0 / 4$[1] $= 1.5$ A[1]

2. 12 V / 3 lamps $= 4$ V[1]

3. It is less than 2 Ω[1]

DIRECT AND ALTERNATING POTENTIAL DIFFERENCE

dc and ac

Cells, batteries and solar cells have two terminals. The electrons always move from the negative terminal to the positive terminal. This means the current is always in the same direction and it is called a **direct current** (**dc**). (Note that conventional current is from positive to negative).

In an **alternating current** (**ac**) the direction of the movement of the electrons is constantly changing. This means that the potential difference is also constantly changing. The potential difference increases to a peak and then decreases to zero. It carries on decreasing to a negative 'peak' and then increases to zero. This cycle repeats continuously, 50 times each second in the UK.

pd-time graphs to show direct and alternating potential difference

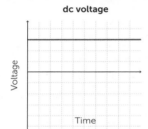

dc voltage

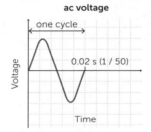

ac voltage

dc

The potential difference-time graph shows a straight line at a constant pd.

ac

The potential difference-time graph shows a curve alternating between positive and negative pd. The positive and negative values indicate the direction of the current.

1. Describe the electricity supply in the UK. Tick **two** boxes. [2]

   ☐ dc supply ☐ Frequency of 50 Hz
   ☐ Frequency of 230 Hz ☐ pd of 110 V
   ☐ pd of 230 V

2. Give **two** examples of devices in the home that use a dc supply. [2]

   1. Frequency of 50 Hz[1] pd of 230 V[1]

   2. Any two devices that use a battery or a cell, not the mains supply. E.g. mobile phone, electric toothbrush, remote control, hand held vacuum cleaner.[1]

MAINS ELECTRICITY

Mains plug

The cable from an electrical appliance to the mains consists of three separately insulated metal wires inside a tube of insulation. It is known as a **three-core cable**.

Each of the three wires inside is colour coded for identification. Inside the appliance plug, the wires are connected as shown. A fuse is also connected to the live wire. Then it is all secured with a cable grip.

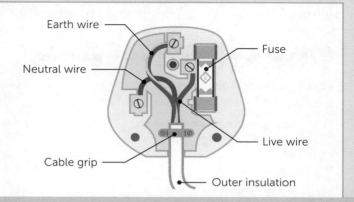

Earth wire — Fuse

Neutral wire — Live wire

Cable grip — Outer insulation

Electrical conductors carry a current e.g. metals.

Electrical insulators do not carry a current e.g. plastic and rubber.

Wire	Colour	Potential difference with the earth	Purpose
Earth	**Green** and yellow stripes	0 V	Safety wire. Only carries a current if there is a fault.
Live	**Brown**	About 230 V	Carries the alternating pd from the supply. Connects to a fuse.
Neutral	**Blue**	At or close to 0 V	Completes the circuit.

Safety features

A **fuse** ▭ contains a wire that will melt if the current gets too high. If this happens, the appliance using the fuse will be at 0 V and not 230 V, so it is made safe.

If a person touches a live wire or appliance, the current travels through them to the ground and they get an electric shock. The **earth** wire is a safety wire to stop the actual appliance becoming live if the live wire touches a conducting part of the appliance, for example, the metal casing of a toaster.

1. Explain why different coloured insulation is used in a three-core cable. [2]
2. Describe **two** dangers of using the mains supply. [2]

   1. *To identify the different wires.[1] To connect wires correctly for safety / for the appliance to work.[1]*
   2. *Electric shock from contact with a live connection[1]. Fire caused by a too much current.[1]*

POWER

The power transfer in any component or appliance is proportional to the potential difference across it and the current through it.

Power, potential difference and current

Power, potential difference and current are linked by the equation:

power = potential difference × current

$$P = V \times I$$

P = power in watts, W

V = potential difference in volts, V

I = current in amperes (amps), A

Remember these definitions of power from **page 251**.

$$\text{power} = \frac{\text{energy transferred}}{\text{time}} = \frac{\text{work done}}{\text{time}}$$

You need to be able to recall and apply both these equations.

Power, current and resistance

Power, current and resistance are linked by the equation:

power = (current)² × resistance

$$P = I^2 R$$

P = power in watts, W

I = current in amperes (amps), A

R = resistance in ohms, Ω

- As $V = I \times R$, substitute $I \times R$ for V in the equation $P = V \times I$
- current² = current × current

Large powers are given in kW.
1 kW = 1000 W

1. Convert 72.61 kW to W. Tick **one** box. [1]

   ☐ 0.07261 W   ☐ 7261 W   ☐ 72 610 W   ☐ 726 100 W

2. Calculate the power, in kW, of an electric iron with a current in it of 11.3 A and a potential difference across it of 230 V. [2]

3. There is a potential difference of 1.5 V across a 7.5 Ω resistor. The current through the resistor is 0.2 A. Show **two** different calculations to confirm that the power of the resistor is 0.3 W. [2]

4. **Higher Tier only:** A 5.0 Ω resistor has a power rating of 12.0 W. Calculate the current, in amps, in the resistor. [3]

1. *72 610 W[1]*

2. *P = V × I*
   = 230 × 11.3[1]
   = 2.60 kW[1]

3. *P = V × I*
   = 1.5 V × 0.2 A = 0.3 W[1]
   P = I²R
   0.2 A × 0.2 A × 7.5 Ω = 0.3 W[1]

4. *P = I²R*
   12 = I² × 5[1]
   I = √(12/5) [1] = 1.5 A[1]

ENERGY TRANSFERS IN EVERYDAY APPLIANCES

Everyday electrical appliances are designed to transfer energy. They transfer energy from cells and batteries (dc supply) or from the mains (ac supply). Appliances with high power ratings need to use the mains supply.

An electrical appliance transfers more energy if it:

- has a **higher power** rating
- is switched on for **more time**

The energy is usually transferred:

- to make electric motors move (**kinetic** energy)
- for use in heating devices (**thermal** energy)

Electrical work

When charge flows in a circuit, energy is transferred by electrical **work**. The amount of energy transferred by electrical work can be calculated using one of these equations.

Energy transferred, power and time are linked by the equation:

energy transferred = power × time

$$E = P \times t$$

E = energy transferred in joules, J

P = power in watts, W

t = time in seconds, s

Energy transferred, charge flow and potential difference are linked by the equation:

energy transferred = charge flow × potential difference

$$E = Q \times V$$

E = energy transferred in joules, J

Q = charge flow in coulombs, C

V = potential difference in volts, V

You need to be able to recall and apply these equations.

Remember: Electrical work done is equal to the energy transferred.

1. Give **one** example of a device that transfers useful energy by electrical work to:
   (a) kinetic energy [1]   (b) thermal energy. [1]

2. (a) A hairdryer has 3 heat settings and 2 speed settings. Explain why it has a power rating of 1650 W to 2000 W. [3]

   (b) Calculate the lowest amount of energy transferred, in kilojoules, when the hairdryer is used for 2 minutes. [3]

   (c) The potential difference across the hairdryer is 230 V. Calculate the charge flow, in coulombs, during the energy transfer in (b). [3]

1. *(a) Any electrical device with a spinning motor e.g. washing machine, hairdryer, food mixer, fan.*[1]

   (b) Any electrical device designed to get hot e.g. hairdryer, tumble dryer, iron.[1]

2. *(a) The power rating varies because the hairdryer transfers different amounts of energy*[1] *depending on the heat / speed settings.*[1] *Hotter / faster settings will have higher power ratings.*[1]

   (b) t = 2 × 60 = 120 s[1]   *(c) E   = Q × V*
   E = P × t   *198 000 = Q × 230*[1]
   1650 × 120[1] *= 198 kJ*[1]   *Q   = 198 000 / 230*[1] *= 861 C*[1]

THE NATIONAL GRID

Electrical power is transferred from power stations to consumers using the National Grid.

The National Grid is the system of cables and transformers in the UK that links power stations to consumers.

Step-up transformers are used to increase the potential difference (pd) from the power stations. The pd is increased from about 25 kV to values up to 400 kV.

Step-down transformers are used to decrease the pd from the transmission cables to a much lower pd of 230 V for use in homes.

The current heating effect means that the current in the transmission cables heats up the cables, dissipating energy to the surroundings. This reduces the amount of useful energy transferred and makes the system less efficient.

The electrical power transferred by the cables is given by the equation:

power = pd × current

So for the same power, when the pd is increased, the current decreases.

The electrical power dissipated by the cables is given by the equation:

power = (current)² × resistance

So by reducing the current, and using wires with a low resistance, the electrical power dissipated by the current heating effect is minimised.

Explain why the National Grid system is an efficient way to transfer energy.   [6]

The power dissipated in the transmission cables is given by the equation $P = I^2R$.[1] In the National Grid, a step up transformer[1] steps up the pd and reduces the current[1]. So, reducing the current, reduces the energy dissipated[1] by the current heating effect[1]. In addition, wires with low resistance are also used to help minimise the energy dissipated.[1]

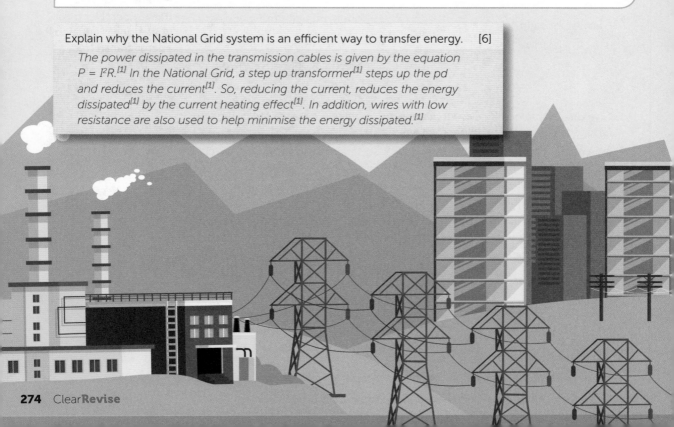

EXAMINATION PRACTICE

01   A student measures the resistance of a lamp.

    01.1   Which **two** components are needed to measure the resistance? Tick **two** boxes.   [2]
        ☐ Ammeter   ☐ Battery   ☐ Voltmeter   ☐ Resistor   ☐ Switch

    01.2   The student uses a battery for a power supply.
        Draw the circuit diagram used to measure the resistance across a lamp.   [3]

    01.3   The current in the lamp is 0.5 A. The potential difference of the battery is 1.5 V.
        Calculate the resistance of the lamp.   [3]

    01.4   A charge of 3.0×10^4 C is transferred by a current of 250 mA.
        Calculate the time taken in seconds for this charge to be transferred.   [4]

02   This question is about understanding components.

    02.1   Name this component.   [1]

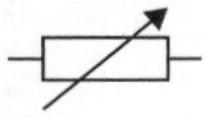

    02.2   Explain how this component works.   [2]

    02.3   Which **one** of these components is used to detect light? Tick **one** box.   [1]

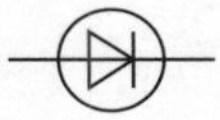

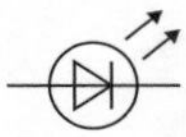

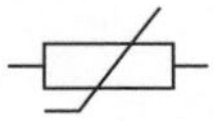

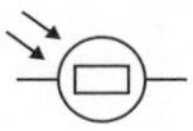

    02.4   Describe how this component is used to control circuits.   [2]

03   The diagram shows an electric circuit.

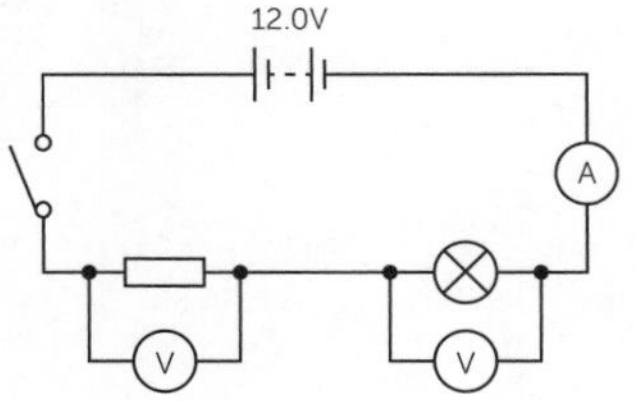

The resistance of the lamp is 2.0 Ω. The ratio of the resistance of the resistor to the resistance of the lamp is 2:1.

    03.1   Calculate the equivalent resistance in the circuit.   [2]

    03.2   Calculate the reading on the ammeter.   [2]

    03.3   Calculate the potential difference across the:
        (a) resistor [1]   (b) lamp [1]

    03.4   The resistance of a wire is proportional to its length. Which graph represents this?
        Tick **one** box.   [1]

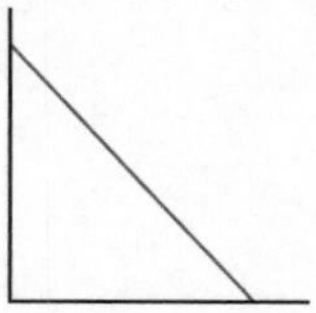

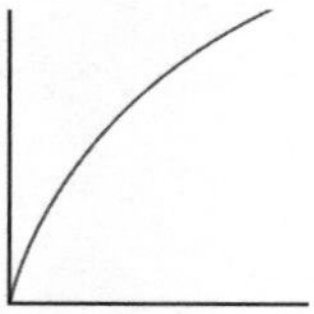

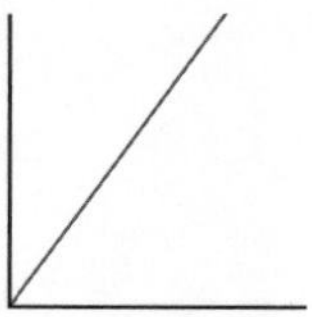

04   Describe the difference between direct and alternating potential difference.   [2]

05   Explain the danger of the live wire touching a conducting part of an electrical appliance.   [2]

06   Describe how transformers are used in the National Grid.   [3]

07   An electric kettle with a power rating of 2.9 kW is connected to the 230 V mains supply.
      07.1   Calculate the current in the kettle.   [3]
      07.2   A different kettle has a lower power rating and is connected to the same mains supply.
             Compare the current in both kettles. Tick **one** box.   [1]
             ☐   The kettle with the higher power rating has a lower current
             ☐   Both kettles have the same current
             ☐   The kettle with the higher power rating has a higher current.

      07.3   Write the equation that links current, power and resistance.   [1]

08   A microwave oven transfers 45 000 J in 70 s when it is connected to the 230 V mains supply.
      08.1   Calculate the power, in watts, of the microwave oven.   [3]
      08.2   Calculate the charge flow through the microwave oven in 70 s.   [3]

DENSITY OF MATERIALS

The **particle model** describes the arrangement of particles in the three different **states of matter**. It is used to predict and explain the behaviour of **solids**, **liquids** and **gases** in different conditions and **densities**.

The particle model

The **particle model** represents the arrangement and movement of particles in the three different states. The particles can be atoms, ions or molecules, but all of them are shown as small, solid spheres.

	Solid	Liquid	Gas
Arrangement of particles	Regular	Random	Random
Relative distance between particles	Close	Close	Far apart
Motion of particles	Vibrate about fixed positions	Move around each other	Move quickly in all directions
Particle diagram			

Calculating density

The density of a material is the mass of a substance per unit volume. **Volume** is how much space a material occupies. For two materials with the same volume, the one with the larger mass will have the higher density. Density, mass and volume are linked by the equation:

$$\text{density} = \frac{\text{mass}}{\text{volume}} \qquad \rho = \frac{m}{V}$$

ρ = density in kilograms per metre cubed, kg/m^3.

m = mass in kilograms, kg

V = volume in metres cubed, m^3

You need to be able to recall and apply this equation.

Density is often given in g/cm^3
1 g/cm^3 = 1,000 kg/m^3

1. Explain the difference in density between the same mass of a substance when it is a solid, a liquid and a gas. Use the particle model. [6]

2. A solid has a mass of 36 g and a volume of 180 cm^3. Calculate the density of the solid. [2]

1. *The density of the same mass of substance will be highest for the solid, a little lower for the liquid and much less for the gas.[1] For the same mass of a substance, there are the same number of particles[1] and the higher the volume, the lower the density.[1] The particles in a solid are packed closely together so have the highest density.[1] When the same mass is in liquid form, there is a little more space between the particles.[1] However, when the substance is in gas form, the particles are spread much further apart so takes up more volume.[1]*

2. *ρ = m/V*
= 36 / 180[1] = 0.2 g/cm^3.[1]

Determining the densities of liquids, regular solids and irregular solids

This practical activity helps you develop your ability to measure length, mass and volume accurately, and then determine densities.

Density, mass and volume

To determine density, the mass and volume of a subject needs to be measured.

Use **density** = $\dfrac{\text{mass}}{\text{volume}}$

Mass is measured using an electronic balance. The mass of a solid can be measured directly by placing it onto a balance. A liquid can be put into a container with a known mass.

The mass of the liquid = (mass of liquid and container) − (mass of empty container)

Volume is measured in different ways depending on both the state and shape of the substance.

Liquids

In a classroom laboratory, the most accurate container for measuring the volume of a liquid is a **measuring cylinder**. For accuracy, choose one with the smallest graduations that will contain the volume of liquid.

Example: For an approximate volume of 4 cm³, use a 5 cm³ cylinder with graduations of 0.1 cm³ instead of a 10 cm³ cylinder with graduations of 0.2 cm³.

Liquid volume is also measured in ml. (1 cm³ = 1 ml)

Regular shaped solids

The volume of a **regular shape** can be found by measuring the quantities that are needed to calculate the volume from the mathematical formula for that shape. This usually includes **length**.

Example:
Volume of a cuboid = length × height × width

Measure the three lengths using a **ruler** or **Vernier callipers**.

Irregular shaped solids

The volume of an irregular shape can be found by displacement. The solid is carefully lowered into a liquid, usually water. The solid will displace the same volume of the liquid and this volume can then be measured. The solid must be fully submerged to obtain its volume.

1. Using a eureka can

The can is filled up to the spout with water and the solid lowered in. The displaced water is collected in a measuring cylinder. The displaced water has the same volume as the irregular solid. Because a liquid takes the shape of a container, the volume can be easily measured.

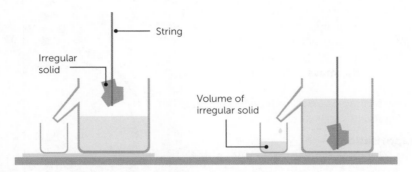

2. Direct measurement in a measuring cylinder

Record the volume of the water in the measuring cylinder. Carefully lower the solid into the water and record the new volume. volume of object = (volume with object) − (volume without object)

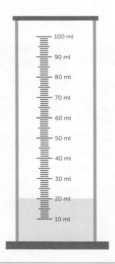

What potential sources of error are there in the two methods described above? [3]

Not lowering in the object carefully so that water splashes out of the measuring cylinder.[1]
Not reading the scale on the measuring cylinder at eye level.[1]
Level of water in the eureka can is below the level of the spout when the object is lowered into it.[1]

CHANGES OF STATE

Changes of state are **physical changes**. This means they can be reversed. The substance can go back to having the original properties by doing the opposite process e.g. heating instead of cooling. This is not the case for a chemical change.

Describing changes of state

Energy is transferred in to, or out of, the substance's thermal energy store by heating or cooling. When a substance changes from one state to another, the arrangement of, movement of, and distance between the particles changes. The **mass** of the substance is **conserved** when it changes state as the number of particles does not change. Even if the particles escape, (e.g. as a gas) the particles still exist.

Boiling and **vaporisation** are the same process as they both describe the change of state from a liquid to a gas. Evaporation is also the process of a liquid changing into a gas, but it is not the same as vaporisation. It takes place at temperatures below the boiling point of the liquid. In evaporation, the particles with the most energy escape from the surface of the liquid to form a vapour. A substance's vapour is a gas that has a temperature below the boiling point of the substance's liquid.

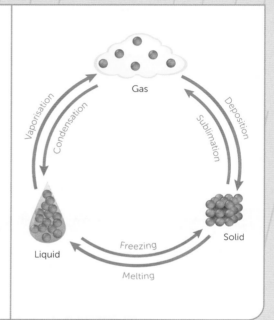

Gas

Vaporisation

Condensation

Deposition

Sublimation

Liquid

Freezing

Melting

Solid

1. Explain what happens to the particles when a substance melts. [3]

2. Define sublimation. [1]

3. Droplets of water form on the inside of a window on a cold day. Explain why this happens. [3]

1. *Energy is transferred to the particles by heating[1] and the particles gain thermal energy[1] so can move away from their fixed positions and start to move around each other as a liquid[1].*

2. *Sublimation occurs when a substance changes state directly from a solid to a gas.[1]*

3. *Energy is transferred from the water particles in the air[1] to the colder glass[1] and water changes from the gas state to the liquid state[1].*

INTERNAL ENERGY

The energy stored in the particles of a system is called **internal energy**. This is the **total kinetic energy** and **potential energy** of all the particles (atoms and molecules) that make up the system.

Energy is stored in particles

The **particle model** shows us that particles move in the liquid and gas states. This means they have a kinetic energy store. Particles also have a potential energy store due to any bonds and intermolecular forces between them.

The energy stored within the system changes when the system is heated or cooled.
- **Heating** increases the internal energy of the particles that make up the system.
- **Cooling** decreases the internal energy of the particles that make up the system.

The effect of this energy change when a system is heated or cooled either changes the temperature of the system, or produces a change of state.

Remember: a system is an object or a group of objects.

1. Give **two** changes that can happen to a liquid when the internal energy is increased. [2]
2. Describe what happens to particles in a gas when the temperature rises. [2]
3. Which form of the same mass of a substance has the most internal energy? Tick **one** box. [1]

        ☐ Gas   ☐ Liquid   ☐ Solid

1. *The temperature of the liquid increases[1] the liquid can change into a gas[1].*
2. *The particles gain kinetic energy[1] and move faster[1].*
3. *Gas.[1]*

TEMPERATURE CHANGES IN A SYSTEM AND SPECIFIC HEAT CAPACITY

When the temperature of a system changes, the change in temperature depends on the mass of the substance heated, the type of material and the energy change in the system. Each substance has an individual specific heat capacity.

Thermal energy

The change in thermal energy stored or released, as the temperature of a system varies, can be calculated using the equation:

change in thermal energy = mass × specific heat capacity × temperature change

$$\Delta E = m \, c \, \Delta\theta$$

ΔE = change in thermal energy in joules, J

m = mass in kilograms, kg

c = specific heat capacity in joules per kilogram per degree Celsius, J/kg °C

$\Delta\theta$ = temperature change in degrees Celsius, °C

> You need to be able to select this equation from the equation sheet and apply it.

★ Know the difference between:

- Specific heat capacity — the amount of energy required to raise the temperature of one kilogram of a substance by one degree Celsius.
- Specific latent heat — the amount of energy required to change the state of one kilogram of a substance at a constant temperature.

A brick has a mass of 0.345 kg. The specific heat capacity of brick is 840 J/kg °C.

(a) Calculate the increase in temperature when 6320 J is transferred to the brick. [2]

(b) How much energy is transferred from the brick as it cools back to the original temperature? [1]

(a) ΔE = $m \, c \, \Delta\theta$
6320 = 0.345 × 840 × $\Delta\theta$
$\Delta\theta$ = 6320 / (0.345 × 840)[1]
= 21.8 °C.[1]

(b) 6320 J is transferred from the brick to the surroundings.[1]

CHANGES OF STATE AND SPECIFIC LATENT HEAT

Energy is needed for a substance to change state from a solid to a liquid, and from a liquid to a gas. This is called the **latent heat**. The same amount of energy is released when the reverse change of state happens as the substance cools.

Changes of state

During a change of state, the energy transferred to the substance changes the energy stored in the substance (the internal energy). The temperature of the substance stays constant and only starts to change when the change of state is complete.

- For a solid changing to a liquid this temperature will be the **melting point** of the substance.
- For a liquid changing to a gas, this temperature will be the **boiling point** of the substance.

Specific latent heat

The specific latent heat of a substance is the amount of energy required to change the state of one kilogram of the substance with **no change** in temperature.

The change in thermal energy, stored or released as a substance changes state, can be calculated using the equation:

$$\text{energy for a change of state} = \text{mass} \times \text{specific latent heat}$$
$$E = m \times L$$

E = energy in joules, J

m = mass in kilograms, kg

L = specific latent heat in joules per kilogram, J/kg

> You need to be able to select this equation from the equation sheet and apply it.

Specific latent heat of fusion	Specific latent heat of vaporisation
for the change of state from solid to liquid	for the change of state from liquid to gas

1. Define:
   (a) Specific latent heat of fusion. [1]
   (b) Specific latent heat of vaporisation. [1]

2. The specific latent heat of vaporisation of nitrogen is 199 200 J/kg.
   Calculate the amount of energy needed to change 0.35 kg of liquid nitrogen into a gas. [2]

   1. *(a) Specific latent heat of fusion is the energy required to change 1 kg of a solid to a liquid[1]*

      (b) Specific latent heat of vaporisation is the energy required to change 1 kg of a liquid to a gas.[1]

   2. *E = m × L*
      = 0.35 × 199 200[1] = 69 720 J[1]

STATE CHANGES IN HEATING AND COOLING CURVES

Heating curve

This **heating curve** shows what happens to the temperature of a substance over time when a solid is heated, changes state to a liquid, and then to a gas. In between the changes of state, the substance is heating up in whatever state it is in.

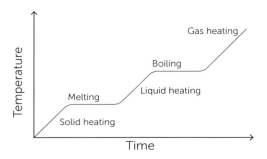

The melting point and boiling point can be found by reading the temperature from the graph when the line is horizontal.

Cooling curve

A **cooling curve** is a graph showing what happens to the temperature of a substance over time when a substance is cooled. The changes of state will be gas to a liquid, and then liquid to a solid.

Remember: During a change of state, the temperature stays constant, so the line is horizontal.

The temperature at which a substance changes from a gas to a liquid is usually still called the boiling point rather than the condensing point. The temperature at which a substance changes from a liquid to a solid is the same temperature as the melting point, but this is sometimes called the freezing point too.

What is represented by the letters A, B, C, D, E, T_1 and T_2 shown on this cooling curve?　　　　[4]

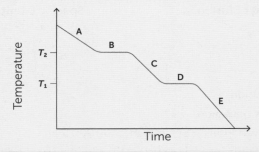

A: A cooling gas; C: A cooling liquid; E: A cooling solid.[1]

B: Changing state from a gas to a liquid (condensing).[1]

D: Changing state from a liquid to a solid (freezing).[1]

T_1 is the melting point and T_2 is the boiling point.[1]

PARTICLE MOTION IN GASES

The particles of a gas move about randomly and continuously, being free to fill up the space of a container. This means they collide with each other and the walls of the container they are in. The collisions exert a force on the container.

Temperature and volume

The **average kinetic energy** of gas particles increases as the temperature of a gas increases.

This can be shown in a frictionless piston at different temperatures. The volume can change because the piston is frictionless. As the gas particles inside the piston are heated, the average kinetic energy increases; the particles move faster, and the piston expands. The opposite happens when the gas particles are cooled.

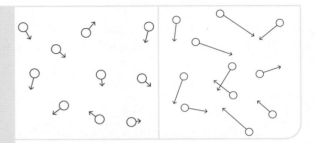

Gas particles are usually molecules. The movement of the gas particles is random.

Temperature and pressure

However, when a gas is kept at a **constant volume**, increasing the temperature of the gas still increases the movement of the particles, but they cannot spread out. This means the pressure exerted by the gas increases.

The pressure increases when temperature increases because:

- the particles are moving faster so the force of each collision increases
- there are more particle collisions each second.

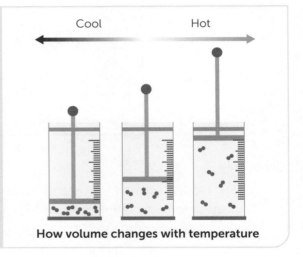

How volume changes with temperature

---

1. What happens to the pressure of a gas in a closed container when the temperature is decreased? Tick **one** box. [1]

   ☐ Decreases   ☐ Stays the same   ☐ Increases

2. Explain how the motion of the particles in a fixed volume of gas is related to its temperature and pressure. [3]

   1. Decreases.[1]

   2. The higher the temperature of the particles, the faster they move[1] and the more frequently they collide with each other and the container walls, increasing the force on the container walls[1]. This means that the pressure of the gas increases as the temperature increases.[1]

EXAMINATION PRACTICE

01   The density of a material is related to its mass and volume.

   01.1   Convert 736.4 g to kg. [1]

   01.2   763.4 g of a liquid has a volume of 50.0 cm³. Calculate the density of the liquid in g/cm³. [2]

   01.3   Calculate the volume of a cube with a side length of 0.02m. [2]

   01.4   A sphere has a density of 5400 kg/m³. Calculate the mass of the sphere in g. [4]

02   This question is about the particle model.

   02.1   Use the particle model to describe what happens to the particles in a substance when it freezes. [3]

   02.2   Define internal energy. [1]

03   This question is about changes of state.

   03.1   Define specific latent heat of fusion. [1]

   03.2   The specific latent heat of fusion for water is 334 000 J/kg.
          Calculate the amount of energy released when 0.250 kg of water freezes. [2]

   03.3   Sketch a cooling curve to show a liquid freezing. [2]

Higher only

04   A 0.5 kg block of solid gold has a temperature of 1000 °C.

   Calculate the amount of energy needed to change this block into liquid gold at 1064 °C. [5]

   The melting point of gold is 1064 °C.

   The specific heat capacity of gold is 130 J/kg °C. The specific latent heat of fusion for gold is 63 000 J/kg.

05   Complete the sentence. Tick **one** box. [1]

   When the temperature of a gas, held at constant volume increases, the pressure exerted by the gas...

   ☐ Increases.

   ☐ Stays the same.

   ☐ Decreases.

THE STRUCTURE OF AN ATOM

The basic structure of an atom is a **positively charged nucleus** surrounded by **negatively charged electrons**.

Atomic particles

Most of the mass of an atom is concentrated in the **nucleus** which contains its **protons** and **neutrons**.

- Protons are positively charged and have the same mass as a neutron.
- Neutrons have no charge so are neutral.

Electrons are about 1/2000 of the mass of a proton or a neutron and are tiny in comparison. Electrons orbit the nucleus in **energy levels**. Each energy level is at a fixed distance from the nucleus. Electrons can move from one energy level to another when electromagnetic radiation is absorbed or emitted.

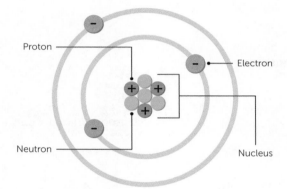

Atom structure

The radius of an atom is about 1×10^{-10} metres. Most of an atom is empty space between the outer electron energy level and the centre of the nucleus.

---

1. How many times bigger is the radius of an atom than the radius of the nucleus?
   Tick **one** box. [1]
   Approximately:
   ☐ 100   ☐ 1000   ☐ 10 000   ☐ more than 10 000

2. Describe what can happen in an atom when electromagnetic radiation is:
   (a) Absorbed by an atom. [1]
   (b) Emitted by an atom. [1]

   1. More than 10 000.[1]

   2 (a) An electron moves into a higher energy level / further away from the nucleus.[1]

   (b) An electron moves into a lower energy level / closer to the nucleus.[1]

MASS NUMBER, ATOMIC NUMBER AND ISOTOPES

The **atomic number** is the **number of protons** in the nucleus of an atom of an element.

Atoms have zero net electrical charge. The number of positive protons in the nucleus equals the number of negative electrons in the energy levels of an atom.

All atoms of a particular element have the same atomic number. This means they all have the same number of protons and electrons.

Each element has a symbol. An atom is represented like this:

The **mass number** is the total number of **protons + neutrons** in an atom.

Isotopes are atoms of the same element that have **different numbers of neutrons**.

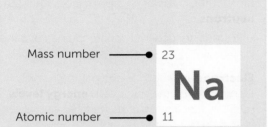

Mass number ⟶ 23

Na

Atomic number ⟶ 11

Ions

An atom forms a positive **ion** when it loses one or more electrons from the outer energy level.

Isotopes of hydrogen

Hydrogen-1	Hydrogen-2	Hydrogen-1
⊖ 1 Electron	⊖ 1 Electron	⊖ 1 Electron
⊕ 1 Proton	⊕ 1 Proton	⊕ 1 Proton
◯ 0 Neutrons	◯ 1 Neutron	◯ 2 Neutrons
mass number = 1 + 0 = 1	mass number = 1 + 1 = 2	mass number = 1 + 2 = 3

---

1. Describe what an atom of lithium $^{7}_{3}$Li is made up of. [3]

2. Complete the sentence. [2]

   Isotopes are atoms of the same element which have the same _____ and a different _____.

   1. *3 protons and 4 neutrons[1] in the nucleus[1] with 3 electrons in energy levels around the nucleus[1].*

   2. *Isotopes are atoms of the same element which have the same atomic number / number of protons[1] and a different mass number / number of neutrons[1].*

THE DEVELOPMENT OF THE MODEL OF THE ATOM

The **atomic model** has changed over time because of new experimental evidence.

Atomic theories

In the early 19th century, atoms were imagined as tiny, solid spheres. The discovery of the **electron** by J.J. Thomson in 1897 led to his **plum pudding model**. The plum pudding model suggested that an atom was a ball of positive charge with negative electrons embedded in it. This model was disproved by a series of results from Rutherford's **alpha particle scattering experiment**.

In this experiment, beams of alpha particles (tiny positively charged particles) were aimed at thin gold foil. The results led to the **nuclear model**. Shortly afterwards, Niels Bohr carried out theoretical calculations showing that electrons orbit the nucleus at set distances. Observations from experiments supported his **electron shell model** and also showed the existence of positively charged **protons**. About 20 years later, James Chadwick demonstrated the existence of **neutrons**.

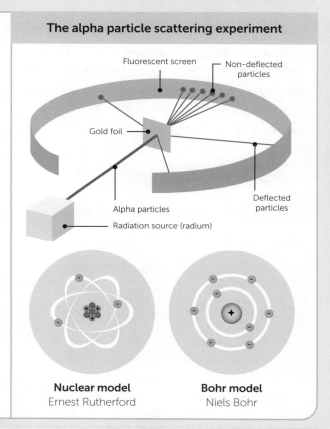

The alpha particle scattering experiment

Fluorescent screen
Non-deflected particles
Gold foil
Deflected particles
Alpha particles
Radiation source (radium)

Nuclear model
Ernest Rutherford

Bohr model
Niels Bohr

---

1. Compare the plum pudding and nuclear models of the atom. [3]
2. Explain why some scientific models have changed over time. [3]
3. Describe how the alpha particle scattering experiment led to the replacement of the plum pudding model by the nuclear model. [4]

1. *Both models have negatively charged electrons.[1] These are embedded in a sphere of positive charge in the plum pudding model[1] but surround a positively charged nucleus in the nuclear model[1].*

2. *Scientists obtain new experimental evidence[1] that does not support the current model[1]. If other respected scientists agree with the new findings the model will be changed.[1]*

3. *All of the fast moving, positively charged alpha particles would have passed straight through the gold foil if the plum pudding model was correct.[1] However, some alpha particles were deflected[1] and a few were repelled straight back[1]. This led to the nuclear model idea that the atomic mass is concentrated in a positively charged centre.[1]*

RADIOACTIVE DECAY AND NUCLEAR RADIATION

Nuclear radiation comes from unstable atomic nuclei.

Radioactive decay

Some atomic nuclei are unstable because they have too many or too few neutrons.

Radioactive decay is a **random** process that occurs when an unstable atomic nucleus (**radioactive isotope**) gives out **nuclear radiation** to become more stable.

Activity is the rate at which a source of unstable nuclei decays. It is measured in **becquerel** (**Bq**). 1 Bq equals one nuclear decay per second.

Count-rate is the number of decays recorded each second by a detector.

Types of nuclear radiation

Name	Symbol	Composition	Range in air	Ionising power
Alpha particle	α	A helium nucleus – two neutrons and two protons	< 5 cm	High
Beta particle	β	A high speed electron, ejected from the nucleus as a **neutron** turns into a proton	~ 1 m	Low
Gamma ray	γ	Electromagnetic radiation which comes from the nucleus	Large distances	Very low

A neutron (**n**) can also be emitted as a form of nuclear radiation.

The **ionising power** is the ability of the radiation to change an uncharged atom into a charged ion.

These different properties determine the choice of the radiation source when using radiation.

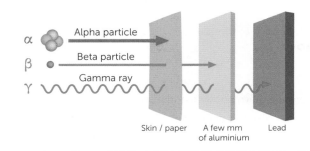

1. Describe how nuclear radiation can be detected and measured. [2]

2. The thickness of paper manufactured in a factory is monitored using a source of beta particles. Explain why sources of alpha particles and gamma rays are not used. [3]

1. *A Geiger-Muller (GM) tube[1] measures the count-rate when connected to a ratemeter.[1]*
   Or: Photographic film[1] turns foggy in the presence of radiation[1].

2. *Alpha particles cannot pass through the paper.[1] Gamma rays pass straight through so would not detect small changes in thickness.[1] Increasing the thickness of paper reduces the count rate of beta radiation so it is the most suitable source.[1]*

NUCLEAR EQUATIONS

Nuclear equations are used to represent radioactive decay.

Alpha particles and beta particles are emitted when unstable nuclei decay. This can cause a change in the charge or the mass of the nucleus which emits the particle.

Balanced **nuclear equations** can be written to show what happens when alpha or beta decay occurs.

An **alpha particle** is a helium nucleus so it is written as: $^{4}_{2}He$.

It has a mass number of 4 and an atomic number of 2. As there are no electrons to balance the protons, the alpha particle has a charge of +2.

Writing nuclear equations

$$^{235}_{92}U \longrightarrow ^{231}_{90}Th + ^{4}_{2}He$$

$231 + 4 = 235$
$90 + 2 = 92$

A **beta particle** is a fast moving electron written as: $^{0}_{-1}e$
It has a mass number of 0 and a charge of −1.

$$^{137}_{55}Cs \longrightarrow ^{137}_{56}Ba + ^{0}_{-1}e$$

$137 + 0 = 137$
$56 + -1 = 55$

1. What happens to the mass number and atomic number of an unstable nucleus when it undergoes alpha decay? [2]

2. Explain why the emission of a gamma ray does not cause the mass or the charge of the nucleus to change. [2]

3. Balance this nuclear equation. [2]

$$^{222}_{86}Rn \longrightarrow \_\_\_\_ Po + ^{4}_{2}He$$

The atomic numbers and mass numbers must be balanced on both sides.

1. *The mass number decreases by 4[1] and the atomic number decreases by 2[1].*

2. *Gamma rays are electromagnetic radiation so have no mass[1] and no charge[1].*

3. *$^{218}_{84}Po$, 218[1] 84[1]*

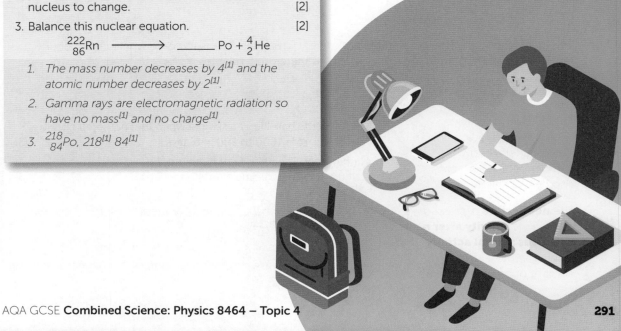

HALF-LIVES AND THE RANDOM NATURE OF RADIOACTIVE DECAY

Radioactive decay is a random process.

Half-life

Radioactive decay is a random process because it is impossible to tell when a particular nucleus in a sample of a radioactive isotope will decay. However, it is possible to predict how many of the nuclei will decay in a certain amount of time.

The **half-life** of a radioactive isotope is:

- the **time** it takes for half of the unstable nuclei to decay.
- the **time** it takes for the count rate, or activity, to halve.

The half-life is different for each **radioactive isotope**.

Calculating half-life

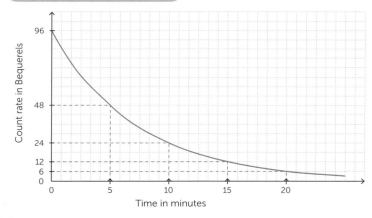

Draw a horizontal line from half of the initial activity level until it touches the graph line. The time given at the vertical line down will be the half-life. This should be the same whatever the starting point on the graph.

The graph shows that the count rate halves every 5 minutes. This is the case whatever the count rate is. The half-life of this radioactive isotope is therefore 5 minutes.

Higher Tier only

The amount, activity or count rate of a radioactive isotope after a known number of half-lives, can also be calculated.

The **net decline** expressed as a ratio is **activity after _n_ half-lives : initial activity** where _n_ represents a number of half-lives.

1. Calculate the half-life of a sample when the measured count rate drops from 1600 to 200 in 6 days. [3]

2. **Higher only:** The half-life of dubnium-270 is 15 hours. Calculate the net decline after 45 hours. [3]

   1. *Half of 1600 is 800, half of 800 is 400, half of 400 is 200.*[1] *So, it takes 3 half-lives to drop from 1600 to 800.*[1] *3 half-lives are 6 days, so the half-life is 6 / 3 = 2 days.*[1]

   2. *45 hours is 3 half-lives.*[1] *½ × ½ × ½ = 1/8.*[1] *Ratio = 1:8.*[1]

RADIOACTIVE CONTAMINATION

Exposure to radioactive materials can be harmful. Measures need to be taken to minimise the risk. There are two types of exposure to consider.

Contamination and irradiation

Radioactive contamination

Contamination involves unwanted direct contact with materials containing radioactive atoms. The exposed object also becomes radioactive. It is the decay of the radioactive atoms that causes the hazard. The level of danger is determined by the type of radiation emitted, its ionising power, and its ability to penetrate air and other materials.

Compare the hazards associated with radioactive contamination and irradiation. [6]

Contamination is the unwanted presence of radioactive particles on other objects after they and a radioactive material have come into direct contact with each other.[√] The object then becomes radioactive as well so will emit radiation[√] until all the particles have decayed[√]. The longer the half-life of the radioactive particles, the longer this takes.[√] Irradiation occurs when an object is exposed to a source.[√] The greater the distance from the source, the weaker the irradiation.[√] The shorter the time spent near the source, the lower the irradiation.[√] Removing the object from the source stops the irradiation.[√] The type of radiation affects the damage it can cause[√]. The more ionising the radiation, the more tissue damage.[√] Contamination with alpha particles inside a body is particularly dangerous[√] because alpha particles are highly ionising[√]. Being exposed to alpha particles outside the body is less harmful as they cannot penetrate skin.[√] Beta particles and gamma rays can penetrate skin.[√] Wearing protective clothing can reduce the risks from both contamination and irradiation.[√]

This question should be marked with reference to the levels of response guidance on page 374.

Irradiation

Irradiation is the process of exposing an object to nuclear radiation. The irradiated object does not become radioactive. For example, fresh fruit is irradiated to destroy bacteria on it and preserve the fruit for longer. The fruit does not become radioactive.

Humans are often irradiated deliberately for medical reasons with a controlled dose of radiation.

Suitable precautions must be taken to protect against any hazard. This includes wearing protective clothing, minimising unnecessary exposure and handling radioactive materials with tongs. Using tongs prevents contamination by stopping hands touching the radioactive material. They also reduce the rate of irradiation by increasing the distance of the hands from the radioactive source.

The effect of radiation on humans is studied globally. It is important to publish the findings so that other scientists can rigorously check them and confirm or challenge the evidence. This process is known as **peer review**.

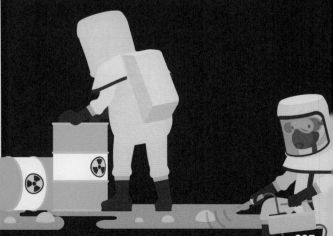

EXAMINATION PRACTICE

01 One measure of the atomic radius of hydrogen is 4.0×10^{-11} m.

01.1 What is the atomic radius of hydrogen in nm? Tick **one** box. [1]

☐ 0.004 nm ☐ 0.04 nm ☐ 4.0 nm ☐ 40 nm

01.2 Calculate the diameter of a hydrogen atom in metres. [1]

02 This represents an atom of fluorine. $^{19}_{9}\text{F}$

02.1 Describe this atom of fluorine in terms of subatomic particles. [3]

02.2 Give the atomic number and mass number of an atom that contains 13 protons and 14 neutrons. [2]

02.3 Describe the difference in atomic particles between isotopes of the same element. [1]

03 This question is about the development of the atomic model.

03.1 Describe the plum pudding model of the atom. [2]

03.2 In the alpha particle scattering experiment, some alpha particles were scattered, some alpha particles were reflected back and most alpha particles passed straight through.
Explain what conclusions can be drawn from each piece of evidence. [3]

04 Radioactive isotopes have different half-lives.

04.1 The count rate of a radioactive substance falls from 240 Bq to 60 Bq after 14 years.
Calculate the half-life of this substance. [2]

04.2 Another isotope has a half-life of 30 minutes.
Calculate the net decline as a ratio after one hour. [2]

04.3 An object is contaminated with radioactive atoms. Explain why the half-life of the radioactive atoms affects the level of hazard. [2]

05 This question is about nuclear radiation.

05.1 Smoke detectors use a source of nuclear radiation to ionise air in the smoke detector.
This allows a small current to flow between electrodes. When smoke gets into the gap, the current reduces, and an alarm sounds.

Radioactive isotope	Radiation emitted	Half-life
Americium-241	Alpha	432 years
Cobalt-60	Gamma	5.3 years
Thorium-228	Alpha	1.9 years
Nickel-63	Beta	100 years

The table shows four sources of nuclear radiation.

Evaluate the suitability of each radioactive isotope for use as a source of radiation in a smoke detector. Include descriptions of the properties in your evaluation. [6]

05.2 Complete the nuclear equation for the alpha decay of Americium-241 [3]

$$^{241}_{95}\text{Am} \longrightarrow \underline{\hspace{1cm}} \text{Np} + \underline{\hspace{1cm}}$$

05.3 The atomic number of nickel is 28.
Write a balanced nuclear equation for the beta decay of nickel-63 to copper. [3]

PHYSICS
TOPICS FOR PAPER 2

Information about Paper 2:

Trilogy 8464:

Written exam: 1 hour 15 minutes
Foundation and Higher Tier
70 marks
16.7% of the qualification grade
All questions are mandatory

Specification coverage

The content for this assessment will be drawn from topics on: Forces; Waves; and Magnetism and electromagnetism.

Questions

A mix of calculations, multiple-choice, closed short answer and open response questions assessing knowledge, understanding and skills.

Questions assess skills, knowledge and understanding of Physics.

SCALAR AND VECTOR QUANTITIES

A **quantity** is an amount or a measure of something. In Physics, these are things that can be measured.

Quantities

There are two types of quantities:

- **Scalar** quantities – which only have a value for size, called a **magnitude**, and
- **Vector** quantities – which have a **magnitude** and also a **direction**.

Vector quantities can be represented by arrows.

- The **length** of the arrow represents the **magnitude** of the vector quantity.
- The **direction** of the arrow represents the **direction** of the vector quantity.

Examples of scalar and vector quantities

Scalar qualities	Vector qualities
Mass	Force
Speed	Velocity
Distance	Displacement
Energy	Acceleration
Temperature	Momentum

Two runners are running at the same speed, but in different directions. Their speeds are the same but their velocities are different. One runner has a velocity of 5 m/s to the left and the other runner has a velocity of 5 m/s to the right.

If we define moving to the right as being in the positive direction, the runner on the left has a velocity of −5 m/s.

1. Explain, giving examples, the difference between speed and velocity. [2]

2. Explain why weight is a vector quantity. [2]

3. An object has a velocity of 4 m/s. What is the velocity of an object moving at the same speed in the opposite direction? [1]

   1. *Speed only has a magnitude, for example 10 m/s[1], velocity also has a direction. for example 10 m/s south[1].*

   2. *Weight is a force[1] which has both magnitude and direction.[1]*

   3. *−4 m/s[1]*

5 m/s

5 m/s

CONTACT AND NON-CONTACT FORCES

Forces are pushes and pulls that act on an object due to an interaction with another object.

Types of force

Force is a vector quantity, so it has size and direction. Forces are represented by arrows, often drawn to scale. The longer the arrow, the larger the force.

The size and direction of a force is determined by the force that is acting and how it is acting.

There are two types of forces:

- **Contact forces** – when objects are touching.
- **Non-contact forces** – when objects are not touching or are separated.

Contact forces

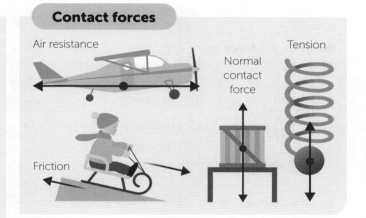

Non-contact forces

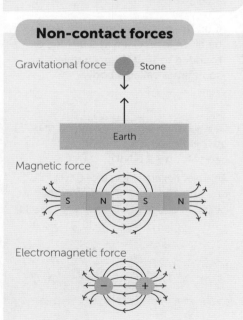

3. Explain the difference between a gravitational force and a magnetic force. [2]

3. *A gravitational force always attracts[1] but a magnetic force can attract or repel.[1]*

Interaction pairs

A book is sitting on a table. The weight of the book is balanced by the reaction force from the table. This force is at right angles to the surface and is called the **normal contact force** from the table. There is an interaction between the two objects which produces a force on each object. 'Normal' means at 90°.

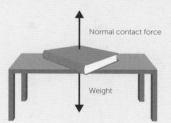

1. Explain why an electrostatic force is a non-contact force. [2]
2. Identify the other force in each of the interaction pairs shown in the contact force diagrams. [4]

1. *There will be a force between two charged objects when they are brought close together but not touching.[1] The objects do not have to be touching for the force to be exerted.[1]*

2. *Tension: weight[1]; Air resistance: pushing force or thrust[1]; Friction: pushing force[1]; Normal contact force: weight[1].*

GRAVITY

Mass and weight

Mass is the amount of matter that an object contains. Mass is constant regardless of the strength of the gravitational field it is in. It is a scalar quantity.

Weight is the **force** acting on a mass due to gravity. Weight changes and depends on the strength of the gravitational field the mass is in.

The weight and mass of an object are directly proportional to each other. The weight of an object can be calculated using the equation:

weight = mass × gravitational field strength

$$W = m\,g$$

W = weight in newtons, N

m = mass in kilograms, kg

g = gravitational field strength in newtons per kilogram, N/kg

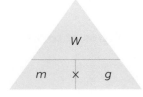

You need to be able to recall and apply this equation.

Force of gravity

The force of gravity near or on the Earth is due to the gravitational field around the Earth.

Forces, including weight, can be measured using a **newtonmeter**, which is a calibrated spring-balance. The spring stretches a fixed distance for each newton.

The weight of an object is considered to act at a single point known as the **centre of mass** of the object. This is useful when drawing diagrams representing forces.

 g on the surface of the Earth will be given in an exam as 9.8 N/kg, or rounded to 10.0 N/kg. An object is sometimes called a body.

1. Calculate the weight of a 0.5 kg mass on the moon.
   Use gMoon = 1.6 N/kg. [2]
2. An object exerts a force of 6.9 N on Earth. Calculate the mass of the object in g. Use *g* = 9.8 N/kg. [4]

   1. W = m × g = 0.5 × 1.6[1] = 0.8 N[1]

   2. W = m × g
   6.9 = m × 9.8[1]
   m = 6.9 / 9.8[1] = 0.7 kg[1] = 700 g[1]

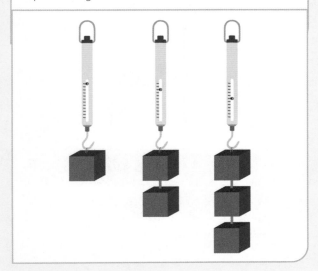

RESULTANT FORCES

There is usually more than one force acting on an object. The different forces can be replaced by a single force that has the same effect as all of the other forces acting together. This single force is called the **resultant force**.

Predicting behaviour

Finding the resultant force simplifies diagrams and calculations. This helps predictions to be made about how an object might behave.

- When a **pair of different sized forces** are in opposite directions, an **unbalanced force** causes a **resultant force**.
- When a **pair of equal sized forces** are in **opposite directions**, this is a **balanced force** and there is no **resultant force**.

Calculating the resultant of parallel forces acting in a straight line

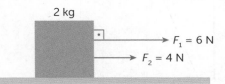

As the forces are in the same direction they can be added together.

$F_1 + F_2 = 6 + 4 = 10$ N

$F_{right} = 10$ N

Remember to give the direction of the force. Forces in opposite directions have opposite values.

F_3 is acting to the left and is in the opposite direction to F_1.

Assume that the positive direction is to the right. So, $F_1 = -3$ N

Add the forces together to find the resultant force.
$F_{resultant} = F_1 + F_3 = 6$ N $+ -3$ N $= 3$ N (to the right).

The diagram shows three blocks in water.

(a) Calculate the resultant force on each block. [3]

(b) Describe the movement of the blocks. [1]

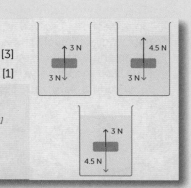

    (a) Block A: resultant force = 3 + −3 = 0 N[1]

        Block B: resultant force = 4.5 + −3 = 1.5 N upwards.[1]

        Block C = resultant force = 3 + −4.5 = −1.5 N, so 1.5 N downwards.[1]

    (b) Block A: stationary, Block B: moves upwards,

        Block C: moves downwards.[1]

RESOLVING FORCES

Vector diagrams and **scale drawings** can be used to solve problems when forces act at angles to each other.

Resolving forces

The arrows are best drawn on graph paper using a ruler, protractor and sharp pencil to show both the magnitude and the direction of the forces.

Choose a suitable scale for the magnitude of each force and add an arrow to show the direction it is in. Any angle is given as a value 'to the horizontal' or 'to the vertical.'

> Remember, force is a vector quantity so has size and direction. Forces are represented by arrows.

Resolving a single resultant force

A single force acting at an angle can be resolved into two components acting at right angles to each other: a **horizontal force** and a **vertical force**. The two forces have the same effect as the single force.

To resolve a force into horizontal and vertical components draw a scale diagram on squared paper. Use a protractor to draw the angle and measure its length using a ruler. Then measure the lengths of the components. The diagram shows a force of 13 N at an angle of 67° to the horizontal. The force has been drawn on squared paper with one large square representing 5 N. Count the number of squares to find the horizontal and vertical components: F_h = 5 N and F_v = 12 N.

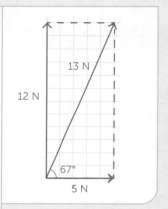

This vector diagram shows how the force in red can be resolved into two forces at right angles to each other. Force F is acting at an angle of $\theta°$ to the horizontal. It can be resolved into a vertical component, F_y and a horizontal component F_x.

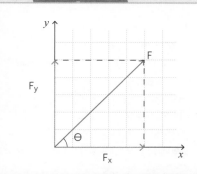

Resolving two perpendicular forces into single resultant force

You can resolve a horizontal and vertical force into a single resultant force by drawing a scale diagram. Draw the horizontal force first. Then draw the vertical force, starting it at the end of the horizontal force. Measure the length of this arrow using a ruler. Measure the size of the angle using a protractor. The scale diagram shows a horizontal force of 6 N and a vertical force of 4 N. The resultant force has a magnitude of 7.2 N acting at 34° to the horizontal.

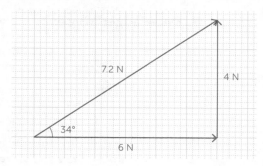

1. Draw a vector diagram to resolve a force of 58 N at 31° to the horizontal into a horizontal force and a vertical force. [2]

2. Use a scale drawing to find the resultant of a force of 20 N acting to the right and a force of 40 N acting upwards. [2]

1. Scale drawing shows:
   horizontal component: 50 N to the right[1];
   vertical component: 30 N upwards[1].

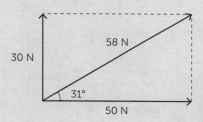

2. Resultant: 45 N[1] at 63° to the horizontal[1].

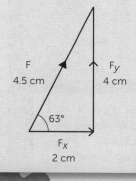

FREE BODY DIAGRAMS

A **free body force diagram** is a simplified diagram showing the forces acting on an object shown as a simple box or a dot. The force arrows act away from the centre of the box or dot.

Balanced forces

The air resistance is equal and opposite to the weight of the parachutist so there is no net force on the parachutist and the forces are balanced.

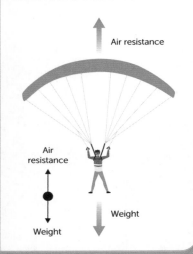

Air resistance

Air resistance

Weight

Weight

Unbalanced forces

The weight of the aeroplane is equal and opposite to the lift of the plane. So there is no net vertical force.

The thrust from the engines of the aeroplane is greater than the air resistance. So the horizontal forces are unbalanced and there is a resultant force to the left.

The same force can sometimes be described using different words. For example, **drag** and **air resistance**; **forward force** and **thrust**. The situation of the object, and the direction of the force arrow, will make what the force represents clearer.

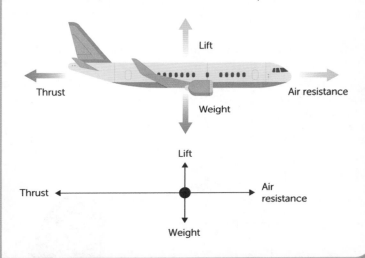

Lift

Thrust

Air resistance

Weight

Lift

Thrust

Air resistance

Weight

Figure X, below, shows a submarine moving forward at a constant depth. Draw a free body diagram of this submarine. [2]

[1] Horizontal
[1] Vertical

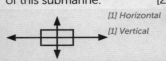

Buoyancy

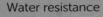
Water resistance

X

Engine force

Weight

WORK DONE AND ENERGY TRANSFER

Mechanical work is done on an object when a force moves the object through displacement. The amount of work done depends on the force used and the distance the object is moved.

Calculating work done

The work done by a force on an object can be calculated using the equation:

$$\text{work done} = \text{force} \times \text{distance}$$

$$W = Fs$$

W = work done in joules, J

F = force in newtons, N

s = distance moved in metres, m

This distance is the **distance moved along the line of action of the force**.

When a force of 1 N causes a displacement of 1 m, 1 J of work is done or 1 J of energy is transferred.

$$1 \text{ joule (J)} = 1 \text{ newton-metre (Nm)}$$

> You need to be able to recall and apply this equation.

Examples of energy transfers when work is done

Here are some energy transfers that happen when work is done:

- A person lifting a mass against gravity transfers energy from their own chemical store to the gravitational potential energy store of the mass.
- Braking in a car transfers energy from the kinetic store of the car to the thermal energy store of the brakes.

Work done against the frictional forces acting on an object causes a rise in the temperature of the object.

> Electrical work also transfers energy and is covered on **page 273**.

---

1. A person with a weight of 600 N goes up a vertical distance of 40 m on a hill. Calculate the work done by the person in kJ. [3]

2. Explain why brake pads get warm when used to stop a bicycle. [3]

1. $W = Fs$
   $= 600 \times 40^{[1]} = 24\,000 \text{ J}^{[1]} = 24 \text{ kJ}^{[1]}$

2. Work is done by the friction force between the brakes and the wheel rims or brake discs.[1] As the bike slows energy is transferred by friction from the kinetic energy store of the bike[1] to the thermal energy store of the brake pads (so their temperature increases)[1].

FORCES AND ELASTICITY

Objects can change shape when forces are applied.

Deformation

Forces can change the shape of a stationary object by **stretching**, **bending** or **compressing** the object. The change of shape is called **deformation**. There are two types of deformation.

- **Elastic** deformation is when the object goes back to its original shape when the force is removed.

- **Inelastic** deformation means that the material does not return to its original shape when the force is removed. There is a permanent change of shape.

To stretch, bend or compress an object, two forces need to be applied to the object. If just one force is applied, the object would just move and would not change shape.

Stretching needs two forces pulling away from each other. Compression needs two forces pushing towards each other. Bending needs one force acting clockwise and the other anticlockwise.

The **extension** of an elastic object (such as a spring) is **directly proportional** to the force applied, provided that the **limit of proportionality** is not exceeded. This means it has not been inelastically deformed. This equation relates the force to the resulting extension (or compression) of an elastic object:

$$\text{force} = \text{spring constant} \times \text{extension}$$

$$F = k\,e$$

F = force in newtons, N

k = spring constant in newtons per metre, N/m

e = extension in metres, m

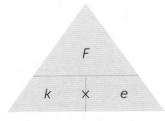

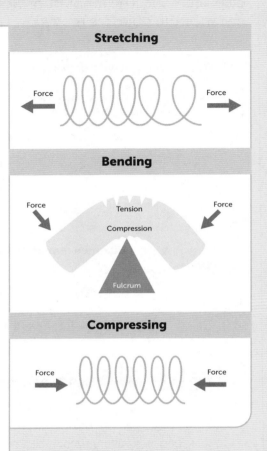

Stretching

Force ← / Force →

Bending

Force | Tension | Compression | Fulcrum | Force

Compressing

Force → / ← Force

You need to be able to recall and apply this equation.

Energy stored in a spring

When an elastic object is stretched or compressed by a force, work is done. This means that an energy transfer takes place and energy is transferred to the elastic potential energy store of the object. As long as the deformation is elastic, the work done on a spring is equal to the elastic potential energy stored.

The work done in stretching (or compressing) a spring (up to the limit of proportionality) can be calculated using the equation:

elastic potential energy = 0.5 × spring constant × (extension)²

$$E_e = \frac{1}{2}k\,e^2$$

> You need to be able to select this equation from the equation sheet and apply it.

1. Complete the diagram to show the force needed to stretch the rubber. [2]

2. A force of 1.8 N is applied to a spring with a spring constant of 12 N/m. The spring goes back to its original shape when the force is removed. Calculate the extension of the spring. [3]

3. The original length of a spring is 10 cm. A force of 20 N is applied to the spring which causes its length to increase to 15 cm. Calculate the spring constant of the spring. [3]

4. A spring is extended by 0.1 m. The spring has a spring constant of 30 N/m. Calculate the work done to stretch the spring. [2]

1. *1 m arrow in opposite directions[1], of the same size[1].*

2. *F = k e   1.8 = 12 × e[1]*
   e = 1.8 / 12[1] = 0.15 m[1]

3. *Extension = extended length – original length = 15 – 10 = 5 cm = 0.05 m[1]*
   F = k e   20 = k × 0.05[1]
   k = 20 ÷ 0.05 = 400 N/m[1]

4. *Work done = $E_e = \frac{1}{2}k\,e^2$*
   = 0.5 × 30 × 0.1²[1] = 0.15 J[1]

Linear and non-linear relationship between force and extension

Overstretched spring

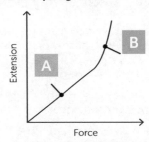

A	Elastic deformation: Straight line so it is a linear relationship.
	The line goes through the origin (0,0) so extension is directly proportional to force.

B	Inelastic deformation:
	Limit of proportionality is exceeded.

Elastic band

Curved line so a non-linear relationship.

Force and extension of a spring

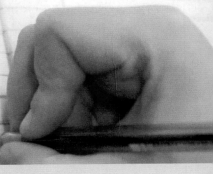

This practical activity helps you to calculate a spring constant and use a graph to find the weight of an unknown object.

> Remember to convert the extension to metres if you measured it in centimetres.

Experiment method

This apparatus is used to measure the extension of a spring as the force is increased by adding masses one at a time. Each force is calculated using $F = mg$.

Make sure the readings on the metre rule are taken horizontally and not at an angle.

Record the original length of the spring when $F = 0$ N. It does not matter if the end of the spring does not line up with zero on the ruler when $F = 0$ N. Take two readings, one at each end of the spring and subtract one from the other to find the original length.

The extension is the total increase in length from the original length of the spring each time the force is increased.

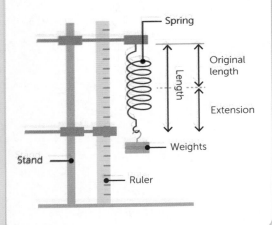

Analysis of results

Plot a line graph of **extension against force**. This means 'Extension of spring in m' is on the y-axis and 'Weight in N' is on the x-axis.

It is a straight-line graph so its equation will be in the form $y = mx + c$.

where m = gradient and c = y-intercept
$c = 0$ as the line goes through the origin.
$y = e$ and $x = F$ so $y = mx$ can be written as
$e = mF$
gradient, $m = e / F$
Using $F = ke$, gradient = e / F = 1 / k

So, the spring constant can be calculated by determining the gradient, and finding the reciprocal of the gradient..

spring constant = 1 / gradient

The graph shows that the extension of a spring is directly proportional to the force applied. This is only for a linear part of the graph before the limit of proportionality has been exceeded. The spring undergoes elastic deformation.

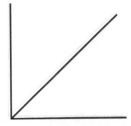

Calculating the spring constant from a graph

Draw a triangle on the line of best fit to determine the gradient.

If you have plotted the extension in centimetres, you will need to convert it to metres when finding the gradient.

Gradient = (0.06 − 0.02) / (4.8 − 1.6)
         = 0.04 / 3.2 = 0.0125

Spring constant   = 1 / gradient
                  = 1 / 0.0125
                  = 80 N/m

You can also use the graph to calculate the elastic potential energy stored in the spring. It is the area under the graph.

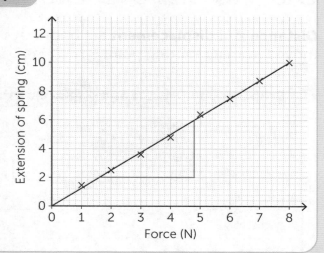

Using the graph to determine the weight of an unknown object

Hang the object on the spring and measure the extension. Use the graph to determine the force at that extension, this will be the weight of the object. It could be converted to mass by dividing it by g, the gravitational field strength.

1. Explain which spring, Spring A or Spring B, is the stiffest. Use the graph in the figure below to help.   [3]

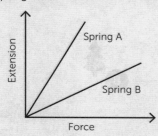

2. Determine the weight of an object which extends the spring in the diagram by 9 cm.
   Show your working on the graph at the top of the page.   [2]

   1. *Spring B is the stiffest spring.[1] A stiffer spring has a higher spring constant, k.[1] The spring constant is inversely proportional to the gradient[1] so the higher the spring constant, the lower the gradient[1].*

   2. *Horizontal line drawn from 9 cm extension.[1] Weight = 7.2 N.[1]*

DISTANCE AND DISPLACEMENT

Distance and **displacement** are both ways of describing how far an object has moved from its starting point.

Properties of distance and displacement

Distance	Displacement
The total length of the path travelled by a moving object from start to finish	The shortest distance between the start and finish points of a moving object
Any shape line	Straight line
Scalar quantity	Vector quantity
No direction	The direction of the straight line is described in terms of an angle or a point of the compass
Measured in metres, m	Measured in metres, m
Will always have a positive value	Values can be positive or negative depending on the direction moved
Will still be total distance moved when an object has moved from one point and returned to the same place	Will be zero when an object starts and finishes at the same place

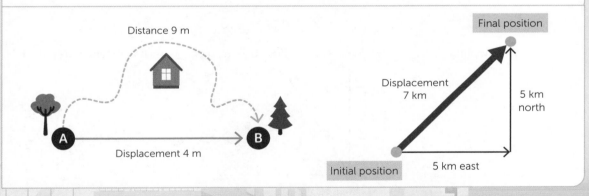

Distance 9 m

Displacement 4 m

Final position

Displacement 7 km

5 km north

Initial position

5 km east

1. Describe the difference between displacement and distance. [1]

2. Describe the displacement of the object above right. [1]

   1. *Distance only has magnitude, displacement has a magnitude and a direction.*[1]

   2. *The displacement is 7 km north east.*[1]

SPEED

Speed is a **scalar** quantity so has size but no direction. Moving objects usually travel with **non-uniform motion** so speed is usually given as a **mean** value over a certain distance or time.

Typical speeds

The speed a person walks, runs or cycles depends on many factors such as the distance travelled, the person's age and fitness, the terrain and the weather.

Vehicle speeds in the UK are usually given in miles per hour (mph). However, in physics, speed is measured in metres per second (m/s).

The table shows some typical values for speed. All can vary.

Example	Typical speed in m/s
Walking	1.5
Running	3
Cycling	6
Car	13–31
Train	50
Plane	250
Sound	330 m/s

Distance travelled, speed and time

For an object moving at constant speed, the distance travelled in a specific time can be calculated using the equation:

distance travelled = speed × time

$$s = v\,t$$

s = distance in metres, m

v = speed in metres per second, m/s

t = time in seconds, s

> You need to be able to recall and apply this equation.

Note that s represents distance here – not speed or seconds. v is also used for velocity.

1. A cyclist travelled at a constant speed of 5 m/s for 25 m. Calculate the time taken. [2]
2. A car journey of 42 km takes 50 minutes. Calculate the mean speed of the car. [4]

   1. $s = v \times t$ $25 = 5 \times t$[1] $t = 25/5$[1] = 5 s[1]

   2. 50 minutes = 50 × 60 seconds = 3000 s[1]
       $s = v \times t$ 42 000 = v × 3000[1]
       v = 42 000 / 3000[1] = 14 m/s[1]

VELOCITY

Velocity is a vector quantity representing speed in a given direction.

Example

Both cars are travelling at the same speed of 20 m/s.

The orange car is travelling at a velocity of 20 m/s to the right. The red car is travelling at a velocity of 20 m/s to the left.

We need to define one direction as positive – this is usually taken to be to the right or upwards.

So we can say that the orange car has a velocity of +20 m/s and the red car has a velocity of −20 m/s.

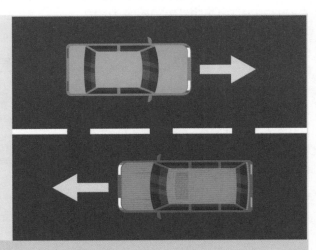

Direction can be described as: right or left, up or down, north, south, east or west

Motion in a circle   Higher Tier

This car is travelling at a constant speed in a circle. The speed is constant because it takes the same time to travel the same distance.

However, the direction of the car changes constantly as it moves in a circle. So the velocity is constantly changing because the direction of the car is constantly changing.

Motion in a circle involves constant speed but changing velocity.

Terminal velocity

An object falling through a fluid initially accelerates due to the force of gravity. As the velocity increases, the drag force increases. Eventually the resultant force will be zero and the object moves at a constant speed in one direction. This constant speed is its **terminal velocity** and is the maximum speed it can reach under those conditions.

1. Describe the difference between speed and velocity. [1]

2. A student takes 5 minutes to walk 510 m from west to east along a straight beach. Calculate the mean velocity of the student. [4]

3. Describe speed and acceleration at terminal velocity. [2]

   1. *Speed only has magnitude and is a scalar, velocity is speed in a given direction and is a vector.*[1]

   2. *t = 5 minutes = 5 × 60 = 300 s*[1]
      s = v × t   510 = v × 300[1]
      v = 510/300[1] *= 1.7 m/s eastwards.*[1]

   3. *Speed is constant*[1]*, acceleration is zero*[1]*.*

THE DISTANCE-TIME RELATIONSHIP

Distance-time graphs

A distance-time graph shows how the distance of an object moving in a straight line changes over time. The **gradient** of the graph equals the **speed** of the object.

   speed = gradient = change in distance ÷ change in time

Uniform motion (top): The car travels the same distance in each equal time interval so has a constant speed.
The gradient is 40 / 4 = 10 so speed = 10 m/s

Non uniform motion (bottom): The car travels a different distance in each equal time interval so the speed changes during the journey shown on this graph.
From A to B the gradient is 20/1 = 20 so the speed = 20 m/s
From B to C the gradient is 10/2 = 5 so the speed = 5 m/s
From C to D the gradient is 10/1 = 10 so the speed = 10 m/s

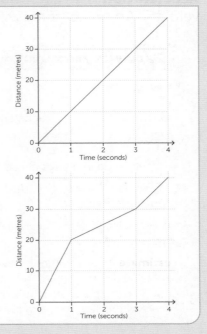

Note that the mean speed for the whole journey is 40 / 4 = 10 m/s as the car has travelled the same total distance in the same time as in the uniform motion graph.

Distance-time graph for an accelerating object   Higher Tier

When an object is **accelerating** its distance-time graph will be curved. The speed of the object at any particular time can be determined by drawing a **tangent** on the graph and finding its **gradient**.

To work out the speed at 15 seconds, draw a **tangent** to the graph.

Gradient = change in distance ÷ **change in time**
= (40 − 0) ÷ (23 − 8) = 40 ÷ 15 = 2.7 m/s

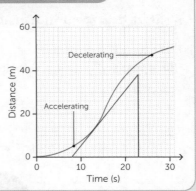

1. Describe what a horizontal line on a distance-time graph represents. [1]
2. A person walks 60 m in 30 s, stops for 5 s, then walks a further 60 m in 20 s. Draw a distance-time graph for this journey. [3]

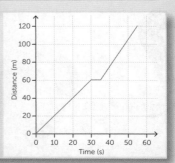

   1. *The object is stationary as it has travelled no distance during the time.*[1]
   2. *Suitable scale.*[1] *Correctly labelled axes.*[1] *Line drawn from (0,0) to (30,60) to (35,60) to (55,120).*[1]

ACCELERATION

Acceleration is a vector quantity representing the rate that an object changes velocity. An object that is slowing down is **decelerating**.

Calculating acceleration

The average acceleration of an object can be calculated using the equation:

$$\text{acceleration} = \frac{\text{change in velocity}}{\text{time taken}}$$

$a = \dfrac{\Delta v}{t}$

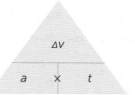

a = acceleration in metres per second squared, m/s²

Δv = change in velocity in metres per second, m/s

t = time in seconds, s

An object falling freely under gravity near the Earth's surface has an acceleration of ~9.8 m/s².

> You need to be able to recall and apply this equation.

To **estimate** the magnitude of **everyday accelerations**, make a sensible approximation. For example, you can estimate the acceleration of a typical car when it accelerates from 0 to 100 km/h (about 30 m/s). Make a sensible approximation of the time taken to accelerate, say 10 s. So an estimate of the car's acceleration is (30 − 0)/ 10 = 3 m/s²
So the acceleration of a typical car is ~3 m/s²

~ means approximately equal to

Equation for uniform acceleration

This equation links uniform acceleration, distance and velocity:

(final velocity)² − (initial velocity)² = 2 × acceleration × distance

$v^2 - u^2 = 2\,a\,s$

v = final velocity in metres per second, m/s

u = initial velocity in metres per second, m/s

a = acceleration in metres per second squared, m/s²

s = distance in metres, m

1. A cyclist increases velocity from 5.0 m/s to 11 m/s with an acceleration of 1.8 m/s². Calculate the time taken for this acceleration. [3]

2. A sprinter accelerates from stationary with an acceleration of 7.5 m/s² for 5.0 m. Calculate the velocity reached by the athlete. [3]

1. $a = \Delta v/t$
   $1.8 = (11 - 5)/ t$ [1]
   $t = 6/1.8$ [1] $= 3.3$ s [1]

2. initial velocity = 0 m/s
   $v^2 - u^2 = 2\,a\,s$
   $v^2 - 0 = 2 \times 7.5 \times 5$ [1]
   $v = \sqrt{75}$ [1] $= 8.7$ m/s. [1]

VELOCITY–TIME GRAPHS

You can use a velocity-time graph to calculate acceleration and distance travelled.

Calculating acceleration using velocity–time graphs

The acceleration of an object can be determined from a velocity–time graph.
- A velocity-time graph shows how the velocity of an object changes over time
- The **gradient** of the graph equals the **acceleration** of the object

Using the gradient
- Gradient = $\Delta v / \Delta t$
- The gradient equals acceleration because $a = \Delta v / t$
- The steeper the line, the greater the acceleration

Deceleration is negative acceleration.

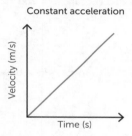
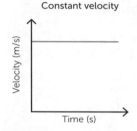
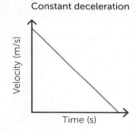

A straight line upwards represents **constant acceleration** as it has a positive gradient.
A horizontal line shows **constant velocity** and zero acceleration as the gradient is zero.
A straight line downwards represents **constant deceleration** as it has a negative gradient.

A train starts from stationary and accelerates at a constant rate for 40 s. It then travels at a constant velocity of 30 m/s for 80s. It slows down to a stop with a constant deceleration for 20 s.

Draw a velocity-time graph for this train journey. [3]

Line (0,0) to (40,30).[1] Line (40,30) to (120, 30).[1] Line (120,30) to (140, 0).[1]

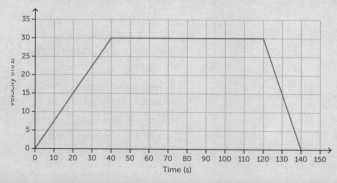

CALCULATING DISTANCE USING VELOCITY-TIME GRAPHS

Velocity-time graphs can also be used to determine the **distance travelled** by (or displacement of) an object on a journey.

Determining distance

The distance travelled over a period of time equals the **area under the velocity-time graph** for that period of time.

The area, and hence the distance travelled, can be calculated by using the values of time and velocity from the graph. Because the graph lines will be horizontal, or sloping up and down, the area under the graph will be a **triangle** or **rectangle** shape, or a combination.

Area of a rectangle = width × height

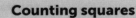

Area of a triangle = ½ × base × height

Area under a velocity-time graph

Calculate the distance travelled shown in the velocity-time graph by calculating the area under the graph.

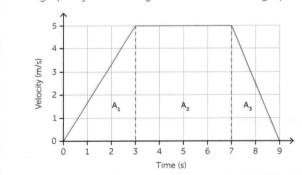

Counting squares

You can also find the distance, or the area under the graph by **counting squares** under the graph line. This is straightforward when they are all whole squares, and more likely to be an estimate when there are several part squares under the line.

This is a velocity-time graph for a train journey.

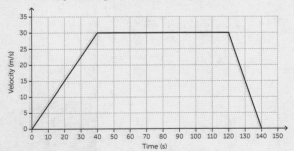

Calculate the total distance travelled by the train. [1]

$Distance = (\frac{1}{2} × 40 × 30) + (80 × 30) + (\frac{1}{2} × 20 × 30)$

$= 600 + 2400 + 300^{[1]} = 3300\ m^{[1]}$

Check that the units match for both axes.

In this case area = s × m/s so the distance travelled is in m.

You could also calculate the distance by finding the area of a trapezium.

NEWTON'S FIRST LAW

Sir Isaac Newton was a mathematician and scientist who studied many things, including forces and motion. He developed three laws of motion which are still used today to understand and describe forces and motion.

The first law of motion

Newton's first law states that:

- if there is **no resultant force** acting on a **stationary object**, then it will remain stationary.

- if there is **no resultant force** acting on a **moving object**, then the object continues to move at the same velocity (same speed, same direction).

A resultant force must act on an object for its velocity to change. When a resultant force acts on a stationary or moving object, the object will:

- stay at the same speed and change direction

- or change speed in the same direction

- or change both speed and direction.

This means when a vehicle travels at a constant speed, the driving force forward is balanced by the **resistive** forces of friction and air resistance acting against the vehicle. For the vehicle to speed up, slow down or change direction, the driving force, or the resistive forces, must change to produce a resultant force.

Higher Tier only

The tendency of an object to stay at rest or continue moving with uniform motion is called **inertia**.

(a) This car is moving at a steady speed. Explain what will happen to the motion of the car. [2]

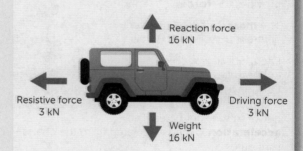

Reaction force 16 kN

Resistive force 3 kN

Driving force 3 kN

Weight 16 kN

(b) The driving force increases to 4 kN. Explain what happens to the car. [2]

(a) There is no resultant force on the car[1] so the car will continue moving at a steady speed[1].

(b) There is a resultant force on the car of 1 kN to the right[1] so the car accelerates to the right[1].

This practical activity helps you to measure the effect on acceleration when force or mass are changed.

Typical equipment

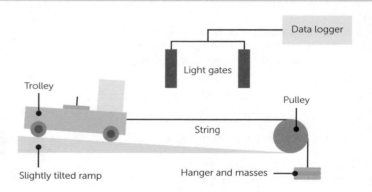

The weight of the hanging masses which is attached to the trolley by string going over the pulley provides the **force**.

The **mass** is changed by adding different masses to the trolley.

Each light gate records the speed of the trolley as the card passes through it. The two light gates record the time taken to travel between the light gates. The **acceleration** is then calculated from the change in speed and the time taken to travel between the light gates.

The string must be horizontal and in line with the trolley.

The hanging masses should just accelerate the car along the ramp – the number should be adjusted so that this happens.

Alternative methods

- Instead of light gates, use a stopwatch to record the time taken for the trolley to travel equally measured distances on the ramp. You can also use ticker-tape and a ticker timer.

- Video the moving trolley so that the film can be slowed down and the distances and times measured more reliably.

- Replace the ramp and trolley with a linear air track and glider. This reduces the opposing friction force on the accelerating object.

Varying the force

You can investigate the effect of varying the force on the acceleration of the trolley with a constant mass.

- The overall mass of the moving items (trolley, masses, string and hanger) must be kept constant.
- Masses are moved from the trolley and placed on the hanger.
- This increases the pulling force on the trolley but keeps the accelerating mass the same.

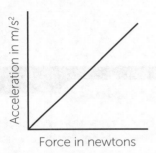

$a \propto F$ so a graph of acceleration against force should be a straight line through the origin.

Varying the mass of an object

You can investigate the effect of varying the mass of an object on the acceleration of the trolley with a constant force.

- The same mass is kept on the hanger on the end of the string.
- The mass and the hangar on the string must give enough force to move the trolley at its greatest mass.
- Add different masses to the trolley.

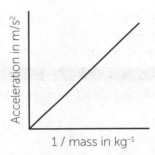

$a \propto 1/m$ so a graph of acceleration against 1/mass should be a straight line through the origin.

1. Explain why two light gates are used in the investigation. [2]
2. How do you calculate the force applied to the trolley? [1]
3. Explain why the ramp needs to be slightly tilted. [2]
4. Give **two** reasons why the length of the string used is important in this investigation. [2]
5. Describe the safety precautions you would need to take when carrying out this investigation. [2]

1. *The velocity of the trolley must be measured at two different points[1] so that acceleration can be calculated using a = Δv/t.[1]*
2. *Use F = mg[1]*
3. *The trolley will experience a friction force as it travels down the ramp[1] so tilting the ramp compensates for this slowing down of the trolley.[1]*
4. *The string needs to be long enough to allow the trolley to stay on the ramp as the hanging masses hit the floor[1] and short enough for the trolley to pass through both light gates before the masses hit the floor.[1]*
5. *Something soft, such as foam, is needed to stop the hanger and mass hitting the floor.[1] This is placed under the hanger.[1]*

NEWTON'S SECOND LAW

The second law of motion

Newton's second law states that the acceleration of an object is:
- proportional to (∝) the resultant force acting on the object
- inversely proportional to the mass of the object.

The equation that links acceleration, force and mass is:

$$\text{resultant force} = \text{mass} \times \text{acceleration}$$

$F = m \times a$

F = force in newtons, N

m = mass in kilograms, kg

a = acceleration in metres per second squared, m/s²

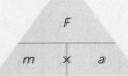

Estimating speed, acceleration and forces for road vehicles

Road vehicles undergo significant acceleration when overtaking or joining a motorway, for example. Typical masses are 30 000 kg for a lorry and 1500 kg for a car. If both vehicles accelerate from 15 m/s to 20 m/s then a greater force will be required to accelerate the lorry because it has a greater mass. Typical accelerations are ~0.4 m/s² for a lorry and ~2 m/s² for a car. The force required can then be estimated using $F = ma$.

So the typical forces are:

F_{lorry} = 30 000 × 0.4 = 12 000 N F_{car} = 1500 × 2 = 3000 N

1. A car with a mass of 1200 kg accelerates at 3 m/s². Calculate the force needed to accelerate the car. [2]

2. A ball has a mass of 420 g and a resultant force of 5.1 N acting on it. Calculate the acceleration of the ball in m/s². [4]

3. **Higher only:** What is the inertial mass of the ball? [1]

   1. $F = m \times a = 1200 \times 3$[1] = 3600 N[1]

   2. 420 g = 0.42 kg[1]
      $F = m \times a$
      5.1 = 0.42 × a[1]
      a = 5.1/0.42[1]
      = 12.1 m/s²[1]

   3. 0.42 kg.[1]

Inertial mass   Higher Tier

Inertial mass is:
- a measure of how difficult it is to change the velocity of an object
- defined as the **ratio** of force over acceleration $\frac{F}{a}$.

An object with a greater inertial mass needs a larger force to produce the same acceleration as an object with a smaller inertial mass.

NEWTON'S THIRD LAW

The third law of motion

Newton's third law states that two objects exert **equal** and **opposite** forces on each other when they interact.

- The forces are sometimes called action-reaction forces.
- The forces act on different objects at the same time.
- Both forces are the same type of force e.g. push, pull.

Be careful! Pairs of equal and opposite forces are not always examples of Newton's third law. The force of gravity acting downwards on a book and the reaction force of the table pushing up on the book, are both acting on the same object – the book. This is not an example of Newton's third law.

Examples of Newton's third law

Newton's third law applies to both contact and non-contact forces. Here are some examples of Newton's third law in equilibrium situations. An equilibrium situation is when the forces are balanced. There is no resultant force in any direction.

Non-contact forces:

- The Earth exerts a gravitational force on the Moon, and the Moon exerts a gravitational force on the Earth.

Contact forces:

- A book exerts a pushing force on a table because of its weight, and the table exerts a pushing force back on the book which has equal value in the opposite direction.
- A person pushes against a wall with a force of 100 N. The wall pushed back against the person with a force of 100 N.

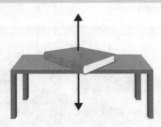

Explain why action-reaction forces are not the same as balanced forces. [1]

Balanced forces act on the same object.[1]
Action-reaction forces act on different objects.[1]

Summary of Newton's laws

Law of inertia: A body will remain at rest of constant velocity unless acted on by an unbalanced force

Law of force and acceleration: The forces experienced by an object are proportional to its mass × the acceleration applied to it.

Law of action and reaction: If two bodies exert a force on each other, the forces are equal in magnitude but opposite in direction.

STOPPING DISTANCE

The stopping distance of a driven vehicle is the total distance that the vehicle travels between the point the driver decides to stop, and the vehicle actually stopping.

Thinking distance and braking distance

The stopping distance of a vehicle is made up of two parts:

stopping distance = thinking distance + braking distance

The **thinking distance** is the distance the vehicle travels during the time the driver is reacting. It is linked to the driver's reaction time. The **braking distance** is the distance the vehicle travels after a force is applied on the brakes.

Given the same braking force, the stopping distance will increase as the speed increases. This is because the vehicle will travel further in the same amount of time.

Describe the relationship between the stopping distance and the speed of a vehicle. [1]

The greater the speed, the greater the stopping distance.[1]

REACTION TIME

Reaction time is the time a person takes to react to a situation.

Factors affecting reaction time

Reaction times are typically in the range **0.2 s** to **0.9 s** and vary between different people.

Factors that affect reaction time include:
- when the driver is tired
- when the driver has taken alcohol or drugs
- when the driver is distracted such as using a mobile phone.

'Longer' can refer to time or distance so use terms carefully. Using 'greater' can be clearer.

Measuring reaction time

Reaction times can be measured in different ways. Because the times involved are small, they are difficult to measure accurately. It is important to record lots of measurements, repeat them and calculate mean values.

Reaction times can be found by measuring the time between a person **detecting a stimulus** and then **responding** to it.

The stimulus could be a light or a sound, and the response can be recorded by pushing a button on a computer.

A typical value of reaction time is 0.7 s. Some people, such as sprinters, can improve their reaction times through training.

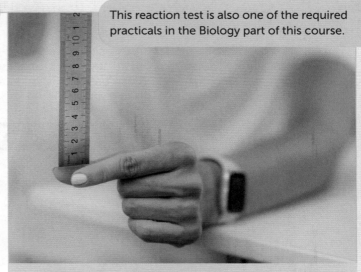
This reaction test is also one of the required practicals in the Biology part of this course.

Reaction time can also be measured by catching a dropped ruler and comparing how far it falls for different people under different situations. The further the ruler falls, the greater the reaction time of the person.

1. Explain the relationship between reaction time and thinking distance. [2]
2. Give **two** disadvantages of using the dropped ruler method to measure reaction time compared with a computer. [1]
3. Suggest why there is a legal limit for alcohol levels when driving. [2]

   1. *The thinking distance is the distance the car travels while the driver reacts to a danger.[1]*
   If the driver has a longer reaction time, the car will have travelled further – the thinking distance is greater.[1]

   2. *The ruler may be dropped from a slightly different height each time due to human error.[1]*
   The time may also be affected by the reaction time of the person using the stopwatch.

   3. *Alcohol can lengthen a person's reaction time[1] which increases the overall stopping distance of a vehicle[1].*

FACTORS AFFECTING BRAKING DISTANCE 1

The **braking distance** of a vehicle is related to the **condition of the vehicle**, the **surface** being driven on and the **weather**.

Braking distance

Wet or icy conditions, and worn tyres reduce friction between the tyres and the road surface. Less friction between the tyres and the surface mean that the vehicle will travel further whilst braking.

Worn brakes require a larger force to be applied so will also increase the braking distance.

The graph below shows values for the stopping distance of a typical car in good condition on dry roads. It can be seen that as speed increases, braking distance has a greater impact than thinking distance on the overall stopping distance.

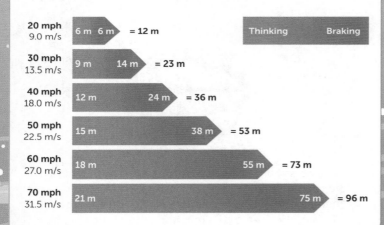

1. A car is travelling at 30 mph in icy conditions.

   (a) Use the graphic opposite to estimate the stopping distance. Tick **one** box. [1]

   ☐ 14 m   ☐ 20 m
   ☐ 23 m   ☐ 28 m

   (b) Explain your answer to part (a). [2]

2. Explain why worn tyres increase braking distance but not thinking distance. [2]

   1. (a) 28m.[1]
      (b) Icy conditions increase the braking distance[1] so the overall stopping distance will be higher than 23 m[1].

   2. The thinking distance is the distance travelled as the driver is responding to a stimulus to stop the vehicle.[1] The condition of the tyres has no impact on the driver's reaction as they are not braking at this time.[1] Less grip on the tyres results in less efficient braking, increasing braking distance.[1]

FACTORS AFFECTING BRAKING DISTANCE 2

A moving vehicle has **kinetic energy** which is dependent on velocity ($E_k = \frac{1}{2}m v^2$). See also **page 247**.

Braking

When a vehicle stops, the energy from the kinetic energy store of the vehicle is transferred to other energy stores. When a force is applied to the brakes of a vehicle, the brakes press against the wheel.

Work is done by the friction force between the brakes and the wheel. Energy is transferred mechanically from the kinetic energy store of the vehicle to the **thermal energy** store of the brakes. This causes the vehicle to slow down and the temperature of the brakes to increase.

The greater the speed of a vehicle, the greater the **braking force** needed to stop the vehicle in a given distance. Greater braking forces cause greater **decelerations**.

A higher speed means that there is more energy in the kinetic energy store. As this energy is transferred to the thermal energy store of the brakes, the brakes can overheat and stop working properly. This can cause the driver to lose control of the vehicle.

Large decelerations can also cause injuries to the occupants of the vehicle.

1. Describe the braking force needed to stop a vehicle in the same distance when the speed doubles. [2]

2. **Higher only:** A driver of a lorry with a mass of 36 000 kg brakes suddenly. The lorry decelerates from 15 m/s to 0 m/s in 20 s. Estimate the braking force on the lorry. [2]

3. **Higher only:** Estimate the force needed to stop a car with a mass of 1500 kg travelling at 60 mph. Use the graphic on **page 322**. [3]

1. *If the speed of the vehicle doubles, the kinetic energy quadruples,[1] so the force needed to stop the vehicle in the same distance quadruples[1].*

2. *$a = \Delta v/t = (15 - 0) / 20$*
   $= 0.75$ m/s^2[1]

   $F = m \times a = 36\ 000 \times 0.75 = 27\ 000$ N.[1]

3. *Stopping distance = 55 m; speed = 27 m/s*
   $Fd = \frac{1}{2}mv^2$
   $F \times 55 = 0.5 \times 1500 \times 27^2$
   $F = (0.5 \times 1500 \times 27^2) / 55$
   $= 9900$ N[1]

Estimating braking forces   Higher only

Braking forces for vehicles under typical conditions can be estimated when the mass and deceleration of the vehicle are known.

Deceleration can be calculated using $a = \dfrac{\Delta v}{t}$ and the force can be calculated using $F = ma$.

When you know the stopping distance, you can also calculate the force using:

work done (F d) = kinetic energy of car ($\frac{1}{2}mv^2$)

CONSERVATION OF MOMENTUM

A moving mass with a velocity also has **momentum**.

Calculating momentum

Note that p is also the symbol used for pressure.

Momentum is a **vector** quantity. A stationary object has zero momentum because v = 0. Momentum is defined by the equation:

momentum = mass × velocity

$p = m\,v$

p = momentum in kilograms metre per second, kg m/s

m = mass in kilograms, kg. v = velocity in metres per second, m/s

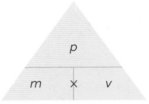

A bird with a mass of 250 g flies at a velocity of 18 m/s. Calculate the momentum of the bird. [3]

$250\ g = 0.25\ kg$[1] $p = m × v = 0.25 × 18$[1] $= 4.5\ kg\ m/s.$[1]

Conservation of momentum

The law of conservation of momentum states that in an event in a **closed system** such as a collision, the total momentum before the event equals the total momentum after the event. A closed system is where no energy escapes or enters from outside the system, and no external forces are acting on the system.

Momentum can have positive and negative values depending on the direction the object moves.

Before collision	Collision	After collision
a) Red ball has momentum p. Blue ball is stationary.		Red ball stops. Blue ball now moves in the same direction with the same momentum p that the red ball had.
b) Red ball has velocity v. Blue ball is moving in the opposite direction with velocity $-v$		Both balls continue to move but in the opposite direction, so red ball now has velocity $-v$, blue ball velocity v.
c) Red ball has velocity v. Blue ball moving in the same direction but with a lower velocity.		Red ball velocity decreases. Blue ball velocity increases.
d) Red ball has momentum p. Blue ball is stationary.		Both balls stick together and move with the same velocity.

EXAMINATION PRACTICE

01   This question is about forces.

01.1 Which quantity is a vector quantity? Tick **one** box.   [1]

☐ displacement   ☐ distance   ☐ mass   ☐ time

The diagram shows the forces on a box, F_X and F_Y.

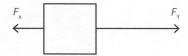

01.2   Compare the two forces.   [2]

01.3   Explain what will happen to the box.   [2]

01.4   **Higher Tier only:** Explain why momentum is a vector quantity.   [2]

01.5   Give **two** examples of non-contact forces.   [2]

01.6   Define weight.   [1]

01.7   Calculate the weight of an object with a mass of 1.6 kg. Use g = 9.8 N/kg   [2]

01.8   Name an instrument used to measure weight.   [1]

02   This question is about resultant forces.

The diagram shows the forces acting on a block.

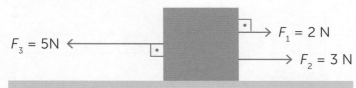

02.1   Calculate the resultant force on the block.   [2]

02.2   **Higher Tier only:** Draw a free body diagram of a cyclist moving to the right along a
        flat road. Label all four forces acting on the cyclist.   [3]

02.3   **Higher Tier only:** A single force of 3 N acts at 48° to the vertical.
        Draw a scale diagram to resolve the force into two forces. Use graph paper.   [2]

02.4   **Higher Tier only:** The diagram represents two forces on a boat. The forces are acting at
        right angles to each other.
        Draw a scale diagram to determine the resultant force on the boat.   [3]

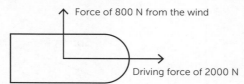

03   This question is about the energy transfer when a force does work.

03.1   Give the equation that links distance, force and work done.   [1]

03.2   Show how the work done in raising a mass through a vertical distance is equal to the
        energy transferred to the gravitational potential energy store of the mass.   [4]

03.3 The work done to stop a car on a flat surface is 6600 J. Calculate the distance travelled, in m, when the braking force is 1200 N. [3]

03.4 Convert 500 Nm into kilojoules (kJ). [2]

04 This question is about elasticity.

04.1 Write the equation that links force, extension and spring constant. [1]

04.2 A spring stretches 0.07 m when a force of 2.1 N is applied. The deformation is elastic. Calculate the spring constant of the spring. Give the unit. [4]

04.3 The sketch shows a small toy that works by compressing a spring.
The compression of the spring is 3.5 cm.
The spring constant is 500 N/m.

Calculate the energy stored in the spring. [3]

04.4 A student investigates the relationship between force and extension for a spring.
The results are shown in the table.

Mass in g	Force in newtons	Length of spring in cm	Extension in cm
0	0	X	0
10	0.1	2.4	0.2
20	0.2	2.6	0.4
30	0.3	2.8	0.6
40	0.4	3.0	0.8
50	0.5	3.2	1.0

Calculate X, the original length of the spring. [1]

04.5 The diagram shows a graph of extension in cm against force in newtons for another spring, Spring A.
Plot a graph of extension against force on the axes above for the spring in the table.
Draw a line of best fit. [3]

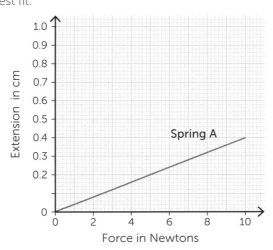

04.6 Determine the spring constant for Spring A. [2]

05　This question is about motion.

05.1　The diagram shows the journey of a car from Town A to Town B
Determine the displacement of the car. [2]

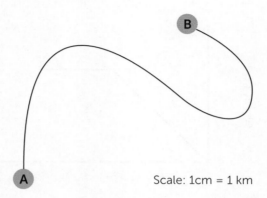

Scale: 1cm = 1 km

05.2　The car travels a further 3.6 km to get home. This journey takes 150 s. Calculate the mean speed of the car for this journey in m/s. [4]

The diagram shows a distance-time graph for the journey of an object.

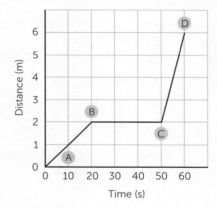

05.3　An object travels 2 m in 20 s.
The object is then stationary for 30 s
The object then travels 4 m in 10 s.
Draw a distance-time graph for the motion of the object. [3]

05.4　Calculate the mean speed of the object for the whole journey. [3]

05.5　**Higher Tier only:** Sketch a distance-time graph for a decelerating object. [1]

06　This question is about acceleration.

06.1　A car accelerates at 1.5 m/s² for 8s.
Calculate the change in velocity in m/s. [3]

06.2　A cyclist increases speed from 5.0 m/s to 9.0 m/s
with an acceleration of 0.5 m/s².
Calculate the distance travelled by the cyclist. [3]

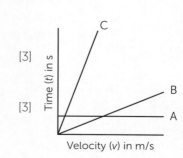

06.3 The diagram shows a velocity-time graph for three objects, A, B and C.
Describe the motion of the three objects. [3]

06.4 The diagram below shows a velocity-time graph for the journey of an object.
Calculate the acceleration of the object for the first 20s of the journey. [2]

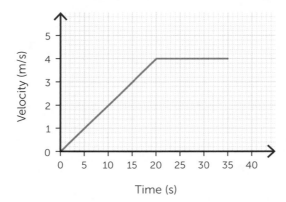

Time (s)

06.5 **Higher Tier only:** Calculate the total distance travelled by the object during the journey. [2]

07 This question is about Newton's Laws of motion

07.1 **Higher Tier only:** What is the tendency of an object to stay at rest or continue moving
with uniform motion called? Tick **one** box. [1]

☐ Acceleration ☐ Deceleration ☐ Inertia ☐ Momentum

07.2 What must happen to make a stationary object move? [1]

07.3 The diagram shows an object moving to the right at a constant speed.

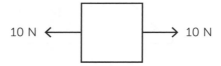

10 N ← → 10 N

Explain what happens to the movement of the object when an upward force of 10 N also
acts on the object. [2]

07.4 A resultant force of 250 N acts on an object. The object accelerates at 12.5 m/s^2.
Calculate the mass of the object in kg. [3]

07.5 Explain the motion of this box. [2]

4 kg → $F = 20$ N

07.6 A student investigates the effect of changing mass on the acceleration of a trolley using this apparatus.

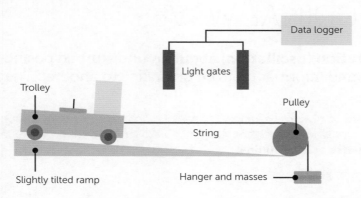

Explain why the number of masses on the hanger is kept constant. [2]

07.7 Give **one** safety factor to consider in this investigation. [1]

07.8 Define Newton's third law of motion. [1]

08 This question is about forces and braking vehicles.

08.1 Define the stopping distance of a vehicle. [1]

08.2 Which of these is correct when a driver is tired? Tick **one** box. [1]
☐ Braking distance increases
☐ Reaction time increases
☐ Stopping distance decreases
☐ Thinking distance decreases

08.3 Give **two** factors that increase the braking distance of a vehicle. [2]

08.4 Explain how a large deceleration can cause brakes to overheat in a vehicle. [3]

09 **Higher Tier only:** This question is about momentum.

09.1 Calculate the momentum of a 5.2 kg ball rolling with a velocity of 0.3 m/s. [2]

09.2 Give the law of conservation of momentum. [1]

TRANSVERSE AND LONGITUDINAL WAVES

A **wave** is a **vibration (oscillation)** about an undisturbed position. Waves transfer **energy** and information from one place to another. They do not transfer **matter**.

Mechanical waves are oscillations (vibrations) of particles and transfer energy through a medium. **Electromagnetic** waves are oscillations of electrical and magnetic fields and also transfer energy. Waves can be **transverse** or **longitudinal**.

Transverse waves

Perpendicular means at right angles.

Water waves, or ripples, are an example of a transverse wave. The water particles vibrate up and down about a fixed point. The vibrations are **perpendicular** to the direction of energy transfer. It is the wave and not the water particles that travel.

Longitudinal waves

Sound waves travelling through air are an example of a longitudinal wave. Air particles vibrate backwards and forwards about a fixed point. The vibration is **parallel** to the direction of energy transfer. Again, it is the wave not the air particles that travel.

Longitudinal waves show areas of **compression** and **rarefaction**. A compression is where the particles are closer together. A rarefaction is where the particles are further apart.

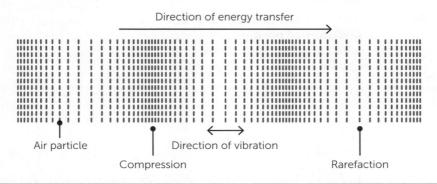

Direction of energy transfer

Air particle

Direction of vibration

Compression

Rarefaction

1. Give **one** similarity and one difference between a longitudinal and a transverse wave. [2]

2. Describe how a slinky spring can be used to show that for both types of wave the particles vibrate about a fixed position. [3]

   1. *Similarity: both transfer energy / both are oscillations; about an undisturbed position.[1] Difference, any one from: Vibration is perpendicular to the direction of travel for transverse waves and parallel to the direction of travel for longitudinal waves / all longitudinal waves need a medium to travel through, some transverse waves do not (EM waves).[1]*

   2. *Mark a point on the slinky and fix it at one end.[1] For longitudinal waves, move the other end back and forth to show that the point also moves back and forth but does not move forward.[1] For the transverse wave, move the end of the slinky up and down to show that the point also moves up and down but does not move forward.[1]*

PROPERTIES OF WAVES

You can describe wave motion in terms of the wavelength, amplitude, frequency and period of the wave.

Amplitude and wavelength

The **amplitude** of a wave is the maximum displacement of a point on a wave away from its **undisturbed position**. The undisturbed position is where there is no vibration. It is also known as the rest position, or the equilibrium position.

> Use a ruler when drawing lines to show the amplitude and wavelength on a wave.

The **wavelength (λ)** of a wave is the distance from a point on one wave to the equivalent point on the adjacent wave.

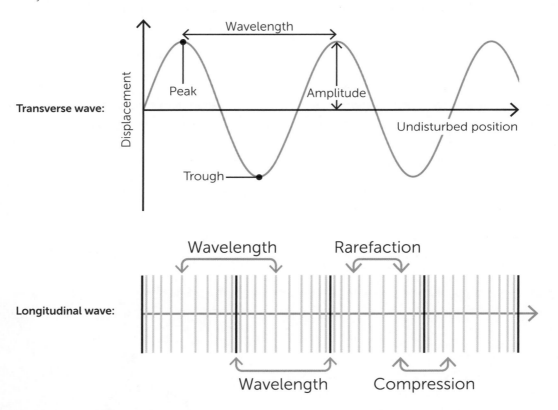

The **peak** and the **trough** are the points usually used on a wave as these are easier to identify on a diagram. Remember that the amplitude is half the distance from a peak to a trough.

In a longitudinal wave, λ is the distance between a compression and a compression, or a rarefaction and a rarefaction.

It is difficult to show amplitude on a longitudinal wave so amplitude is usually shown on a transverse wave.

Frequency

The **frequency (f)** of a wave is the number of waves passing a point each second.

The **period (T)** of a wave is the time taken for one full cycle of the wave.

period = $\dfrac{1}{frequency}$

$T = \dfrac{1}{f}$

T = period in seconds, s

f = frequency in hertz, Hz. **1 Hz = 1 wave per second.**

> You need to be able to select this equation from the equation sheet and apply it

Wave speed

The **wave speed (v)** is the speed at which the energy is transferred, or the speed that the wave moves, through a medium.

All waves obey the **wave equation**:

wave speed = frequency × wavelength

$v = f\lambda$

v = wave speed in metres per second, m/s

f = frequency in hertz, Hz

λ = wavelength in metres, m

> You need to be able to recall and apply this equation.

A **medium** is a something that a wave can be transmitted through.

1. Label the amplitude and the wavelength on this diagram of a wave. [2]

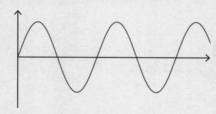

2. Describe the relationship between wavelength and frequency. [1]
3. Calculate the period of a wave with a frequency of 0.5 Hz. [2]
4. A wave travels with a frequency of 7.0 Hz and a wavelength of 0.4 m. Calculate the speed of the wave. [2]

   1. *Labelled: Amplitude as the vertical height from horizontal line to peak.[1] Wavelength as the distance between any point and the same point in the next wave.[1]*
   2. *The longer the wavelength, the lower the frequency.[1]*
   3. *T = 1/f = 1 / 0.5[1] = 2s[1]*
   4. *v = f λ = 7.0 × 0.4[1] = 2.8 m/s[1]*

MEASURING WAVE SPEED

There are various methods for measuring the speed of sound in air and the speed of ripples on the surface of water.

Measuring the speed of sound in air

One possible method is:

- A person makes a loud noise (e.g. banging together blocks of wood) and indicates as they make the sound.
- A second person stands 100–200 m away and starts a stopwatch when they see the visual sound indication.
- They stop the watch when they hear the sound and record the time taken.

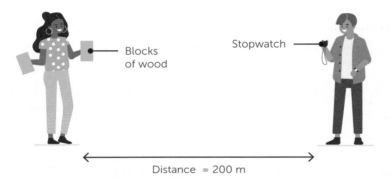

Blocks of wood

Stopwatch

Distance = 200 m

The sound is seen being made before it is heard because light travels much faster than sound in air (3×10^8 m/s, and 330 m/s)

An alternative method is to record the time taken to hear an echo from a wall (the sound travels to the wall and back). This has the advantage of a greater distance, so the time taken for the sound wave to travel will be longer and easier to measure. Also, the person with the stopwatch stands next to the person making the sound which may mean the stopwatch is started more accurately.

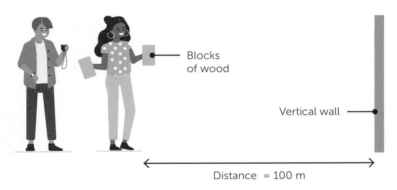

Blocks of wood

Vertical wall

Distance = 100 m

Measuring the speed of waves and ripples on a water surface

One possible method is to time how long it takes a water wave to travel a known distance such as along a sea wall. A ripple tank can also be used for this.
See the next page for more details. **RPA 20**.

1. Explain how the reaction time of the person timing affects the measurement of the speed of sound using a wall. [2]

2. You are going to measure the speed of a water wave by timing how long it takes to travel a known distance along a sea wall.
   Describe how you would take your measurements. [3]

   1. *The reaction time of the person timing introduces an error which makes the time longer than the true value.[1] This has the effect of calculating the speed to be lower than its true value.[1]*

   2. *Identify two obvious features or points on the sea wall that you can use for timing the waves.[1] Measure the distance between these two points.[1] Time how long it takes several waves to travel between these two points.[1]*

You can see a lightning flash before you hear the clap of thunder. The light from the flash travels at the speed of light, whereas the sound from the clap of thunder travels at the speed of sound. You can use this information to calculate how far away the lightning flash was.

REQUIRED PRACTICAL 20

Investigating waves

This practical activity helps you make observations of waves in fluids and solids to identify the suitability of apparatus to measure speed, frequency and wavelength.

Both investigations involve measurements of wavelength, and measurements or recording of frequency. You can calculate speed using: **wave speed = frequency × wavelength**.

Water waves in a ripple tank

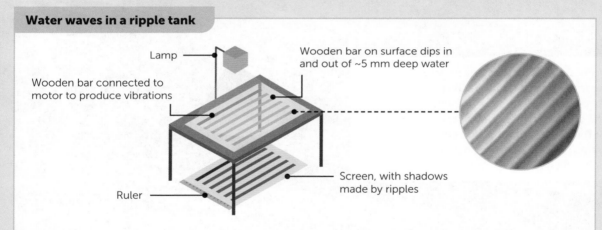

Lamp

Wooden bar connected to motor to produce vibrations

Wooden bar on surface dips in and out of ~5 mm deep water

Screen, with shadows made by ripples

Ruler

Light from the lamp shines through the water and an image of the waves can be observed on white card below the tank. The image will look clearer in a darker room.

To measure wavelength: lay a metre ruler at right angles to the waves. Measure the total length the waves travel. Divide the total length by the number of waves to find the wavelength.

To measure frequency: set the vibration generator to a low frequency so it is possible to count the number of waves passing a point in 10 seconds. Record observations from as many people as possible and calculate a mean.

The **number of waves** is needed for both values and can be difficult to measure. The waves can be video recorded and slowed down, or paused, so that the number of waves can be measured more accurately. The timer must also be recorded so that the real time taken is known to determine the frequency.

1. Explain how the frequency is calculated from the number of waves passing in 10 seconds. [1]
2. 18 waves pass a point in 10 seconds. Calculate the frequency of the waves. [1]
3. The length of the tank is 27 cm. 9 waves are seen on the tank.
   Calculate the wavelength of the waves [2]
4. Calculate the speed of the waves in the ripple tank. [2]

1. *Frequency = number of waves / 10.*[1]
2. *Frequency = 18/10 = 1.8 Hz*[1]
3. *2.7 cm = 0.27 m*[1] *wavelength = 0.27 / 9 = 0.03 m*[2]
4. *v = 1.8 × 0.03*[1] *= 0.054 m/s*[1]

Waves on a stretched spring

When a string is vibrated at certain frequencies, **standing waves** are generated on the string. This is a wave that appears to be stationary so it is easier to measure the wavelength than from a travelling wave.

A standing wave is generated on a string by vibrating the string at one end. The tension in the string is adjusted to get a visible standing wave.

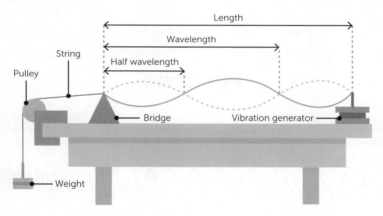

Wavelength: Use a metre ruler to measure the total length of the standing wave produced. Count the number of **half wavelengths** and determine the wavelength. Vary the power supply to the vibration generator to produce different wave frequencies. Measure the wavelengths for standing waves with different numbers of half wavelengths.

$$wavelength = \frac{total\ length}{number\ of\ half\ wavelengths} \times 2$$

Frequency: This is the frequency supplied to the vibration generator so it is recorded rather than measured.

5. What is the purpose of the hanging mass? [1]
6. Calculate the wavelength of this standing wave in m. [3]

21.0 cm

5. *To keep the string tight.*[1]
6. *21 cm = 0.21 m*[1] *0.21 / 3 × 2*[1] *= 0.14 m*[1]

TYPES OF ELECTROMAGNETIC WAVES

Electromagnetic waves are transverse waves that transfer energy from the source of the waves to an absorber.

Properties of electromagnetic waves

- Transverse waves
- Transfer energy from the **source** of the waves to an **absorber**.
- Form a continuous spectrum.
- Do not need a medium to travel through.
- All travel at the same velocity through a **vacuum** (3×10^8 m/s).
- Are grouped in terms of their wavelength and their frequency.

A vacuum means there are no particles at all.

The electromagnetic spectrum

Electromagnetic waves form a continuous spectrum.

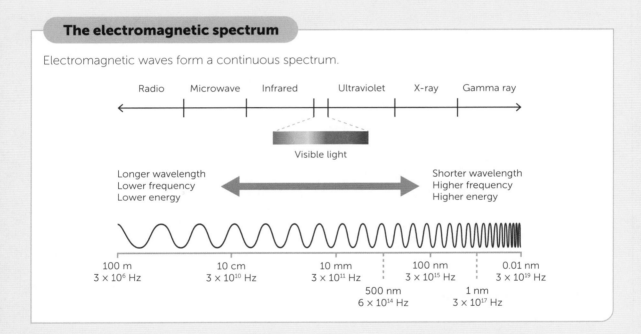

1. Which of the electromagnetic waves can be detected by the human eye? [1]
2. Which electromagnetic wave has the longest wavelength? [1]
3. Which electromagnetic wave has the highest frequency? [1]
4. Give **two** electromagnetic waves that have lower energy than visible light. [2]

1. *Visible light.* [1]
2. *Radio waves.* [1]
3. *Gamma rays.* [1]
4. *Any two from: infrared / microwave / radio.* [2]

PROPERTIES OF ELECTROMAGNETIC WAVES 1

Waves change speed when they move from one substance into another. This can also cause a change in direction. This change in direction is called **refraction**.

Ray diagrams

When the angle of incidence of light at a boundary is less than 90°:

- the light is refracted towards the normal when it goes from air to another material like glass or water
- the light is refracted away from the normal when it goes from a material such as glass or water to air

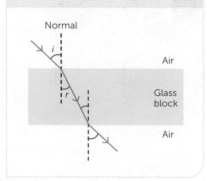

Light travels faster in air than in water. Describe what happens to light as it passes from water to air at an angle of incidence of 40°.   [3]

The light refracts and changes direction[1] and bends away from the normal[1] at an angle of refraction greater than 40° and less than 90°.[1]

Explaining refraction   Higher Tier

Different substances may absorb, transmit, refract or reflect electromagnetic waves in ways that vary with wavelength.

Refraction is caused by the difference in velocity of the wave in different substances. For example, light travels more slowly in air and glass than it does in water. The shorter the wavelength of an electromagnetic wave , the more it is refracted.

A **wave front** is a line that joins all the points on a wave that are moving up and down together at the same time. As the wavefronts meet the boundary between air and glass, they slow down as they enter the glass. The left-hand end of the wavefront meets the boundary and slows down before the right-hand end, and this causes the direction of the wavefront to change. The wavelength gets shorter, but the frequency remains the same.

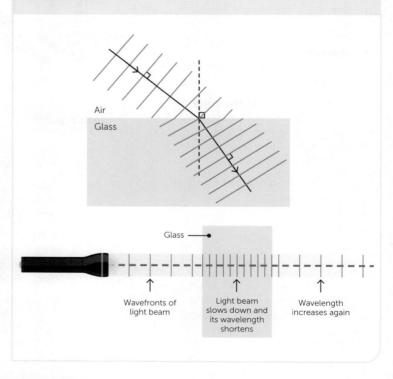

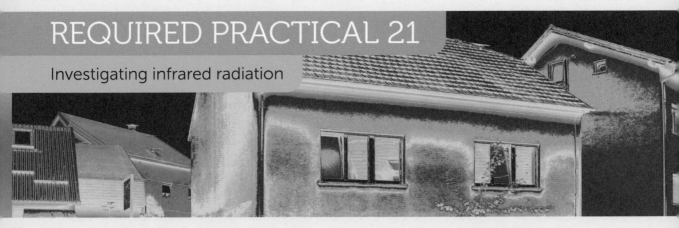

This practical activity helps you to record accurate temperatures and observe the effects of infrared radiation interacting with different surfaces.

Emission

All objects emit infrared (IR) radiation. The hotter the surface, the more radiation is emitted. The amount of IR radiation emitted also depends on the type of surface. We cannot see IR radiation so we use an **IR detector** or observe temperature changes near the surface.

An IR detector measures the amount of infrared radiation emitted from each surface. A thermometer can also be used to measure the temperature of the air near the surface. A detector has a higher **resolution** than a thermometer so will be better for detecting small differences in the amount of radiation emitted.

A metal box called a **Leslie cube** is used for this practical. The bottom of the box is placed on a heat proof mat and the top has a hole with a lid so that very hot water can be placed inside.

The box has different surfaces on the four vertical sides, typically:
- matt black
- shiny black
- shiny silver
- matt white.

As it is a cube, all surfaces have the same surface area. The amount of infrared emitted is recorded and compared for each surface. The higher the value the greater the emission.

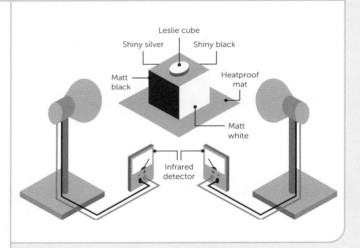

Absorption

We can also use a Leslie cube to investigate how well the different surfaces absorb IR radiation.

Fill the Leslie cube with a known volume of cold water. Put a thermometer in the top of the cube and plug the hole with a bung.

Place a radiant heater a known distance (about 10 cm) from one surface. Switch the heater on and record the temperature of the water every 30 seconds for about 10 minutes.

Repeat for the other three faces, replacing the water each time with the same volume of cold water and making sure that the radiant heater is the same distance from the face.

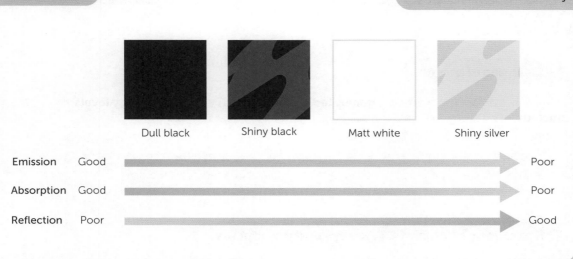

		Dull black	Shiny black	Matt white	Shiny silver	
Emission	Good					Poor
Absorption	Good					Poor
Reflection	Poor					Good

1. (a) Give **two** control variables for the absorption investigation. [2]

   (b) Explain why you need to control the variables you have named. [2]

2. A student uses a Leslie cube to observe IR emission.

   (a) The Leslie cube is left for a short while after the hot water is added before recording the emission. Explain why. [2]

   (b) Suggest the best type of graph to present the results. Explain your choice. [3]

   (c) Describe how standard laboratory thermometers can be used to investigate the emission from the Leslie cube. [4]

1. *(a) Any two from: distance of the IR source from the surface / intensity of IR source / volume of water used each time.[1]*

   (b) To keep them constant[1] so that it is a fair test[1]

2. *(a) It allows the surfaces to heat up to the same temperature as the water[1] so that the surfaces all have the same starting temperature.[1]*

   (b) A bar chart[1] because the independent variable is a categoric variable[1] so a line graph cannot be plotted.[1]

   (c) Paint four[v] thermometer bulbs matt black.[v] Place a thermometer near each surface[v] and at the same distance[v] from each surface. Make the distance small[v] to increase the IR radiation incident on the thermometer bulb.[v] Record the temperature rise for each surface[v] over a set time[v]. The thermometer reading that reaches the highest temperature[v] indicates the surface with the greatest emission[v].

   This question should be marked in accordance with the levels based mark scheme on page 374.

PROPERTIES OF ELECTROMAGNETIC WAVES 2

Changes in atoms

When electromagnetic waves are **generated** or **absorbed**, by atoms the **energy levels**, or **nucleus**, of an atom changes. This happens over a wide frequency range.

- When an electromagnetic wave is absorbed, it can cause an electron to move to a higher level.
- When an electron returns to its original energy level it can generate an electromagnetic wave.
- Gamma rays are generated by a change in the nucleus of a radioactive atom.

Hazardous effects of UV, X-rays and gamma rays

The hazardous effect that some electromagnetic waves can have on human body tissue depends on both the **type** of radiation and the size of the **dose**.

Radiation dose is a measure of the risk of harm resulting from exposure of the body to a source of radiation. It is measured in sieverts (Sv).

1000 millisieverts (mSv) = 1 sievert (Sv)

Ultraviolet waves, X-rays and gamma rays have higher energy, so exposure carries more risk. **Ultraviolet waves** can cause skin to age prematurely and increase the risk of skin cancer. The use of sunscreen to block UV waves is recommended, and tanning beds need to be used with caution. **X-rays** and **gamma rays** are both **ionising** radiation. This means they remove electrons from atoms. This can cause mutations in cells and genes, increasing the risk of cancer.

Radio waves   Higher Tier

Radio waves can be produced by **oscillations of electrons** in electrical circuits. Radio waves can be absorbed by a conductor such as an aerial. This can induce the electrons in the conductor to oscillate and generate an alternating current with the same frequency as the absorbed radio waves. This is how terrestrial TV and radio is broadcast and received.

---

1. Complete the sentences. [2]

   An electron can move to a higher energy level when electromagnetic radiation is _____.

   When an electron moves to a lower energy level electromagnetic radiation can be _____.

2. The radiation dose from a single X-ray is about 0.1 mSv. A fatal dose of radiation is about 5 Sv. Explain why hospital staff working with X-rays need to be shielded when a patient is having an X-ray. [2]

   1. *absorbed[1] emitted[1]*

   2. *As hospital staff could see many patients each day, this could lead to a high dose of radiation over a period of time.[1] So staff shield to minimise the dose of X-rays they are exposed to, to reduce the potential damage that could be caused.[1]*

USES AND APPLICATIONS OF ELECTROMAGNETIC WAVES

The electromagnetic spectrum

Electromagnetic waves have very many applications. Some examples are shown in the table.

Wave	Uses	**Higher Tier only** Suitability of each electromagnetic wave
Radio waves	Television and radio, communication with ships, aeroplanes and satellites	• Travel long distances before being absorbed. • Reflect off the ionosphere in the atmosphere so can be sent very long distances around the earth.
Microwaves	Satellite communications, mobile phones, cooking food	• Pass through the Earth's atmosphere without reflection or refraction. • Food contains a lot of water which absorbs microwaves and heats up.
Infrared	Electrical heaters, cooking food, infrared cameras, TV remote controls	• Emitted by hot objects and easily absorbed by surfaces. • Check for heat being emitted from animals or heat loss from buildings.
Visible light	Fibre optic communications	• Thin strands of glass which reflect visible light inside to carry pulses of light in cables. • Short wavelength means a lot of information can be transmitted.
Ultraviolet	Energy efficient lamps, sun tanning	• Short wavelength UV converted to visible light inside the bulb. • Tanning beds use UV light to reproduce the effect of the Sun on skin.
X-rays and gamma rays	Medical imaging, medical treatments	• X-rays and gamma rays have high energy and are very penetrating. • Both pass through tissue but X-rays are absorbed by bones, so bones show up on photographic film. • Gamma rays can also be used to target and destroy cancers.

1. Give **two** types of electromagnetic waves used for cooking. [1]
2. Give **two** types of electromagnetic waves used for communicating. [1]
3. **Higher Tier only:** Describe why X-rays and not gamma rays are used to diagnose broken bones. [2]

   1. *Microwave and infrared.[1]*
   2. *Any two from: radio waves, microwaves, visible light.[1]*
   3. *X-rays can pass through surrounding tissue but are absorbed by bone.[1] Gamma rays pass straight through both tissue and bone.[1]*

EXAMINATION PRACTICE

01 This question is about the properties of waves.

01.1 The diagram represents a longitudinal wave. Label a **wavelength**, a **compression** and a **rarefaction**. [3]

01.2 The period of a wave is 0.4 s. Calculate the frequency of the wave. Include the unit. [4]

01.3 A wave with a frequency of 160 Hz travels at a speed of 8.0 m/s.
Calculate the wavelength of the wave. [3]

01.4 Describe a method for two students to measure the speed of sound in air using echoes.
Include steps to reduce the effect of errors. [6]

02 This question is about measuring properties of waves in liquids and solids.

02.1 A ripple tank has a total length of 0.56 m. 28 waves are counted on the ripple tank.
Calculate the wavelength of the waves. [2]

The table shows the frequency of a wave in a ripple tank observed by four students A, B, C and D.

Student	Frequency in Hz
A	2.9
B	3.3
C	3.2
D	X
Mean	**3.2**

02.2 Calculate value X. [2]

02.3 The wave has a wavelength of 0.075 m. Calculate the speed of the wave. [2]

02.4 Explain why taking a video of the waves on a ripple tank is better than taking a photograph of the waves. [3]

The diagram shows a standing wave generated on a string. The wave has a frequency of 12.0 Hz.

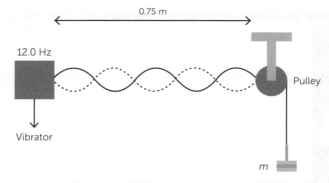

02.5 Calculate the speed of the wave in the string. [4]

03   This question is about electromagnetic waves.

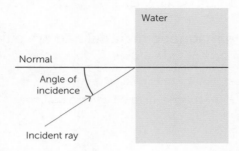

03.1 Give **two** features of all electromagnetic waves.   [2]

03.2 Complete the ray diagram to show what happens as light travels from air to water.
      Include labels.   [2]

03.3 **Higher Tier only:** The diagram shows wavefronts travelling from deep water to
      shallow water.
      Explain what happens to the wavefronts as they travel from deep water to shallow water. [3]

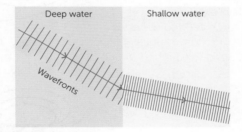

A student uses a Leslie cube to investigate the emission of infrared radiation from different types
of surface.

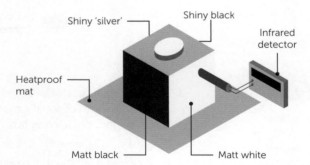

03.4 Give the independent and dependent variables for this investigation.   [2]

03.5 Give **one** control variable.   [1]

03.6 Give the type of electromagnetic wave that is the most damaging to human tissue.   [1]

03.7 **Higher Tier only:** Describe how radio waves can be generated.   [3]

03.8 Give **two** uses of infrared radiation.   [2]

03.9 **Higher Tier only:** Explain why radio waves are used for transmitting terrestrial
      television signals.   [2]

POLES OF A MAGNET

Magnets are made from a magnetic material and are either permanent or induced:

- A **permanent** magnet produces its own magnetic field.
- An **induced** magnet is a temporary magnet created when a magnetic material becomes a magnet while in a magnetic field.

Permanent magnets

Bar and horseshoe magnets are examples of permanent magnets.

The ends of a magnet are called the **poles** and are described as **north (N)** or **south (S)**. The poles are the places where the magnetic forces are strongest.

A permanent magnet will always exert a force on another magnet. The type of force exerted depends on the interacting poles. These are non-contact forces as the magnets do not need to touch to experience the force.

When two magnets are brought close together, they exert a force of attraction or repulsion on each other.

- two **like** poles will **repel** each other (N/N; S/S).
- two **unlike** poles will **attract** each other (N/S).

Permanent magnets also attract magnetic materials.

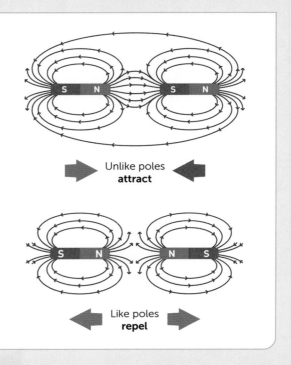

Induced magnets

When a magnetic material (that is not a permanent magnet) is placed in a magnetic field, a magnetic field is **induced**, and the material becomes a **temporary** magnet. When the material is removed from the magnetic field the induced magnet quickly loses nearly all of its magnetism.

An induced magnet will always cause a force of attraction.

1. Magnetic materials are affected by magnets and attracted to both poles of a magnet. Which material is **not** magnetic? Tick **one** box. [1]

   ☐ Cobalt ☐ Copper ☐ Iron ☐ Nickel ☐ Steel

2. How can it be shown that a magnetic material is a permanent magnet? [1]

3. Describe the forces on Magnet 2. [2]

Magnet 1   Magnet 2   Magnet 3

1. *Copper[1]*
2. *It will attract and also repel another magnet.[1]*
3. *The forces are non-contact forces[1] Magnet 2 experiences a force of attraction from Magnet 1[1] and a force of repulsion from Magnet 3.[1]*

MAGNETIC FIELDS

A magnetic field is the region around a magnet where a non-contact magnetic force acts on either:

- another magnet
- or a magnetic material (iron, steel, cobalt and nickel)

Strength and direction of the magnetic field

The **strength** of the field depends on the **distance** from the magnet. The field is strongest at the **poles** of the magnet.

The **direction** of the magnetic field at any point is given by the direction of the force that would act on another north pole placed at that point. The direction of a magnetic field line is from the north (seeking) pole of a magnet to the south (seeking) pole of the magnet.

Remember the force between two magnets can be attractive or repulsive. The force between a magnet and a magnetic material is always attractive.

Magnetic field pattern of a bar magnet

A magnetic **compass** contains a small bar magnet. The north needle of a compass is attracted to the south pole of a magnet. The magnetic field lines of a magnet can be plotted using a compass.

When a compass is away from other magnets it always points towards the same place on Earth. This shows that the **Earth** must have a **magnetic field**. The compass needle points in the direction of the Earth's magnetic field and provides evidence for a **magnetic core** in the Earth.

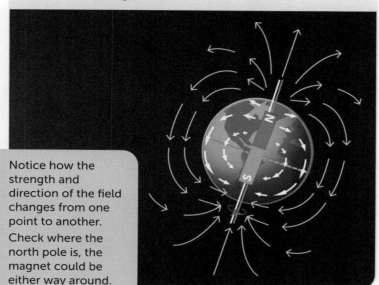

Notice how the strength and direction of the field changes from one point to another.
Check where the north pole is, the magnet could be either way around.

These six steps describe how to plot the magnetic field pattern of a bar magnet on a piece of paper using a plotting compass. Put the six steps in the correct order. [3]

A Repeat this until the south pole of the magnet is reached;

B Join the crosses to show the field lines;

C Move the compass so the south pole of the needle touches this cross;

D Place the compass near the N pole of the magnet;

E Mark a small cross where the compass needle is pointing;

F Repeat for different positions from the N pole of the magnet.

The correct order is: D, E, C, A, B, F. 2 marks for 5 in correct order, 1 mark for 4 in correct order.

ELECTROMAGNETISM

Electromagnetism is the effect where a magnetic field is produced by result of an electric current flowing through a wire. It is a fairly simple effect yet has many useful applications which affect the way we live.

Current in a wire

A **magnetic field** is produced around a conducting wire when a **current** flows through the wire. The **strength** of the magnetic field depends on both the **current** and the **distance** from the wire.

- As the current in the wire increases, the strength of the magnetic field increases: they are directly proportional.

- As the distance from the wire increases, the strength of the magnetic field decreases: they are inversely proportional.

The **direction** of the magnetic field depends on the **direction** of the current in the wire.

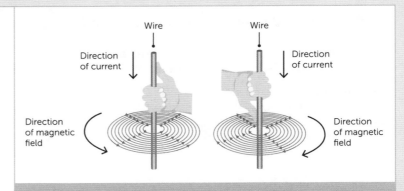

Right hand grip rule for a wire:
- thumb points in the direction of the current
- fingers point in the direction of the field

Conventional current direction is from the positive terminal to the negative terminal of the supply.

Solenoid

Making a coil of wire increases the strength of the magnetic field produced by a current through the wire. A **coil** of wire used in electromagnetism is called a **solenoid**.

- The magnetic field **inside** a solenoid is **strong** and **uniform**.

- The magnetic field **around** a solenoid has a similar shape to the magnetic field around a **bar magnet**

To increase the **strength** of the magnetic field of a solenoid:

- increase the **current** in the wire
- increase the **number of turns** of wire in the coil
- add an **iron core** inside the coil.

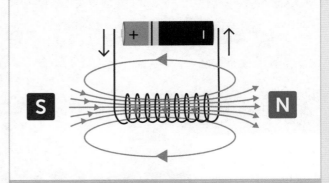

Right hand grip rule for a solenoid:
- fingers curl in the direction of the current
- thumb points in the direction of the north pole

Electromagnet

An **electromagnet** is a **solenoid** with an **iron core** inside the coil.

Electromagnets have advantages over permanent magnets:

- they can be switched on and off which is useful for moving magnetic items.

- their strength can be altered by varying the current.

Electromagnets have very many useful applications.

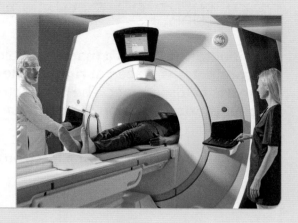

1. Describe how the magnetic effect of a current can be demonstrated. [2]

2. Give **three** ways to increase the strength of a magnetic field around a solenoid. [3]

3. (a) A student investigates the effect of current on the strength of an electromagnet as shown.

   Give the independent, dependent and control variables. [3]

   (b) Describe how the student could carry out this experiment using the equipment shown in the diagram. [6]

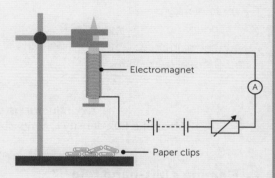

1. *Plotting compasses or iron filings can be placed near a current carrying wire.[1] The effect on the direction of the compass needle/pattern made by the iron filings gives evidence for a magnetic field.[1]*

2. *Increase the current in the wire[1] increase the number of turns in the solenoid[1], or add an iron core[1].*

3. (a) *Independent variable: current in the coil[1]; dependent variable: number of paper clips picked up[1]; control variables: number of coils of wire; size/material of paper clips; distance the electromagnet is held from the paper clips[1].*

   (b) *Adjust the power supply / variable resistor to choose a suitable value for the current e.g. 0.1A[√] Hang as many paper clips as possible from the electromagnet.[√] Record the number of paperclips in a table with the current.[√] Turn off the electromagnet.[√] Repeat the test at this current[√]. Adjust the power supply/variable resistor to increase the current to e.g. 0.2A.[√] Repeat the method to hang paper clips.[√] Collect readings for at least five different values of current.[√] Calculate a mean.[√] Check for any anomalous results.[√] The more paper clips the electromagnet picks up, the stronger the electromagnet is.[√]*

   This question should be marked in accordance with the levels based mark scheme on page 374.

FLEMING'S LEFT-HAND RULE

A conductor produces a magnetic field when a current flows through it. When the conductor is placed at **right angles** to another magnetic field the two magnetic fields interact. The conductor exerts a force on the magnet and the magnet exerts a force on the conductor, and this is known as the **motor effect**.

Calculating the size of the force on a conductor

For a conductor carrying a current at right angles to a magnetic field:

force = magnetic flux density × current × length

$F = BIl$

F = force in newtons, N

B = magnetic flux density in tesla, T

I = current in amperes (amps), A

l = length of wire in metres, m

You need to be able to select this equation from the equation list and apply it.

A conductor usually means a metal wire.
Magnetic flux density is a measure of the strength of the magnetic field.

Fleming's left-hand rule

You can use Fleming's left-hand rule to determine the direction of the force on a conductor in a magnetic field. This is a handy way to represent the relative orientations of the force or movement (thu**m**b), the magnetic field (**f**irst finger) and the **c**urrent in the conductor (se**c**ond finger).

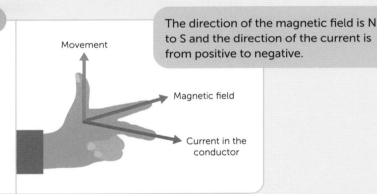

The direction of the magnetic field is N to S and the direction of the current is from positive to negative.

1. Calculate the magnetic flux density of a field that exerts a force of 0.2 N on a wire that is 1.0 m long and carrying a current of 0.8 A.   [3]

2. Explain what will happen when a wire carrying a current is placed south to north in a magnetic field.   [2]

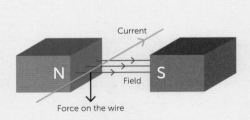

1. $F = BIl$ $0.2 = B × 0.8 × 1.0$[1]
    $B = 0.2 / (0.8 × 1.0)$[1] $= 0.25$ T[1]

2. Nothing[1] because the current is parallel to the magnetic field and so they will not interact as they are not at right angles to each other[1].

ELECTRIC MOTORS

An **electric motor** is an application of the **motor effect**. When a coil of wire carrying a current is placed at right angles to another magnetic field, the force experienced by the coil causes the coil to rotate. Electric motors have multiple uses and it is very useful for an appliance to produce a turning force.

A simple electric motor

- When the current flows in the direction shown in the diagram through the loop of wire in the magnetic field, the force acts upwards on the left side and downwards on the right side of the loop.

- This produces a clockwise moment. The left side of the loop moves up and the right side moves down.

- When the loop is vertical, the force on the loop is zero and there is no moment. At this point there is no connection to the battery because of the gap in the split-ring commutator.

- The loop has momentum and keeps moving. The split-ring commutator swaps the connections with the battery so the current starts flowing again, but now in the opposite direction through the loop. This means that the force on the loop is still upwards on the left and downwards on the right so the loop keeps moving in the same direction.

- Note that the current still flows from positive to negative. The loop has switched sides so the current flows the opposite way through the wire.

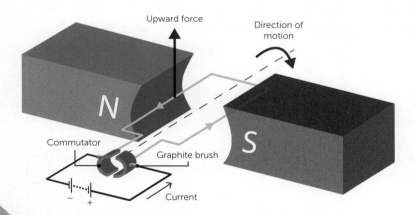

1. Which rule is used to determine the direction of the force on each side of the loop of wire? [1]

2. Explain the purpose of a split ring commutator in an electric motor. [2]

   1. *Using Fleming's left-hand rule.[1]*

   2. *The split-ring commutator keeps the motor rotating in the same direction[1] by allowing the current in the coil to change direction each half-turn as the coil rotates[1].*

EXAMINATION PRACTICE

01 This question is about magnets and magnetic fields.

01.1 A bar magnet is hung on a thread so it can move freely.
Explain what will happen when another bar magnet is brought towards the south pole
of the hanging magnet as shown. [2]

01.2 Name **two** magnetic materials. [2]

01.3 Permanent and induced magnets are both made of magnetic materials.
Describe the differences between a permanent and an induced magnet. [2]

01.4 Draw the magnetic field pattern of a bar magnet. [3]

02 This question is about electromagnetism.

02.1 Draw the magnetic field pattern for a straight wire carrying a current. [2]

02.2 Draw the magnetic field pattern for the wire when it is coiled into a solenoid. [3]

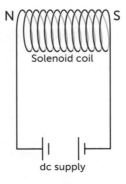

Solenoid coil

dc supply

A student investigated factors that affect the strength of an electromagnet by measuring the number of pins the electromagnet picked up each time.
The table shows the number of pins picked up under different conditions.

| Number of coils | Number of cells | | | | | |
| | 1 | | 2 | | 3 | |
	No core	Iron core	No core	Iron core	No core	Iron core
5	1	5	4	8	5	10
10	3	7	2	12	9	17
15	5	9	7	15	11	20
20	7	15	9	19	13	25

02.3 Which result is anomalous? [1]

02.4 Write three conclusions about the factors that affect the strength of an electromagnet from the results. [3]

03 **Higher Tier only:** The diagram shows a wire carrying a current at right angles to a magnetic field.

03.1 The wire is 1.8 m long. The current is 1.2 A. The magnetic flux density of the magnetic field is 0.03 T.
Calculate the force exerted on the wire. [2]

03.2 Draw an arrow on the diagram to show the direction of the force. [1]

03.3 A coil of wire carrying a current is placed in a magnetic field.
Explain how the force on the wire causes the coil to rotate. [2]

EXAMINATION PRACTICE ANSWERS

01.1 Award one mark for each correct column. [2]

Sub-cellular structure	Eukaryotic cell	Prokaryotic cell
Nucleus	✓	✗
Cell membrane	✓	✓

01.2 In a single large DNA loop [1] and in smaller circular plasmids [1]. [2]

02 Chloroplasts [1], large vacuole filled with sap [1], cell wall [1]. [3]

03.1 Without the nucleus there is more space for haemoglobin [1], so the red blood cell can carry more oxygen [1]. [2]

03.2 Correct conversion and standard form i.e. 8.0×10^{-6} m. [2] Correct answer without standard form 0.000008 m is [1] only. [2]

03.3 6 cm = 60,000 μm. [1] 60,000 ÷ 8 = magnification of ×7,500. [1] [2]

03.4 Mitochondria carry out aerobic respiration [1], which provides the energy needed for movement/muscle contraction [1]. [2]

03.5 They contain protein fibres that can shorten, causing the muscle to contract. [1]

04.1 Ribosomes are too small to see using a light microscope [1]. The electron microscope has better resolution [1] and allows greater magnification [1] than the light microscope. [3]

04.2 So that colourless or transparent structures can be more easily distinguished. [1]

04.3 The coarse focus knob is used to quickly bring the cells into view. [1] The fine focus makes smaller adjustments to bring cells into sharp focus especially at higher magnifications. [1] [2]

05.1 The agar contains nutrients that the micro-organisms feed on. [1] It provides a surface for colonies to grow on [1]. [2]

05.2 *This question should be marked with reference to the levels-based mark scheme on* **page 374**. Indicative content: [6]
Contamination from the **air** in the laboratory [✓]. Prevented by keeping the culture covered/taping the lid on the petri dish. [✓] Using a Bunsen burner to generate upward air currents. [✓]
Contamination from the **equipment** used [✓]. Prevented by flaming the inoculation loop [✓] to kill microorganisms on it [✓] by heat sterilising / autoclaving equipment [✓].
Contamination from the **person** making the culture [✓]. Prevented by not touching the surface of the culture medium [✓], taking care not to breathe on the culture [✓], incubating at temperatures below body temperature/25 °C [✓] to avoid culturing microorganisms that can grow in the body [✓].

06.1 Differentiation. [1]

06.2 Three marks from: Cell division and differentiation for growth continues in mature plants but not in mature animals [1]. In mature animals, cell division is mainly restricted to repair of damaged tissues [1] and replacement of worn out cells [1]. [3]

07 *This question should be marked with reference to the levels-based mark scheme on* **page 374**. [6]
Indicative content: Disinfectant A is more effective at killing bacteria at all concentrations [✓]. But the difference in effectiveness is small at medium and high concentrations [✓]. Disinfectant A is 5 times more expensive than B [✓]. Disinfectant B may be more cost effective / can be used at a higher concentration for a lower cost [✓]. The human toxicity of B is lower than A, so it would be safer to use [✓]. The effectiveness of both disinfectants increases as concentration increases [✓]. There is a big difference in effectiveness between the low and medium concentrations [✓]. The difference between effectiveness of medium and high concentrations is small [✓]. Manipulation of data to support reasoning e.g. Medium concentration of B is more than four times as effective than the low concentration [✓]. A conclusion is made that is consistent with the reasoning (e.g. a medium concentration of disinfectant B should be used based on effectiveness, dilution factor, cost and toxicity) [✓].

08.1 Mitosis. [1]

08.2 Risk of cell rejection as donated cells are not genetically identical to the patient. [1] Donated stem cells do not use embryos so fewer ethical or religious issues. [1] [2]

09.1 Mass would decrease. [1]

09.2 Water would move out of the solution in the bag by osmosis [1] through the partially permeable membrane [1] from the more dilute solution to the more concentrated sugar solution [1]. [3]

09.3 The higher the temperature the faster the rate of diffusion/osmosis [1] because (water) particles have more energy [1]. The size of the bag would affect the surface area for diffusion / osmosis [1]. A larger bag would allow faster diffusion of water /osmosis.   [4]

09.4 Rate of water uptake is 12 / 2 [1] = 6 grams per hour. [1]   [2]

Topic 2

01.1 Highly folded / villi / epithelial cells have microvilli to increase surface area for absorption. [1] Thin wall to reduce diffusion distance. [1] Good blood supply increases concentration gradient. [1]   [3]

01.2 The pancreas.   [1]

01.3 Fatty acids / glycerol.   [1]

01.4 Proteins.   [1]

01.5 The optimum temperature of the lipase may be 30 °C so the enzyme will be most effective at removing stains at this temperature [1]. At lower temperatures, the reaction rate of the enzyme will be slow [1] at higher temperatures the enzyme will denature and no longer work [1].   [3]

02   Protein [1] and starch [1] were present (deduct a mark for each additional substance named).   [2]

03.1 See graph to right. Axes have suitable linear scales and are labelled [1] (linear scale means that numbers go up in equal increments). Six points correctly plotted. [1] Suitable line of best fit. [1]   [3]

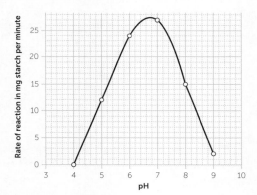

03.2 Optimum pH is 6.7± 0.2. (Give credit if pH at peak of graph stated.)   [1]

03.3 Rate at pH 7.5 is 22 mg of starch per minute.   [1]

04.1 Blood returns to the heart [1] in the pulmonary vein [1] it passes through the left atrium and ventricle [1], then leaves the heart via the aorta to travel to body cells [1].   [4]

04.2 A group of cells found in the right atrium [1] act as a pacemaker causing regular contractions of the heart [1].   [2]

04.3 $(\frac{60}{15}) \times 35 = 140$ beats per minute.   [1]

04.4 Pulmonary artery.   [1]

04.5 Rearrange equation. $\frac{cardiac\ output}{heart\ rate}$ = stroke volume or $\frac{2.9}{70}$. [1] Stroke volume is 0.04 dm³. [1]   [2]

05.1 *This question should be marked with reference to the levels-based mark scheme on **page 374**. Indicative content:*   [6]
Many alveoli provide large surface area to maximise exchange [✓]. Alveoli have thin walls to allow for easy diffusion to cells [✓]. Capillary wall thin /only 1 cell thick [✓] which gives a short diffusion pathway [✓]. Surrounded by many capillaries to give a large surface area for exchange [✓]. Good blood supply increases concentration gradient for diffusion [✓]. Muscles of diaphragm and between ribs allow ventilation by breathing [✓] which increases concentration gradient for diffusion [✓]. Moist surface of alveoli allows gases to dissolve [✓]. Cartilage rings in trachea/bronchi keep airways open to reduce resistance to flow [✓].

05.2 Three marks from: They contain haemoglobin that combines with oxygen. [1] They have no nucleus so they can carry more oxygen. [1] They are small and flexible so that they can fit through narrow blood vessels. [1] Their shape gives them a large surface area (in relation to their volume) to increase diffusion. [1]   [3]

06   Drugs such as statins can be taken to reduce blood cholesterol levels [1] and prevent build-up of fatty material in the coronary arteries [1]. A stent can be inserted into an affected artery to keep it open.[1]   [2]

07.1 Lowest number of deaths in September. [1] Highest number of deaths in January. [1]   [2]

07.2 *This question should be marked with reference to the levels-based mark scheme on **page 374**.* Indicative content: [5]
Generally there is a higher number of cases in winter when temperatures are lowest [✓] but the lowest temperature month does not have highest number of deaths [✓]. No evidence is provided for viruses [✓] but the incidence of flu and colds is usually higher in winter so a link is possible [✓]. More evidence is needed to determine the relationship between temperature and asthma [✓] as a correlation between temperature and asthma deaths does not determine the cause [✓]. Other factors that have a seasonal pattern may be involved. [✓]

08 *This question should be marked with reference to the levels-based mark scheme on **page 374**.* Indicative content: [6]
Epidermis is thin/transparent allowing light to pass through [✓], Epidermis has stomata to allow the gas exchange required for photosynthesis [✓], guard cells in epidermis control gas exchange [✓], xylem supplies water required for photosynthesis [✓], xylem provides support holding leaf up to light [✓], palisade mesophyll and spongy mesophyll cells have chloroplasts to trap light energy [✓], shape of palisade mesophyll cells allows many cells to be packed in a small area to trap light efficiently [✓], spongy mesophyll has air spaces between cells to allow diffusion of gases / CO_2 and O_2 [✓].

09.1 Leaf area is $0.5 \times 0.5 = 0.25$ mm². [1] Stomata per mm² $= \frac{5}{0.25} = 20$. [1] [2]

09.2 Sunlight heats the upper surface more than the lower surface. [1] So having fewer stomata on the upper leaf reduces water loss. [1] [2]

Topic 3

01.1 Bacteria reproduce independently by binary fission. [1] Viruses rely on the infected cell for replication. [1] [2]

01.2 HIV infects and damages cells of the immune system / white blood cells. [1] These cells defend the body against infectious diseases and cancer. [1] Over time, more of these cells are destroyed and the person becomes ill with other infectious diseases or rare cancers. [1] [3]

01.3 Antiretroviral drugs. [1]

02.1 Protist. [1]

02.2 Virus. [1]

02.3 Through sexual contact / it is an STD. [1]

02.4 A course of antibiotics. [1]

03 The disease is caused by infection with the malaria protist. [1] This occurs when a person is bitten by an infected mosquito. [1] Mosquitoes become infected when they feed on the blood of an infected person. [1] [3]

04.1 $91,823 - 19,278 = 72,545 / 91,823 = 79\%$ [2]

04.2 The biggest reduction in cases was in the years immediately following introduction of the programme. [1] The number of cases did fluctuate before the programme [1]; but numbers were never as low as after the vaccination programme [1]. [2]

04.3 It is unlikely that every case of smallpox globally would have been reported. [1]

04.4 Smallpox is caused by a virus. Antibiotics will only kill bacteria. [1]

04.5 *This question should be marked with reference to the levels-based mark scheme on **page 374**.* [4]
The person would have memory cells for the pathogen. [✓] White blood cells / lymphocytes would recognise the pathogen [✓]. They would produce the correct / specific antibodies. [✓] Antibodies are produced very quickly. [✓] The pathogen would be destroyed before the person became ill. [✓]

05 Lining of the trachea and bronchi produces a sticky mucus that traps pathogens in inhaled air. [1] Cells lining the bronchi have cilia that sweep the mucus and pathogens out of the lungs [1] to the throat to be swallowed [1]. The passages of the nose are lined with hairs that trap pathogens on inhaled dust particles. [1] [4]

06.1 Aspirin [1] is a painkiller / extracted from willow [1]. [2]

06.2 *This question should be marked with reference to the levels-based mark scheme on **page 374**.* [6]
Pre-clinical tests / trials of the new drug on cells / tissues / live animals [✓] to test toxicity, dosage and efficacy. [✓] Clinical trials on healthy volunteers at very low doses [✓] to monitor for safety / side effects [✓]. Only if safe, do further clinical trials on patients [✓] to find the optimum dosage and test for efficacy [✓]. Carry out double blind trials using a placebo which does not contain the new drug. [✓] Patients should be randomly allocated to each group [✓] so no one knows who has placebo or the new drug. [✓] Data should be published after peer review to help prevent false claims. [✓]

1.1   Carbon dioxide + **water** → **glucose** + oxygen   [1]

1.2   In the chloroplasts.   [1]

1.3   Light intensity is limiting photosynthesis up to point A. [1] Increased light gives increased rate of photosynthesis because more energy is available for the reaction. [1] As light intensity increases above point A, the rate of photosynthesis levels off [1] because some other factor / temperature is now limiting [1].   [4]

1.4   A higher temperature increases the rate of photosynthesis [1] because it increases the rate of the collisions between enzymes and substrate molecules [1].   [2]

1.5   Raising the temperature would only increase growth if light levels are high enough. [1] At high light levels, raising the temperature by 10 °C could increase photosynthesis significantly / by 1/3. [1] Fuel for heating has an economic cost. [1] The farmer would need to maximise photosynthesis / growth while still making a profit. [1]   [4]

1.6   Length of time that the plants are illuminated for / wavelength of light.   [1]

2.1   Place the tube of pondweed in a water bath / beaker of water which absorbs heat. [1] The temperature of the water can be monitored and controlled. [1]   [2]

2.2   One from: The size of bubbles may vary. / Bubbles may be difficult to count if released quickly.   [1]

2.3   Changing the amount of pondweed would change the number of cells / chloroplasts available to carry out photosynthesis [1] which would affect the rate of photosynthesis. [1]   [2]

2.4   Use the same piece of pondweed / the same number of leaves / the same mass.   [1]

2.5 $25 - 14 = 11$ [1]. $\frac{11}{25} \times 100 = 44$ %. [1]   [2]

2.6   Prediction – 35 bubbles per minute. [1] Because the rate of photosynthesis has not increased from 200 to 250 Watts. [1] Some other factor is probably limiting the rate of photosynthesis. [1]   [3]

2.7   Glucose is produced in photosynthesis. [1] Cellulose is made from / is a polymer of glucose. [1]   [2]

3.1   Carbon dioxide.   [1]

3.2   Ethanol.   [1]

3.3   To keep out / exclude oxygen from the air [1] so that yeast will respire anaerobically [1].   [2]

3.4   Count the bubbles / measure gas volume produced [1] in a stated time using a stopwatch or timer [1].   [2]

3.5   **glucose** + oxygen → water + **carbon dioxide**   [1]

4.1   Heart rate is highest after running and lowest after standing. [1] Mean heart rate after running is approximately twice that for standing. [1] Stated difference e.g. heart rate after walking is 55 bpm more than standing and running is 75 bpm more than standing. [1]   [3]

4.2   *This question should be marked with reference to the levels-based mark scheme on* **page 374**.   [5]
      Choose two groups of people based on the amount of exercise they do. [✓] Use large groups of people (at least 5). [✓] Record heart rate for each person immediately after exercise. [✓] Calculate means and compare results for each group. [✓] Control variables for people - same gender / same age / no health issues/other factors such as smoking or diet. [✓] Other control variables same type of exercise / same time / same conditions for exercise. [✓]

4.3   Muscles had become fatigued [1] so they would no longer contract properly. [1]   [2]

4.4   Synthesis / formation of glycogen.   [1]

Higher Tier only:

4.5   Oxygen debt caused faster and greater volume of breathing. [1] Extra oxygen is needed to remove the lactic acid from anaerobic respiration [1] to convert lactic acid to glucose in the liver [1].   [2]

01.1 The lining of the uterus is breaking down. [1]

01.2 The lining of the uterus thickens [1] in preparation for implantation of an embryo [1]. [2]

02.1 X = sclera. [1] Y = iris.[1] Z = pupil. [1] [3]

02.2 Decrease in light intensity. [1]

02.3 Muscles [1] contract [1]. [2]

03 A root will always turn and grow downwards [1] towards the pull of gravity [1] whereas a shoot will always turn and grow upwards [1] towards light [1]. [4]

04.1 Contraceptive pill. [1]

04.2 Because a person might forget to take it. [1]

04.3 A diaphragm is a barrier (to sperm). [1]

05 This question should be marked with reference to the levels-based mark scheme on **page 374.** [6]
Receptors detect stimulus/pain [✓]. Receptor generates electrical impulses [✓]. Electrical impulses pass to spinal cord along sensory neurone [✓]. Sensory neurone connects to relay neurone [✓] via synapses [✓] and then to motor neurone [✓]. Electrical impulses cause muscles to contract [✓] moving arm [✓].

06 C: Lower side. [1]

7.1 0.01s. [1]

7.2 Hormones are carried in the blood [1] so they have to travel around the circulatory system to reach their target [1]. [2]

Higher Tier only:

08 If blood glucose concentration is high, the pancreas secretes insulin [1] which causes body cells to take up glucose and the liver to convert excess glucose into glycogen [1]. When blood glucose concentration is low, the pancreas secretes glucagon [1] which causes the liver to convert stored glycogen to glucose [1]. [4]

Higher Tier only:

09.1 10 days. [1]

09.2 After 12 days. [1]

01.1 The insect became trapped in the resin where conditions prevented it from decaying. [1]

01.2 The hard parts of plants [1] decay more slowly [1] that the softer parts so are more likely to fossilise. [2]

01.3 When rootlets decay [1], they leave a tunnel [1] in the soil which becomes fossilised. [2]

02.1 Carl Linnaeus to created the binomial system. [1] Carl Woese created the three-domain system. [1] [2]

02.2 The binomial system is a naming system [1] using the genus and the species [1] as a two-part name. [2]

02.3 Kingdom, phylum [1] family, genus [1]. [2]

03.1 A dominant allele. [1]

03.2 dd [1]

03.3 [1]

	D	d
d	Dd	dd
d	Dd	dd

03.4 Probability = 0.5. [1]

04.1 Circle around a single g on top row or left column. [1]

04.2 Gg [1]

04.3 25% [1]

05. Fertilisation → restores the normal number of chromosomes [1], mitosis → preserves the normal number of chromosomes [1], meiosis → halves the normal number of chromosomes [1]. [3]

06. Species C and D are more closely related to each other than to species B. [1]

Topic 7

01.1 Tertiary consumer → leopard seal. [1]   Primary consumer → krill. [1]   Apex predator → orca. [1] [3]

01.2 When silverfish eat krill, some of this food is not absorbed so it is egested [1]. Materials absorbed from the gut may enter the body cells of the silverfish. Some of these materials are respired [1] which breaks them down into carbon dioxide and water. The carbon dioxide and other waste materials from the body cells are excreted [1]. [3]

02.1 Transect. [1]

02.2 Species B only lives on the small sand dunes [1] whereas species C lives on both the large and the small sand dunes [1]. [2]

02.3 Species A [1] because its habitat at the top of the beach would be the first to be flooded [1] by the sea [1]. [2]

03 This question should be marked with reference to the levels-based mark scheme on **page 374**. [6]
Dead organisms contain carbon-containing molecules [✓]. Microorganisms decay these materials [✓]. Microorganisms release carbon dioxide [✓] to the atmosphere [✓] which is then available for plants to use for photosynthesis [✓]. Without microorganisms, carbon would remain locked in dead materials / would not be recycled for re-use by plants [✓].

04 Maximising food production requires the use of cheap compost to support plant growth and peat-based compost is cheap and readily available. [1] However, the peat bogs from which peat is extracted are rare habitats [1] and their destruction results in a loss of biodiversity [1]. After peat extraction, the decay of remaining peat releases carbon dioxide into the atmosphere, increasing carbon dioxide emissions. [1] [4]

05.1 Any one from: Same type of dead leaves, same mesh bag, buried at same depth, left for same time. [1]

05.2 $21.5 - 7.4 = 14.1$ rate $= 14.1/365$ [1] rate $= 3.9 \times 10^{-2}$ g day^{-1} [1] [2]

05.3 Decay slower under the trees because less sunlight reaching the ground would give cooler temperatures.[1] Less rain would reach the dead leaves under the trees, giving drier conditions.[1] [2]

06.1 $2360 - 1920 = 440$; $\frac{440}{2360} \times 100$ [1] $= 18.64 / 18.6$ % [1] [2]

06.2 This question should be marked with reference to the levels-based mark scheme on **page 374**. [6]
As the size of the deer population increases, the amount of food for wolves increases [✓] so the wolf population increases [✓]As the wolf population increases, they eat more deer [✓] so fewer deer are present to reproduce [✓], fewer offspring result [✓] and the size of the deer population decreased [✓].

07 Increased carbon dioxide emissions has caused more carbon dioxide to dissolve in the sea [1], reducing the pH. [1] Acidic seawater dissolves and softens the skeletons [1] of aquatic animals and can kill them. [3]

08.1 Use a random number table[1] to generate coordinates [1] at which to place each quadrat. [2]

08.2 Area of lawn is $10 \times 10 = 100$ m^2 [1] quadrat fits into lawn area $\frac{100}{(0.2 \times 0.2)} = 2500$ times. [1] Estimated population size $= 2500 \times 7 = 17\,500$. [1] [3]

08.3 Used more quadrats. [1]

08.4 Soil moisture level. [1] Soil pH. [1] [2]

09.1 Any two of: Placed the origin at the corner of the graph paper [1] increased the scale of both axes [1] limited the oxygen level axis to 5–20 mg dm^3. [1] [2]

09.2 1.5 mg dm^3 [2]

09.3 Biological material entering the stream [1] between 150 and 200 metres [1] and then bacteria in the water used oxygen for aerobic respiration [1] as the digested / decayed the material. [1] [4]

10. Introduce policies to reduce the amount of meat farmed [1] increased taxes on meat products [1] provide subsidies for farmers who convert land to arable with protein rich crops [1] change food labelling policies to encourage consumers to eat food made with protein rich crops [1] [4]

EXAMINATION PRACTICE ANSWERS

01 Any two from: [2]
- Compounds formed in chemical reactions, but mixtures not formed in chemical reactions.
- Compounds contain two or more elements chemically combined together, but substances in a mixture are not chemically combined together.
- Elements in compounds are present in fixed proportions, but substances in mixtures are in variable proportions.
- Chemical properties of elements in a compound are changed, but chemical properties of substances in mixtures are unchanged.
- Compounds are only separated into their elements by chemical reactions, mixtures can be separated by physical methods.

02 $2Fe_2O_3 + 3C \rightarrow 4Fe + 3CO_2$ Correctly balanced. [1]

03.1 It showed that atoms could be divided [1] leading to the plum pudding model [1] in which atoms are positive balls with negative electrons inside [1]. [2]

03.2 1 mark for each correct row to 3 marks: [3]

Name of subatomic particle	Relative charge	Relative mass
Neutron	0	1
Electron	−1	very small
Proton	+1	1

03.3 7 protons [1] 8 neutrons [1] 10 electrons [1]. [3]

03.4 2.8×10^{-10} m. [1]

04.1 relative atomic mass$= \dfrac{(69 \times 63) + (31 \times 65)}{(69 + 31)}$ [1] $= \dfrac{4347 + 2015}{100}$ [1] $= 63.6$ [1] 3 marks for correct answer without working. [3]

04.2 The chemical properties of an element are determined by the number of electrons [1]; the two isotopes have the same number of electrons / 29 electrons [1]. [2]

05.1 He left gaps for undiscovered elements [1]; he changed the order of some elements [1]. [2]

05.2 Group 5 [1] period 3 [1] (Allow top right for 1 mark only.) [2]

05.3 (Positively charged ions because) metals / transition metals are placed between groups 2 and 3 [1] and metals are elements that react to form positive ions / non-metal elements do not form positive ions [1]. [2]

06 Indicative content: [6]

Filtration
- to separate carbon particles from the ink
- as a residue

Fractional distillation
- of the filtrate
- to separate the propanol from the water

Chromatography
- of the filtrate / ink
- to separate the coloured substances from one another

Safety precautions relevant to the experiment with reasons
- eye protection because of solvent / hot liquid
- care with hot apparatus to avoid burns
- heat flask with an electrical heater / care with naked flames because ethanol is flammable

07.1 The atoms of Group 0 elements have full outer shells / stable arrangements of electrons [1], so they have little tendency to share electrons [1]. [2]

07.2 The boiling point increases going down the group / increases as the relative atomic mass increases. [1]

07.3 Answer in the range 3.65–3.90 g/m³ [1] because the relative atomic mass of krypton is about half-way between those of argon and xenon [1] and its density should be about half-way between their densities [1]. [3]

07.4 Oganesson is predicted to be a solid / a metal rather than a non-metal. [1]

08.1 $2Na(s) + 2H_2O(l) \rightarrow 2NaOH(aq) + H_2(g)$ 1 mark for correct formulae, 1 mark for correct balancing, 1 mark for state symbols. [3]

08.2 The outer electron of potassium is further from the nucleus [1] so the potassium nucleus has a weaker force of attraction for the outer electron [1] and the outer electron is lost more easily [1]. [3]

09.1 A halogen cannot displace itself from its salts. [1]

09.2 The order of decreasing reactivity is chlorine, bromine, iodine [1] because chlorine displaces bromine and iodine from their salts [1] and bromine displaces iodine but not chlorine [1]. [3]

09.3 $Br_2 + 2I^- \rightarrow 2Br^- + I_2$ [1] for correct formula, [1] for correctly balanced. [2]

Topic 2

01.1 MgS. [1]

01.2 Electrons are transferred from the outer shell of a magnesium atom to the outer shell of a sulfur atom [1] forming Mg^{2+} ions [1] and S^{2-} ions [1]. [3]

01.3 (Strong) electrostatic forces of attraction [1] between oppositely charged ions [1]. [2]

01.4 Two from: high melting point [1], high boiling point [1], does not conduct electricity when solid [1], conducts electricity when molten/dissolved [1]. [2]

02.1 4 shared pairs of electrons [1] no other dots or crosses shown [1]. [2]

02.2 Methane exists as small molecules [1] which have weak forces between them [1] and which need relatively little energy to overcome [1]. [3]

03.1 Both contain carbon atoms joined by covalent bonds [1] and both forms giant structures [1]. Each carbon atom forms four covalent bonds in diamond but only three covalent bonds in graphite [1]. Graphite forms layers (of hexagonal rings of atoms) but diamond does not [1]. [4]

03.2 They have delocalised electrons [1] which can move and carry charge through the structures [1]. [2]

Answer for 02.1

04.1 Atoms in metals are held together by strong (metallic) bonding [1] which takes a lot of energy to overcome [1]. [2]

04.2 Its atoms are in layers [1] that can move over one another [1]. [2]

04.3 The lithium atoms distort the layered structure of the metal [1] which makes it more difficult for layers of atoms to move over one another [1]. [2]

04.4 Energy is transferred [1] by delocalised electrons [1]. [2]

05.1 Simplest whole number ratio [1] of atoms of each element in a substance [1]. [2]

05.2 FeO. [1]

06.1 (Addition) polymer. [1]

06.2 A covalent bond. [1]

06.3 They have relatively strong intermolecular forces [1] which need a lot of energy to overcome [1]. [2]

07 Indicative content: [6]

Section X
- dodecanoic acid is in solid state
- particles in regular/lattice arrangement
- vibrate about fixed positions
- particles gain energy
- vibrate more rapidly

Section Y
- dodecanoic acid changes state from solid to liquid / melts
- particles gain energy
- energy used to overcome some of the molecular forces
- regular/lattice structure broken
- arrangement of particles becomes random

Section Z
- dodecanoic acid is in liquid state
- transition metals typically have high melting points
- particles gain energy
- particles move around each other
- particles move randomly
- movement increases

01.1 $(3 \times 24) + (2 \times 14) = 72 + 28 = 100$ (1 mark for correct answer without working).   [1]

01.2 $56 + (2 \times 14) + (2 \times 3 \times 16) = 56 + 28 + 96 = 180$ (1 mark for correct answer without working).   [1]

02   77.4% [2] Evidence of working [1] e.g. $\frac{3 \times 16}{62} \times 100$ [2]

03   (0.6 g of) carbon dioxide escaped from the boiling tube.   [1]

04   Volume = $\frac{50}{1000}$ = 0.05 dm³ [1] Mass = $8.0 \times 0.05 = 0.4$ g [1] Correct answer scores 2 marks.   [2]

Higher Tier only:

05    $44 \times 0.25 = 11$ g.   [1]

06   M_r of NaCl = $(23 + 35.5) = 58.5$ and M_r of Cl_2 = $(2 \times 35.5) = 71$ [1].
Amount of NaCl = $\frac{7.25}{58.5}$ = 0.124 mol [1]
From the balanced equation, mole ratio NaCl : Cl_2 = 2 : 1 so 0.062 mol of Cl_2 [1]
Mass of Cl_2 = $71 \times 0.062 = 4.40$ g [1]   [4]

07   M_r of Fe_2O_3 = $(2 \times 56) + (3 \times 16) = 160$ [1].
Amount of Mg = $\frac{960}{24}$ = 40 mol [1]   Amount of Fe_2O_3 = $\frac{2.0 \times 1000}{160}$ = 12.5 mol [1]
Amount of Mg needed = $3 \times 12.5 = 37.5$ mol which is less than the 40 mol added [1]   [4]

08.1   Mass of O_2 = $(6.85 - 6.69) = 0.16$ g.   [1]

08.2   Amount of PbO = $\frac{6.69}{223}$ = 0.03 mol.   Amount of O_2 = $\frac{0.16}{32}$ = 0.005 mol.   Amount of Pb_3O_4 = $\frac{6.85}{685}$ = 0.01 mol [1]
Simplest whole number ratio is 6 : 1 : 2 [1]
$6PbO + O_2 \rightarrow 2Pb_3O_4$ [1]   [3]

01.1   Z, X, Y.   [1]

01.2   A metal cannot displace itself from its compounds.   [1]

01.3   Magnesium was oxidised because it gained oxygen [1]. Copper(II) oxide was reduced because it lost oxygen [1].   [2]

02.1   Nitric acid.   [1]

02.2   Copper does not react with dilute acids.   [1]

02.3   Two from: copper(II) oxide [1], copper(II) hydroxide [1], copper(II) carbonate [1].   [2]

03.1   The pH of hydrochloric acid was less than 7 but the pH of the sodium hydroxide was more than 7.   [1]

03.2 $H^+(aq) + OH^-(aq) \rightarrow H_2O(l)$ 1 mark for correct reactants correctly balanced, 1 mark for state symbols.   [2]

04   Aluminium sulfide: Al_2S_3 [1], sodium phosphate Na_3PO_4 [1].   [2]

05.1   Aluminium is more reactive than carbon.   [1]

05.2   Aluminium oxide is insoluble in water so it must be melted [1]. The molten mixture has a much lower melting point than aluminium oxide alone [1] which reduces the amount of energy needed for electrolysis to happen [1].   [3]

05.3   The positive electrode is made from carbon [1]. Oxygen is given off at the positive electrode [1] which reacts with the carbon (to form carbon dioxide) and wears it away [1].   [3]

06.1   It contains zinc ions and chloride ions [1] which are free to move [1].   [2]

06.2   Cathode: zinc [1] Anode: chlorine [1] .   [2]

06.3   Chlorine is still given off at the anode [1] but hydrogen is given off [1] (instead of zinc).   [2]

Higher Tier only:

07.1 $Cu^{2+} + 2e^- \rightarrow Cu$ [1]

07.2 $2Cl^- \rightarrow Cl_2 + 2e^-$ [1]

08.1   (Magnesium) because it loses electrons [1] $Mg \rightarrow Mg^{2+} + 2e^-$ [1].   [2]

08.2   Reduction and oxidation reactions happen at the same time.   [1]

09.1   It is only partially ionised/dissociated [1] in aqueous solution [1].   [2]

09.2   pH 5.   [1]

01 A reaction which transfers energy to the surroundings [1] causing the temperature of the surroundings to increase [1]. [2]

02.1 The process is endothermic[1] because the temperature goes down [1]. [2]
02.2 Named application that needs low temperatures, e.g. sports injury pack, ice pack for food. [1]

03 Indicative content: [6]
- X is the activation energy
- X is the energy needed for the reaction to happen
- X is the energy needed to break bonds in $CaCO_3$
- X is an endothermic process
- X is 2191 kJ/mol
- Y is the overall energy change
- Y is 179 kJ/mol
- Y is positive, so the reaction is endothermic
- more energy needed to break bonds in reactants than is released when bonds form in the products
- Z is the energy released when bonds form in CaO and CO_2
- Z is an exothermic process
- Z = (2191 − 179) = 2012 kJ/mol

Higher Tier only:

04 Energy in to break bonds: (H–H) + 151[1]   Energy out when bonds form: (2 × 298) = 596 [1].
Overall energy change = (H–H) + 151 − 596 = −9 [1] (H–H) = −9 − 151 + 596 = 436 kJ/mol [1]. [4]

01 Increase the surface area / crush the lump to make a powder. [1]
02.1 One from: use a gas syringe [1] / use an inverted measuring cylinder. [1] [1]
02.2 Mean rate = $\frac{48}{60}$ [1]=0.8 [1] cm³/s.[1] [3]
02.3 Line B [1] because calcium was the limiting reactant / hydrochloric acid was in excess [1], 1.5 times more calcium was used so 1.5 times more hydrogen would be produced [1]. [3]
02.4 **Higher Tier only:** Evidence of drawing a tangent at 40 s on line A and reading off values [1]. Calculations, for example:
Rate of reaction = $\frac{(100 \text{ cm}^3 - 70 \text{ cm}^3)}{(55 \text{ s} - 15 \text{ s})} = \frac{30 \text{ cm}^3}{40 \text{ s}}$ [1] = 0.75 [1] (cm³/s) [3]

03 The minimum amount of energy that particles need to react. [1]
04.1 The concentration of reactants decreases [1] so the frequency of collisions decreases [1]. [2]
04.2 Reactant particles gain energy / move faster [1], the frequency of collisions increases [1], a greater proportion of collisions are successful / a greater proportion of reactant particles have the activation energy [1]. [3]

05.1 Vanadium(V) oxide [1] because it is present in the reaction mixture but does not appear in the balanced equation [1]. [2]
05.2 Catalysts increase the rate of reaction [1] by providing a reaction pathway with a lower activation energy [1]. [2]

06.1 It shows that the reaction is reversible. [1]
06.2 The colour would change from blue to pink. [1]
06.3 Heat it up. [1]
06.4 The reverse reaction is endothermic [1] and transfers the same amount of energy as the forward reaction [1]. [2]

07.1 The reacting substances must not escape for the reaction to reach equilibrium [1] but reactants/products can escape from an open boiling test tube [1]. [2]
07.2 It will be 0.50 g/s (the same) [1] because at equilibrium the rate of the forward reaction is exactly the same as the rate of the reverse reaction [1]. [2]

8   Indicative content: [6]

Variables to control
- mass of magnesium ribbon
- volume of hydrochloric acid
- surface area of magnesium ribbon
- temperature of reaction mixture

Method
- connect a bung and delivery tube to a gas syringe / upturned measuring cylinder in a trough of water
- use measuring cylinder
- to place a known volume of hydrochloric acid in a conical flask
- add a known mass of magnesium ribbon
- immediately start stopwatch and connect the flask to the delivery tube
- record the volume of gas at regular intervals

Analysis
- calculate the mean rates of reaction at each concentration of hydrochloric acid
- plot a graph of mean rate against concentration

Higher Tier only:

09   They increase the rate of reaction / reduce the time taken to reach equilibrium. [1]

10.1 (It increases the equilibrium yield) because there are fewer molecules of gas on the product side of the equation [1]
      so the equilibrium position moves to the right [1]. [2]

10.2 (It decreases the equilibrium yield) because increasing the temperature shifts the equilibrium position in the
      direction of the endothermic change [1] and this is away from dichloroethane [1]. [2]

11   The colour of the mixture will turn dark red [1] because the equilibrium position will move to the right [1] to form more
      products / away from the additional reacting substance [1]. [3]

Topic 7

01.1 It contains carbon and hydrogen only. [1]

01.2 Its name ends in ane [1] and its formula fits the general formula C_nH_{2n+2} [1] [2]

02   1 mark for correct number of each atom [1], 1 mark for all single bonds [1]: [2]

03.1 Two from: liquefied petroleum gases / LPG [1], petrol [1], diesel oil [1], kerosene [1],
      heavy fuel oil [1].
      [2]

03.2 Crude oil is heated to boil/evaporate it [1]; vapours pass up the fractionating column [1], they cool and condense [1]
      at different heights/temperatures [1]. [4]

04.1 It is a measure of how easily a substance flows. [1]
04.2 (As the molecules become larger) the boiling point increases [1] and the flammability decreases [1]. [2]

05    $C_5H_{12} + 8O_2 \rightarrow 5CO_2 + 6H_2O$. 1 mark for correct formulae, 1 mark for correct balancing. [2]

06   Mix with bromine water [1], bromine water changes from orange-brown to colourless with hexene but stays
      orange-brown / no change with hexane [1]. [2]

07.1 One from: to match supply of fractions with demand for them [1], to produce more fuels [1], to produce starting
      materials for polymers [1], to produce starting materials for other substances / petrochemicals [1]. [1]
07.2 Steam cracking needs higher temperatures [1] and higher pressures [1] than catalytic cracking. [2]
07.3 **2**$C_8H_{18} \rightarrow C_7H_{16} + C_3H_8 + $**3**$C_2H_4$. 1 mark for each balancing number. [2]

01   A **glowing** splint [1] held in the gas relights if the gas is oxygen [1].   [2]

02   Shake the gas with limewater / bubble the gas through limewater [1]. The limewater turns cloudy white / milky [1].   [2]

03   Collect a sample of each gas in a test tube [1]. (For the gas from the cathode) hold a burning splint near the test tube [1]. It ignites the gas with a pop sound if the gas is hydrogen [1]. (For the gas from the anode) hold **damp** litmus paper in the gas [1]. It turns white if the gas is chlorine [1].   [4]

04.1   In chemistry a pure substance only contains one element or compound / is not mixed with any other substance [1] In everyday use, a pure substance has nothing added to it / is in its natural state [1].   [2]

04.2   The components of an alloy are in measured quantities.   [1]

05   $R_f = \dfrac{74\ mm}{120\ mm}$ [1] = 0.6166 [1] = 0.62 to 2 significant figures [1].   [3]

06   Measure its melting point [1], if the dodecanoic acid is pure it should melt at 43.8 °C / it will not melt at 43.8 °C if it is not pure [1].   [2]

07   Fertilisers are a mixture of components [1] in carefully measured quantities [1] each with a purpose / to make sure that the fertiliser has the required properties [1].   [3]

08   There is a mobile phase [1] and a stationary phase [1]; different substances distribute between the two phases by different amounts [1] so they move up the paper at different rates [1].   [4]

09   Indicative content: [6]

Apparatus
• chromatography paper
• beaker or other suitable container
• pencil and ruler
• propanol

Method
• draw a pencil line near the bottom of the paper
• write a spot of pen ink on the line
• place the paper in the solvent / propanone
• make sure the solvent is below the pencil line
• allow the solvent to move through the paper
• remove and dry the paper before the solvent reaches the top

Analysis
• examine the results
• ink containing one coloured substance produces one spot
• ink containing more than one coloured substance produces more than one spot

01.1   Both atmospheres contain carbon dioxide [1] and nitrogen [1].   [2]

01.2   One from: Venus has more carbon dioxide [1] less nitrogen [1] less oxygen.   [1]

02   Nitrogen four-fifths / 80% / 78% [1]; oxygen one-fifth / 20% / 21% [1].   [2]

03   Volcanoes gave off water vapour [1] which cooled and condensed [1].   [2]

04   carbon dioxide + water → glucose + oxygen. Reactants correct [1], products correct [1].   [2]

05   Photosynthesis [1] reduced the proportion of carbon dioxide [1] and increased the proportion of oxygen [1].   [3]

06.1   One from: methane [1], water vapour [1].   [1]

06.2   One from: use of hydrocarbon/fossil fuels [1], cement manufacture [1].   [1]

06.3   Greenhouse gases in the atmosphere let radiation from the Sun reach the Earth's surface [1], absorb radiation emitted by the Earth's surface [1] let short wavelength through but absorb long wavelength radiation [1].   [3]

07   Two from: more heat waves / fewer cold periods [1], more droughts / more rainfall [1], rising sea levels / coastal flooding [1], Melting ice caps / melting glaciers [1], Changes in seasons / farming practices [1].   [2]

08.1 The total amount of carbon dioxide (and other greenhouse gases) given off [1] over the life cycle of a product / service / event [1]. [2]

08.2 One sensible way to reduce individual carbon footprint, e.g. walk rather than take the car [1], turn off the lights when leaving a room [1], turn down heating [1]. [1]

09 One from: carbon dioxide [1], carbon monoxide [1], carbon [1]. [1]

10 Carbon monoxide [1], carbon [1], water (vapour) [1]. [3]

11 It is odourless [1] and colourless [1]. [2]

12 Carbon / pollution particulates. [1]

13 Indicative content: [6]

Apparatus	**Method**	**Analysis**
• fuels may contain sulfur	• in hot engines/furnaces	• sulfur dioxide and oxides cause acid rain
• sulfur reacts with oxygen	• nitrogen and oxygen	• effect of acid rain, e.g. damage to trees / rivers / aquatic life / buildings
• when the fuel is used	• in the air	• sulfur dioxide and oxides cause respiratory problems
• balanced equation, e.g. $S + O_2 \rightarrow SO_2$	• react together	
	• balanced equation, e.g. $N_2 + 2O_2 \rightarrow 2NO_2$	

Topic 10

01.1 It removes insoluble solids from the water. [1]
01.2 It kills microbes / sterilises the water [1] which prevents disease caused by drinking water [1]. [2]
01.3 One from: distillation / simple distillation [1], reverse osmosis [1]. [1]
01.4 One from: treating sea water uses large amounts of energy [1], the UK has a lot of rainfall / fresh water [1]. [1]

02.1 The waste water is screened / grit is removed [1]. [2]
02.2 One from: bacteria carry out anaerobic digestion [1] of sewage sludge [1]; bacteria carry out aerobic biological treatment [1] of effluent [1]. [2]

03.1 Aluminium ore is no longer being made / is made very slowly [1] so it will run out if we keep using it [1]. [2]
03.2 Two from: less energy is used [1], reduces pollution (e.g. from mining) [1], landfill sites for waste material are running out [1], recycled aluminium is cheaper [1], recycling is more sustainable [1]. [2]

04.1 Meeting the needs of people today [1] without compromising the ability of people in the future to meet their own needs [1]. [2]
04.2 Two from: Polyester comes from a limited resource / crude oil is a limited resource [1], polyester is not biodegradable [1], cotton is a renewable resource [1], mixing cotton into the shirt reduces the amount of polyester needed [1]. [2]

05 Indicative content: [6]

Raw materials
• wood is a renewable resource
• plastics are made from crude oil
• crude oil is a limited resource
• cutting down trees harms the environment
• fewer trees can absorb less carbon dioxide
• drilling for oil harms the environment

Manufacturing
• paper bags need more energy
• more energy may cause more pollution
• carbon dioxide may be a pollutant
• carbon dioxide is a greenhouse gas

Transport
• paper bags are heavier
• more fuel needed to transport paper bags
• larger lorries may be needed

Use
• paper bags may break more easily
• more paper bags may be needed
• paper bags cannot easily be reused

Disposal
• paper bags decompose
• plastic bags do not decompose
• both types of bag may be recycled to make other products

06.1 Plants absorb gold / gold compounds [1], plants are harvested [1], plants are burned to produce ash containing gold / gold compounds [1]. [3]
06.2 Two from: less damage to the environment / less waste produced [1], less expensive / less energy needed [1], can extract smaller amounts of gold / gold from less concentrated sources [1], burning plants release useful energy [1], plants absorb carbon dioxide from the atmosphere as they grow [1]. [2]

EXAMINATION PRACTICE ANSWERS

Topic 1

01    $E_k = \frac{1}{2}mv^2$ [1]

02.   0.05m [1]

03. $E_k = \frac{1}{2}mv^2 = 0.5 \times 0.4 \times 8^2$ [1] = 12.8 [1] J [1] [3]

04.1 $E_p = mgh = 85 \times 9.8 \times 2$ [1] = 1666 J [1] [2]

04.2 As the cyclist falls, energy is transferred from the gravitational potential energy store to the kinetic energy store, [1] and as the cyclist hits the ground the kinetic energy is transferred to the thermal energy store of the air in the tyre, and the surroundings and/or elastic potential energy store of the tyres [1]. [2]

04.3 **Higher Tier only:** $E_p = E_k$ so $E_k = 0.5 \times 85 \times v^2 = 680$ J [1] $v = \sqrt{[680 / (0.5 \times 85)]}$ [1] = 4.0 m/s^2 [1] [3]

05   Trend: The number of diesel cars sales went down 2018-19. [1]

      Reason: Any one from: diesel is a known air pollutant / some places are banning diesel cars from city centres / Government targets to ban production of new fossil fuel cars by 2030. [1]

      Trend: The number of electric car sales went up 2018-19. [1]

      Reason: Any one from: becoming more available / prices are starting to come down / they are becoming cheaper to run / increase in electric car charging points ; no car tax / people are encouraged to buy them as they contribute less to carbon dioxide emissions. [1] [4]

06   Advantage: (any one from) does not emit carbon dioxide / sulfur dioxide / contribute to global warming. [1]

      Disadvantage: (any one from) hugely expensive to build/decommission the power station; disposal of radioactive waste; risk of catastrophic accident. [1] [2]

07   Air is a better insulator than brick [1] so having a layer of air reduces the rate that thermal energy can transfer through the walls [1] increasing the thickness of the wall by having two layers also reduces the rate of energy transfer [1]. [3]

08   Efficiency = useful power output / total power input; 0.45 = useful power output / 30; useful power output = 0.45 × 30 [1] = 13.5W [1] $E = Pt = 13.5 \times 20 = 270$ J [1] [3]

09   The same amount of energy needs to be supplied by both heaters (to warm up the room) [1] and the more powerful heater transfers that energy more quickly [1]. [2]

10.1 The amount of energy in joules needed to raise the temperature of 1 kg of water by 1°C. [1]

10.2 The time the heater is on. [1]

10.3 Thermal energy from the heater was also transferred to heat up the air/beaker/surroundings [1] so reduce this loss by increasing the insulation of the container [1]. [2]

10.4 **Higher Tier only:** 250 g = 0.25 kg [1]; $\Delta E = mc\Delta\theta$; 2100 = 0.25 × 4200 × $\Delta\theta$ [1]; $\Delta\theta$ = 2100 / (0.25 × 4200) [1] = 2 °C [1] [4]

Topic 2

01.1 ammeter, voltmeter. [2]

01.2

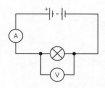

      All symbols correct [1] battery, ammeter and lamp in series [1] voltmeter across lamp [1]. [3]

01.3 $V = IR$; 1.5 = 0.5 × R [1] R = 1.5 / 0.5 [1] = 3 Ω [1] [3]

01.4 250 mA = 0.25 A [1]; $Q = It$; 3.0 × 10^4 = 0.25 × t [1]; t = 3.0 × 10^4 / 0.25 [1] = 1.2 × 10^5 s [1] [4]

02.1 Variable resistor. [1]

02.2 Changes length of wire connected. [1] Longer wire, higher resistance. [1] [2]

02.3 LDR 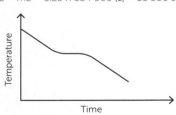 [1]

02.4 As the light intensity increases, the resistance decreases. [1] This means the current changes and circuits can be switched on and off. [1] [2]

03.1 $R_r = 2 \times 2.0 = 4.0\ \Omega$ [1] $R_{total} = 2.0 + 4.0 = 6.0\ \Omega$ [2]

03.2 $V = IR$; $12.0 = I \times 6.0$ [1]; $I = 12 / 6 = 2\ A$ [2]

03.3 Ratio of pd resistor : pd lamp is 2:1. (a) resistor 8 V [1] (b) lamp 4 V [1] (*Could also use V = I R*) [1] / [1]

03.4 Graph C. [1]

04 Direct pd is in one direction only. [1] Alternating pd changes direction.[1] [2]

05 The appliance also becomes live. [1] A person touching it would receive an electric shock because the current goes through them. [1] [2]

06 Step-up transformers increase the pd in the transmission cables [1] and this reduces the current so less energy as dissipated as wasted heat [1]. A step down transformer then reduces the pd back to a safe value for domestic use. [1] [3]

07.1 $P = VI$; $2900 = 230 \times I$ [1] $I = 2900 / 230$ [1] $= 12.6\ A$ [1] [3]

07.2 The kettle with the higher power rating has a higher current. [1]

07.3 power = (current)^2 × resistance [1]

08.1 $E = Pt$; $45\,000 = P \times 70$ [1] $P = 45\,000 / 70$ [1] $= 643\ W$ [1] [3]

08.2 $E = QV$; $45\,000 = Q \times 230$ [1] $Q = 45\,000 / 230$ [1] $= 196\ C$ [1] [3]

Topic 3

01.1 0.7364 kg [1]

01.2 $\rho = m / V = 763.4/50.0$ [1] $= 15.3\ g/cm^3$ [1] [2]

01.3 $V = 0.02 \times 0.02 \times 0.02$ [1] $= 8 \times 10^{-6}\ m^3$ [1] [2]

01.4 $\rho = m / V$; 5400 mass $/ 8 \times 10^{-6}$ [1] mass $= 5400 \times 8 \times 10^{-6}$ [1] $= 4.32 \times 10^{-2}\ kg$ [1] $= 43.2\ g$ [1] [4]

02.1 Energy is transferred from the liquid. [1] The particles lose energy [1] and stop moving freely so just vibrate around fixed positions as a solid. [1] [3]

02.2 The total kinetic and potential energy that the particles in a substance have. [1]

03.1 The amount of energy needed to melt 1kg of a solid to a liquid at a constant temperature. [1]

03.2 $E = mL = 0.25 \times 334\,000$ [1] $= 83\,500\ J$ [1] [2]

03.3

Labelled axes [1], correct shape cooling curve [1]. [2]

04 **Higher only:**
To heat the solid gold to melting point: $\Delta E = mc\Delta\theta = 0.5 \times 130 \times 64$ [1] $= 4160\ J$ [1]
To melt the gold: $E = 0.5 \times 63\,000$ [1] $= 31\,500\ J$ [1]
Total energy is $4160 + 31\,500 = 35660\ J$ [1] [5]

05 Increases. [1]

01.1 0.04 nm [1]

01.2 $2 \times 4.0 \times 10^{-11}$ m $= 8.0 \times 10^{-11}$ m [1]

02.1 9 protons [1] 10 neutrons [1] 9 electrons[1] [3]

02.2 Atomic number 13 [1]   Mass number 27 [1] [2]

02.3 They have different numbers of neutrons. [1]

03.1 The atom is a ball of positive charge [1] with negative electrons embedded in it. [1] [2]

03.2 Some alpha particles repelled: the nucleus is positively charged [1]; some alpha particles were reflected back: the atomic mass is concentrated in the centre [1]; most alpha particles passed straight through: an atom is mainly empty space [1]. [3]

04.1 240 / 2 = 120; 120 / 2 = 60 so 2 half-lives is 14 years [1] so one half life is 7 years [1]. [2]

04.2 1 hour is 2 half-lives. $\frac{1}{2} \times \frac{1}{2} = \frac{1}{4}$ kg [1] ratio = 4.0 : 16.0 = 1:4 [1]. [2]

04.3 The longer the half-life is, the longer the radioactive decay will continue at a higher level [1] so will therefore be more of a hazard for longer. [1] [2]

05.1 The source for the smoke detector needs to be ionising [✔] and safe [✔]. Nickel-63 is a beta source which is less ionising than alpha [✔] and also passes through skin so is unsafe. [✔] The gamma source cobalt-60 is only weakly ionising [✔] and penetrates a long way so is unsafe. [✔] Alpha radiation is strongly ionising [✔]. It does not penetrate outside the smoke alarm [V]. So, the best choice is one of the two alpha sources [✔] The one with the longest half-life [✔] will be more appropriate as it will last longer [✔]. Half of the thorium-228 is decayed after just under 2 years [✔] and will need to be changed too often [✔]. So the most appropriate source for the smoke alarm is Americium-241 [✔] as it is an ionising source, has a long half life and can be used safely in the home as long as it is not ingested. [✔]

   This extended response question should be marked holistically in accordance with the levels-based mark scheme on page 374. [6]

05.2 $^{241}_{95}$Am $\longrightarrow$ $^{237}_{93}$Np $+$ $^{4}_{2}$He [1] [3]

05.3 $^{63}_{28}$Ni $\longrightarrow$ $^{63}_{29}$Cu $+$ $^{0}_{-1}$e [1] [3]

01.1 Displacement. [1]

01.2 F_y is in the opposite direction to F_x [1] and has twice the magnitude F_x. [1] [2]

01.3 There is resultant force acting to the right [1] so the box will accelerate to the right. [1] [2]

01.4 Momentum is mass multiplied by velocity. [1] Velocity a vector quantity so momentum will have both magnitude and direction/also be a vector quantity. [1] [2]

01.5 Any **two** from: magnetic, electrostatic, gravitational. [2]

01.6 Weight is the force acting on a mass due to gravity. [1]

01.7 $W = m \times g = 1.6 \times 9.8$ [1] $= 15.7$ N [1] [2]

01.8 Newtonmeter. [1]

02.1 Resultant force = 2 + 3 − 5 [1] = 0 N [1] [2]

02.2 Forces labelled correctly. [1] Right arrow longer than left arrow. [1] Weight and reaction arrows equal. [1] [3]

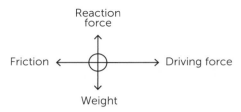

02.3 Horizontal force 2.2 N [1] vertical force 2 N [1] [2]

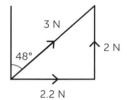

02.4 Scale drawing of original forces [1] resultant force 2150 N [1] at 22° to the horizontal [1]. [3]

03.1 Work done = force × distance. [1]

03.2 ($W = F_s$) $E_p = mgh$ [1]
Work done is against gravity so $F = mg$ [1]
The distance moved along the line of action of the force, so $s = h$ [1]
so work done (= F_s)= $m \times g \times h = E_p$ gained by the mass. [1] [4]

03.3 $W = F_s$; 6600 = 1200 × s [1]; s = 6600/1200 [1] = 5.5 m [1] [3]

03.4 500 Nm = 500 J [1] = 0.5kJ [1] [2]

04.1 Force = spring constant × extension. [1]

04.2 $F = ke$; 2.1 = k × 0.07 [1]; k = 2.1/0.07 [1] = 30 [1] N/m [1] [4]

04.3 3.5 cm = 0.035 m [1]; E_e= 0.5ke^2 = 0.5 × 500 × 0.035 × 0.035 [1] = 0.31 J [1] [3]

04.4 e.g. 2.4 − 0.2 = 2.2 cm [1]

04.5 5 Correctly plotted points.[2] Line of best fit.[1] [3]

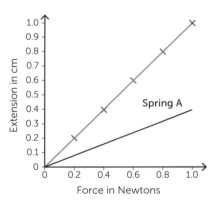

04.6 (k = 1 / gradient). gradient = 0.4/10 [1] k = 25 N/cm [1] [2]

05.1   Displacement is 5.5 km [1] at 45° to the horizontal (or north east) [1]   [2]

05.2   3.6 km = 3600 m [1]; $s = vt$; 3600 = v × 150 [1]; v = 3600/150 [1] = 24 m/s [1].   [4]

05.3   Graph drawn with distance on vertical axis and time on the horizontal axis, with sensible scale [1].
       Points plotted correctly. [1] Point joined with a straight line [1].   [3]

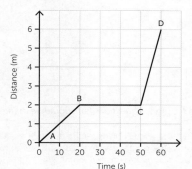

05.4 $s = vt$; 6 = v × 60 [1]; v = 6/60 [1] = 0.1 m/s [1]   [3]

05.5   [1]

06.1 $a = \Delta v/t$; 1.5 = Δv /8 [1]; Δv = 1.5 × 8 [1] = 12 m/s [1]   [3]

06.2 $v^2 - u^2 = 2as$; $9^2 - 5^2$ = 2 × 0.5 × s [1]; s = (81 − 25) / 1.0 [1] = 56 m [1]   [3]

06.3   A is travelling at a constant velocity with zero acceleration. [1]
       The velocity of B is increasing, with constant acceleration. [1]
       The velocity of C is also increasing, but with a greater constant acceleration than B. [1]   [3]

06.4   acceleration = gradient = 4/20 [1] = 0.2 m/s² [1]   [2]

06.5   distance = area under graph = (0.5 × 20 × 4) + ((35 − 20) × 4) = 40 + 60 [1] = 100 m [1]   [2]

07.1   Inertia.   [1]

07.2   A resultant force must act on the object.   [1]

07.3   The object will continue to move to the right at a constant speed.[1] There will be a resultant force upwards so the
       object will also accelerate upwards.[1]   [2]

07.4 $F = ma$; 250 = m × 12.5 [1]; m = 250/12.5 [1] = 20 kg [1].   [3]

07.5   There is a resultant force (of 20 N) to the right [1], so the box will accelerate to the right [1].   [2]

07.6   The force on the trolley is provided by the hanging masses. [1] As the effect of mass is being investigated,
       the force must be kept constant. [1]   [2]

07.7   Any **one** from: protect people/floor from falling masses; take care with any electrical equipment used.   [1]

07.8   Two objects exert equal and opposite forces on each other when they interact.   [1]

08.1   Stopping distance = thinking distance + braking distance.   [1]

08.2   Reaction time increases.   [1]

08.3   Any **two** from: wet/icy road surface; worn tyres ; worn brakes.   [2]

08.4   A large deceleration means a greater braking force [1] so more work is done transferring the kinetic energy of the
       vehicle to thermal energy of the brakes [1] so the temperature of the brakes rises rapidly with no time to cool [1]. [3]

09.1 $p = mv$ = 3.5 × 0.4 [1] = 1.4 kg m/s.[1]   [2]

09.2   For a closed system, the total momentum before an event equals the total momentum after an event.   [1]

01.1 [3]

01.2 $T = 1/f$; 0.4 = 1/f [1]; f = 1/0.4 [1] = 2.5 [1] Hz[1]   [4]

01.3 $v = f\lambda$; 160 = 8 × λ [1]; λ = 160/8 [1] = 20 m [1]   [3]

01.4   ***Suggested method: This is an extended response question that should be marked in accordance with the levels based mark scheme on page 374.***

      The students stand together in front of a large wall.[✓] Measure the distance from the wall [✓] using a long tape measure [✓] and record the distance in metres. [✓] Make the distance as large as possible [✓] to increase the time taken for the sound to travel. [✓] One student makes a sound by banging two blocks together. [✓] The other student starts timing with a stopwatch as the sound is made [✓] stops timing when the echo is heard and records the time in seconds. [✓] Repeat this several times.[✓] Remove any anomalies [✓] and calculate a mean time.[✓] Swap the students around and repeat the experiment [✓] to reduce the effect of their different reaction times. [✓] The distance the sound has travelled is two times the distance to the wall. [✓] Calculate the speed of sound using speed = distance/time. [✓]   [6]

02.1   wavelength = length of tank/number of waves = 0.56/28 [1] = 0.02 m. [1]   [2]

02.2   (2.9 + 3.3 + 3.2 + X)/4 = 3.2[1]; X = (3.2 × 4) – (2.9 + 3.3 + 3.2) = 3.4 (Hz). [1]   [2]

02.3 $v = f\lambda$ = 3.2 × 0.075 [1] = 0.24 m/s. [1]   [2]

02.4   A video can be used to determine both frequency and wavelength [1] as long as the timer is also recorded [1]. A photograph can help determine wavelength but not frequency. [1]   [3]

02.5   λ = 2 × 0.75/5 [1] = 0.3 m [1]
      $v = f\lambda$ = 12 × 0.3 [1] = 3.6 m/s [1]   [4]

03.1   Any **two** from: Transverse waves/ transfer energy from a source to an absorber / can travel through a vacuum / travel at the same speed through air or a vacuum.   [2]

03.2   Ray labelled refracted ray with arrow as shown [1] angle of refraction labelled which is less than the angle of incidence [1].   [2]

Air   Water

Refracted ray

Normal

Angle of incidence

Angle of refraction

Incident ray

03.3   **Higher Tier only:** The waves slow down as they move from deep to shallow water (shorter wavelength) [1]. As the wavefront touches the boundary the speed of that part slows down whilst the rest of the wavefront continues at the faster speed until it also meets the boundary and slows down. [1] This causes the wavefront to change direction and travel through the shallow water at a different angle. [1]   [3]

03.4   Independent variable: the type of surface. [1] Dependent variable: the amount of infrared radiation emitted. [1]   [2]

03.5   Any one from: the distance of the IR detector from each surface / the temperature of the surface / the time taken to measure the amount of radiation emitted.   [1]

03.6   Gamma rays.   [1]

03.7   **Higher Tier only:** An electric field is created around a wire when a current flows through it.[1] When the current changes, the electric field changes.[1] The changing current produces radio waves in the electric field.[1].   [3]

03.8   Any two from: Electrical heaters / cooking food / infrared cameras.   [2]

03.9   **Higher Tier only:** Radio waves are not easily absorbed in the atmosphere so can travel long distances. [1] Their energy is not high enough to cause any harm or other effects.[1]   [2]

01.1 The hanging magnet will move away when the south pole is brought towards the south pole [1] and move towards it when the north pole is brought towards the south [1]. [2]

01.2 Any **two** from: cobalt, iron, nickel, steel. [2]

01.3 A permanent magnet produces its own magnetic field whereas an induced magnet becomes a magnet when it is placed in a magnetic field.[1] A permanent magnet can attract and repel whereas an induced magnet can only attract. [1] [2]

01.4 At least 2 loops above and below equally spaced [1] field lines closer together at both poles [1] arrows from N to S [1]. [3]

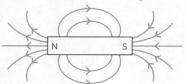

02.1 Concentric circles [1] clockwise direction [1]. [2]

02.2 At least 2 loops above and below equally spaced [1] field lines closer together at both poles [1] field direction labelled [1]. [3]

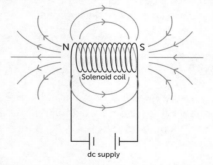

02.3 2 (pins) (10 coils, 2 cells, no core) [1]

02.4 The strength of the electromagnet increases when: there are more coils [1] there is an iron core [1] there are more cells/is a greater current/voltage [1]. [3]

Higher Tier only:

03.1 $F = BIl = 0.03 \times 1.2 \times 1.8$ [1] $= 0.065$ N [1]. [2]

03.2 Upwards arrow from conductor labelled F. [1]

03.3 The current travels in opposite directions on each side of the coil of wire [1] so the force acting in each side will be in opposite directions producing a moment on the coil. [1] [2]

LEVELS BASED MARK SCHEME FOR EXTENDED RESPONSE QUESTIONS

What are extended response questions?

Extended response questions are worth 4, 5 or 6 marks. These questions are likely to have command words such as 'compare', describe', 'explain' or 'evaluate'. You need to write in continuous **prose** when you answer one of these questions. This means you must write in full sentences, not bullet points, and organise your answer into paragraphs.

You may need to bring together skills, knowledge and understanding from two or more areas of the specification. To gain full marks, your answer needs to be logically organised, and ideas linked to give a sustained line of reasoning.

Some extended response questions may involve calculations. These need two or more steps that must be done in the right order. These questions will include the command words 'calculate' or 'determine'.

Marking

Written answers are marked using 'levels of response' mark schemes. Examiners look for relevant points (indicative content) and also use a best fit approach. This is based on your answer's overall quality and its fit to descriptors for each level. Extended response calculations give marks for each step shown.

Example level descriptors

Level descriptors vary, depending on the question being asked. Level 3 is the highest level and Level 1 is the lowest level. No marks are awarded for an answer with no relevant content. The table gives examples of the typical features that examiners look for.

Level	Marks	Descriptors for a method	Descriptors for an evaluation
3	5–6	The method would lead to a valid outcome. All the key steps are given, and they are ordered in a logical way.	The answer is detailed and clear. It includes a range of relevant points that are linked logically. The answer uses relevant data that may be given in the question. A conclusion is made that matches the reasoning in the answer.
2	3–4	The method might not lead to a valid outcome. Most of the key steps are given, but the order is not completely logical.	The answer is mostly detailed but not always clear. It includes some relevant points with an attempt at linking them logically. Data may not be used fully. A conclusion is given that may not fully match the reasoning given.
1	1–2	The method would not lead to a valid outcome. Some key steps are given, but they are not linked in a clear way.	The answer gives separate, relevant points. Uses little or no data that may be given in the question. The points made may be unclear. If a conclusion is given, it may not match the reasoning given in the answer.

COMMAND WORDS

A **command word** in a question tells you what you are expected to do.

The structure of a question

You should see one command word per sentence, with the command word coming at the start. A command word might not be used, however, if a question is easier to follow without one. In these cases, you are likely to see:

- What ...?
- Why ...?
- How ...?

Command word	What you need to do
Balance	Add correct balancing numbers to an equation, chemical equation or nuclear equation.
Calculate	Use the numbers given to work out an answer.
Choose	Select from a range of options.
Compare	Write about **all** the similarities and/or differences between things.
Complete	Complete sentences by adding your answers in the spaces provided.
Define	Give the meaning of something.
Describe	Recall a fact, event or process accurately.
Design	Describe how something will be done, such as a practical method.
Determine	Use the data or information given to you to obtain an answer.
Draw	Produce a diagram, or complete an existing diagram.
Estimate	Work out an approximate value.
Evaluate	Use your knowledge and understanding, and the information supplied, to consider evidence for and against something. You must include a reasoned judgement in your answer.
Explain	Give the reasons why something happens, or make something clear.
Give, name, write	Only write a short answer, commonly just a single word, phrase or sentence.
Identify	Name or point out something.
Justify	Support your answer using evidence from the information given to you.
Label	Add the correct words or names to a diagram.
Measure	Use a ruler or protractor to obtain information from a photo or diagram.
Plan	Write a method.
Plot	Mark data points on a graph.
Predict	Write a likely outcome of something.
Show	Give structured evidence to come to a conclusion.
Sketch	Make an approximate drawing, such as a graph without axis units.
Suggest	Apply your knowledge and understanding to a new situation.
Use	You **must** base your answer on information given to you, otherwise you will not get any marks for the question. You might also need to use your own knowledge and understanding.

USEFUL BIOLOGY EQUATIONS

You need to know how to carry out some calculations that are specific to biology topics.

Magnification

You should learn the equation for magnification and be able to use it. At higher tier you may need to rearrange the equation to calculate the size of an image or object. Don't forget to convert image and object measurements to the same units, for example multiply measurements in mm by 1000 to convert to μm.

$$\text{magnification} = \frac{\text{size of image}}{\text{size of real object}}$$

Cross-sectional areas of colonies or clear areas around colonies

The size (area) of a colony of bacteria and the area of a clear zone where bacteria have been killed (e.g. by an antibiotic) approximate to the area of a circle. The area of a circle = πr^2. Remember that r is the radius of the circle, which is half the diameter.

The number of bacteria in a population

An exam question might state the number of bacteria at the start of an experiment and give you the mean division time. This is the average time taken for the number of bacteria to double. You should be able to calculate the number of bacteria in the population after a certain time. Often, answers can be worked out using simple maths. For more difficult problems this formula can be used:

$$\text{Final population of bacteria} = \text{Number at start} \times 2^{\text{number of divisions}}$$

First work out the number of divisions per hour. Multiply this by time in hours to get the number of divisions.

Surface area to volume ratio

A cell or structure with a large surface area to volume ratio is better adapted to exchange materials with the surroundings. Divide the surface area by the volume to find how many units of surface area there are to a single unit of volume.

USEFUL CHEMISTRY EQUATIONS

Mathematical skills account for 20% of the marks in the exams.
Foundation Tier students may be given equations like the ones below.
All students may be given an unfamiliar equation if a question needs one.

Shown in the specification

$$\text{mean rate of reaction} = \frac{\text{quantity of reactant used}}{\text{time taken}}$$

$$\text{mean rate of reaction} = \frac{\text{quantity of product formed}}{\text{time taken}}$$

$$R_f = \frac{\text{distance moved by substance}}{\text{distance moved by solvent}}$$

$$\% \text{ yield} = \frac{\text{mass of product actually made}}{\text{maximum theoretical mass of product}} \times 100$$

$$\% \text{ atom economy} = \frac{\text{relative formula mass of desired product from equation}}{\text{sum of relative formula masses of all reactants from equation}} \times 100$$

Not shown in the specification

relative formula mass (M_r) = sum of relative atomic masses (A_r) of atoms shown in the formula

relative atomic mass (A_r) using abundances of two isotopes, A and B =

$$\frac{(\text{mass number} \times \text{percentage}) \text{ of A} + (\text{mass number} \times \text{percentage}) \text{ of B}}{100}$$

mass of solute (g) = concentration of solution (g/dm^3) × volume of solution (dm^3)

mass (g) = amount (mol) × M_r

$$\text{concentration of solution (mol/dm}^3\text{)} = \frac{\text{amount of solute (mol)}}{\text{volume of solution (dm}^3\text{)}}$$

volume of gas at room temperature and pressure (dm^3) = amount of gas (mol) × 24

overall energy change of a reaction =
(energy in to break bonds in reactants) − (energy out when bonds form in products)

$$\text{gradient of a graph} = \frac{\text{change in vertical axis}}{\text{change in horizontal axis}}$$

USEFUL PHYSICS EQUATIONS

Mathematical skills account for 20% of the marks in Trilogy exams.

Calculation questions in physics usually involve an equation. Most equations need to be learnt and then recalled and applied in the exam. Others need to be selected from a sheet given in the exam.

Some equations are **higher tier only**.

All the equations you are expected to learn, select and apply are given here.

Select and apply

Word equation	Symbol equation	
(final velocity)^2 − (initial velocity)^2 = 2 × acceleration × distance	$v^2 - u^2 = 2\,a\,s$	
elastic potential energy = 0.5 × spring constant × (extension)^2	$E_e = \frac{1}{2}\,k\,e^2$	
change in thermal energy = mass × specific heat capacity × temperature change	$\Delta E = m\,c\,\Delta\theta$	
period = $\dfrac{1}{\text{frequency}}$	$T = \frac{1}{f}$	
force on a conductor (at right angles to a magnetic field) carrying a current = magnetic flux density	$F = B\,I\,l$	H
thermal energy for a change of state = mass × specific latent heat	$E = m\,L$	
potential difference across primary coil × current in primary coil = potential difference across secondary coil × current in secondary coil	$V_p I_p = V_s I_s$	H

Recall and apply

Word equation	Symbol equation
weight = mass × gravitational field strength (g)	$W = m\,g$
work done = force × distance (along the line of action of the force)	$W = F\,s$
force applied to a spring = spring constant × extension	$F = k\,e$
distance travelled = speed × time	$s = v\,t$
acceleration = $\dfrac{\text{change in velocity}}{\text{time taken}}$	$a = \dfrac{\Delta v}{t}$
resultant force = mass × acceleration	$F = m\,a$
momentum = mass × velocity	$p = m\,v$
kinetic energy = 0.5 × mass × (speed)^2	$E_k = \dfrac{1}{2}m\,v^2$
gravitational potential energy = mass × gravitational field strength (g) × height	$E_p = m\,g\,h$
power = $\dfrac{\text{energy transferred}}{\text{time}}$	$P = \dfrac{E}{t}$
power = $\dfrac{\text{work done}}{\text{time}}$	$P = \dfrac{W}{t}$
efficiency = $\dfrac{\text{useful output energy transfer}}{\text{total input energy transfer}}$	
efficiency = $\dfrac{\text{useful power output}}{\text{total power input}}$	
wave speed = frequency × wavelength	$v = f\,\lambda$
charge flow = current × time	$Q = I\,t$
potential difference = current × resistance	$V = I\,R$
power = potential difference × current	$P = V\,I$
power = (current)^2 × resistance	$P = I^2 R$
energy transferred = power × time	$E = P\,t$
energy transferred = charge flow × potential difference	$E = Q\,V$
density = $\dfrac{\text{mass}}{\text{volume}}$	$\rho = \dfrac{m}{V}$

H

MATHS SKILLS FOR SCIENCE

Standard form

Standard form is a way of writing very large or very small numbers and is written as:

$$A \times 10^B$$

- A is a number greater than or equal to 1 and less than 10
- B is any integer (negative or positive whole number)

Examples

Number	Standard form
0.0050 61	5.61×10^{-3}
170 000 000	1.7×10^8
0.012 03	1.203×10^{-2}
8 040 000	8.04×10^6

Rounding to *n* decimal places

When rounding to n decimal places (dp):

- look at the nth decimal place
- if the next digit is 5 or more, round up by increasing the preceding digit by one
- if it is 4 or less, keep the preceding digit the same

For example, 0.365 rounded to 2 dp is 0.364 is 0.36.

0.8675 rounded to 1 dp is 0.9

Rounding to *n* significant figures

- Look at the first non-zero digit, go $n - 1$ digits to the right, and follow the rules for rounding to n decimal places
- Fill any places after it with a zero and stop when you reach the decimal point.

Examples

	1 sf	2 sf	3 sf
9375	9000	9400	9380
56.27	60	56	56.3
0.003684	0.004	0.0037	0.00368

Units and conversions

Quantities in one unit can be converted into a different unit and the size of the measurement will still be the same.

Units need to be correct when using equations, it is usually the SI unit.

A common conversion is to change time from hours and minutes to the SI unit of seconds:

seconds = hours × 60 × 60

seconds = minutes × 60

Other conversions depend on the prefix of the unit

Common prefixes

	10^B	Prefix	Symbol
1 000 000 000	10^9	Giga-	G
1 000 000	10^6	Mega-	M
1 000	10^3	Kilo-	k
0.01	10^{-2}	Centi-	c
0.001	10^{-3}	Milli-	m
0.000 001	10^{-6}	Micro-	μ
0.000 000 001	10^{-9}	Nano-	n

Example conversions for the SI unit of metres (m)

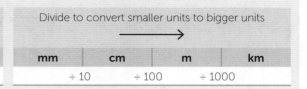

km	m	cm	mm
× 1000	× 100	× 10	

mm	cm	m	km
÷ 10	÷ 100	÷ 1000	

Multiply to convert bigger units to smaller units

Divide to convert smaller units to bigger units

Using data

The **mean** is an average of a set of values.

To calculate the mean, add up all the values and divide this total by the number of values there are.

$$\text{mean} = \frac{\textbf{total sum of the values}}{\textbf{how many values there}}$$

Graphs

The most commonly used graph in physics is a **line graph** although you could see **bar** charts and **pie** charts too.

A line graph shows the relationship between two **continuous** variables.

- the **independent** variable is plotted on the horizontal **x-axis**.
- the **dependent** variable is plotted on the vertical **y-axis**.

Plotting a line graph	Drawing a line of best fit
• Look at the **range** of values you need to plot on each axis. • Choose appropriate **scales** for the small and large squares. • Intervals such as 1, 2, 5, and multiples such as 10 or 100 are good to use. Do not use other intervals such as 3, 6 or 9. • Make sure each axis uses at least half of the height or width of any given grid. • Label both axes with the correct variable and unit. • To plot (x, y) find the value on the x-axis, then go up to the value on the y-axis. • Use a sharp pencil to plot each point as a small x, accurate to ± 1 small square.	A line of best fit is an indication of the relationship between two variables from experimental data. Lines of best fit can be **straight** or **curved**. • Ignore any clearly **anomalous** points. • Use a sharp pencil. • Draw the line of best fit through most of the points with equal numbers of points above and below the line. • Use a transparent ruler for **straight** lines so you can see all the points. • Draw a **curved** line free hand as a smooth curve, not dot to dot with a ruler. • Avoid drawing double lines.

Information from graphs

Directly proportional

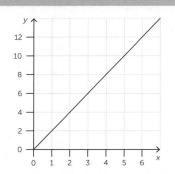

- Straight line graph
- Through the origin
- When x doubles, y doubles
- So x and y are directly proportional to each other

Inversely proportional

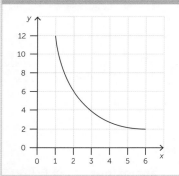

- Curved graph
- When x doubles, y halves
- So x and y are inversely proportional to each other.

Determining a gradient from a straight line graph

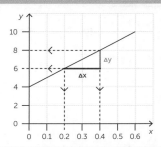

- Choose 2 points on the line which have easy to read values.
- Draw a right-angled triangle from these 2 points
- Determine the values of Δy and Δx
- Gradient = Δy / Δx
- $y = mx + c$ where m is the gradient and c is the point the line intercepts the y axis

Example

$\Delta y = 8 - 6 = 2$ $\Delta x = 0.4 - 0.2 = 0.2$

gradient = $2 / 0.2 = 10$ $y = 10x + 4$

Determining a gradient from a curved graph

Higher Tier only

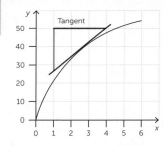

- A **tangent** is a straight line that touches a curve at one point
- Draw a tangent at the point (e.g. $x = 2$) on the line where you want to find the gradient
- Draw a right-angled triangle from these 2 points
- Determine the values of Δy and Δx
- Gradient = Δy / Δx = $(50 - 28)$ / $(3.6 - 1)$
- = $22 / 2.6$
- = 8.5 (1 dp)

Rearranging equations

The **subject** of an equation is the quantity that is on its own. The **equations** that you need to recall, or choose from a sheet, often need to be **rearranged** to make a different quantity the **subject**.

- Whatever is done to one side of the equation needs to be done to the other
- **Inverse operations** are used, usually **×** and **÷ which are the opposite of each other**.

Examples

$V = I \times R$

V is the subject. Rearrange the equation to make I the subject. You need to get I on its own.

I is multiplied by R, so to remove R from the right-hand side, both sides of the equation need to be divided by R

$\dfrac{V}{R} = \dfrac{I \times R}{R}$, so $I = \dfrac{V}{R}$

$P = \dfrac{E}{t}$

P is the subject. Rearrange the equation to make t the subject. You need to get t on its own.

E is divided by t, so first make E the subject by multiplying both sides of the equation by t

$P \times t = \dfrac{E \times t}{t}$ so $E = P \times t$

Now get t on its own by dividing both sides by P

$\dfrac{P \times t}{t} = \dfrac{E}{P}$, so $t = \dfrac{E}{P}$

Sometimes knowing the units can help you remember an equation, or you can work out a unit from an equation.

It is much better to understand and learn how to rearrange an equation properly. However, using a **formula triangle** can help as long as you use the correct triangle. Cover the quantity you want to find and then see if you need to multiply or divide the other quantities.

	$A = B \times C$	$A = B / C$
General equation type		
Example	potential difference = current × resistance $V = I \times R$	power = $\dfrac{energy\ transferred}{time}$ $P = E / t$
Formula triangle	$V = I \times R$ $I = \dfrac{V}{R}$ $R = \dfrac{V}{I}$	$P = \dfrac{E}{t}$ $E = P \times t$ $t = \dfrac{E}{P}$

THE PERIODIC TABLE

1	2												3	4	5	6	7	0
																		4 **He** Helium 2
7 **Li** Lithium 3	9 **Be** Beryllium 4												11 **B** Boron 5	12 **C** Carbon 6	14 **N** Nitrogen 7	16 **O** Oxygen 8	19 **F** Fluorine 9	20 **Ne** Neon 10
23 **Na** Sodium 11	24 **Mg** Magnesium 12												27 **Al** Aluminium 13	28 **Si** Silicon 14	31 **P** Phosphorus 15	32 **S** Sulfur 16	35.5 **Cl** Chlorine 17	40 **Ar** Argon 18
39 **K** Potassium 19	40 **Ca** Calcium 20	45 **Sc** Scandium 21	48 **Ti** Titanium 22	51 **V** Vanadium 23	52 **Cr** Chromium 24	55 **Mn** Manganese 25	56 **Fe** Iron 26	59 **Co** Cobalt 27	59 **Ni** Nickel 28	63.5 **Cu** Copper 29	65 **Zn** Zinc 30	70 **Ga** Gallium 31	73 **Ge** Germanium 32	75 **As** Arsenic 33	79 **Se** Selenium 34	80 **Br** Bromine 35	84 **Kr** Krypton 36
85 **Rb** Rubidium 37	88 **Sr** Strontium 38	89 **Y** Yttrium 39	91 **Zr** Zirconium 40	93 **Nb** Niobium 41	96 **Mo** Molybdenum 42	[98] **Tc** Technetium 43	101 **Ru** Ruthenium 44	103 **Rh** Rhodium 45	106 **Pd** Palladium 46	108 **Ag** Silver 47	112 **Cd** Cadmium 48	115 **In** Indium 49	119 **Sn** Tin 50	122 **Sb** Antimony 51	128 **Te** Tellurium 52	127 **I** Iodine 53	131 **Xe** Xenon 54
133 **Cs** Caesium 55	137 **Ba** Barium 56	139 **La** Lanthanum 57	178 **Hf** Hafnium 72	181 **Ta** Tantalum 73	184 **W** Tungsten 74	186 **Re** Rhenium 75	190 **Os** Osmium 76	192 **Ir** Iridium 77	195 **Pt** Platinum 78	197 **Au** Gold 79	201 **Hg** Mercury 80	204 **Tl** Thallium 81	207 **Pb** Lead 82	209 **Bi** Bismuth 83	[209] **Po** Polonium 84	[210] **At** Astatine 85	[222] **Rn** Radon 86
[223] **Fr** Francium 87	[226] **Ra** Radium 88	[227] **Ac** Actinium 89	[261] **Rf** Rutherfordium 104	[262] **Db** Dubnium 105	[266] **Sg** Seaborgium 106	[264] **Bh** Bohrium 107	[277] **Hs** Hassium 108	[268] **Mt** Meitnerium 109	[271] **Ds** Darmstadtium 110	[272] **Rg** Roentgenium 111	[285] **Cn** Copernicium 112	[286] **Nh** Nihonium 113	[289] **Fl** Flerovium 114	[289] **Mc** Moscovium 115	[293] **Lv** Livermorium 116	[294] **Ts** Tennessine 117	[294] **Og** Oganesson 118

1 **H** Hydrogen 1

Notes

The Lanthanides (atomic numbers 59–71) and the Actinides (atomic numbers 90–103) are omitted.

Relative atomic masses for Cu and Cl have not been rounded to the nearest whole number.

Key

Relative atomic mass
Atomic symbol
Name
Atomic (proton) number

KEY TERMS IN PRACTICAL WORK

Experimental design

Key term	Meaning
Evidence	Measurements or observations collected using a valid method
Fair test	When the dependent variable is only affected by the independent variable
Hypothesis	A suggested explanation for observations or facts
Prediction	A reasoned statement that suggests what will happen in the future
Valid	A valid method involves fair testing and is suitable for an investigation
Valid conclusion	A discussion of a valid experiment and what it shows

Variables

A variable is a characteristic that can be measured or observed.

Type of variable	Meaning
Categoric	It has names or labels rather than values
Continuous	It has values rather than names or labels
Control	It affects the dependent variable, so it must be kept the same or monitored
Dependent	It is measured or observed each time the independent variable is changed
Independent	It is deliberately changed by the experimenter

Measurements and measuring

Key term	Meaning
Accurate	Close to the true value
Calibrated	A device is calibrated when its scale is checked against a known value
Data	Measurements or observations that have been gathered
Interval	The measured gap between readings
Precise	Very little spread about the mean value
Range	The values between the measured maximum and minimum values
Repeatable	When the same results are obtained using the same method and apparatus
Reproducible	Someone else gets the same results, or when different apparatus and methods are used
Resolution	The smallest change a measuring device can show
True value	The value you would get in an ideal measurement
Uncertainty	An interval in which the true value will be found

Errors

Type of error	Meaning
Anomaly	Anomalous results lie outside the range explained by random errors
Measurement	The difference between the true value and a measured value
Random	Unpredictably different readings – their effects are reduced by repeats
Systematic	Readings that differ from true values by the same amount each time
Zero	A type of systematic error where a device does not read 0 when it should

INDEX

H

Haber process 210
habitat 104, 123
half equations 130, 188
half-life 292
halide salts 147
halogens 146
hardness 161
heart 29
heating 281
 curves 284
heavy fuel oil 217
herd immunity 51
heterozygous 86
HIV 45
homeostasis 68
homologous series 216
homozygous 86
hormones 72
host cells 45
HPV 34
human
 endocrine system 72
 genome 85
 immune system 50
 nervous system 70
 reaction time 71
 reproduction 74
hydrocarbons 216, 218
hydrochloric acid 179, 181, 186
hydroelectricity 258
hydroelectric power 257
hydrogen, test for 226
hypothesis 203

I

indicator (pH) 184
induced magnets 346
inelastic deformation 304
inertia 315
inertial mass 318
infectious diseases 44
infertility 77
infrared 343
infrared radiation 340
inherited disorders 88
insoluble reactant 182
insulators 255, 263
insulin 73
interdependent species 102

internal energy 281
intrauterine devices 76
investigating F = ma 316
investigating waves 336
In Vitro Fertilisation (IVF) 77
iodine solution 27
ion 138
ionic
 bonds 150
 compounds 146, 152, 158,
  180, 187, 188
 equations 130
 structures 152
ions 177, 180, 288
irradiation 293
irregular shaped solids 279
isotopes 138, 288
IVF 77

J

James Chadwick 136, 289
J.J. Thomson 136, 289

K

kerosene 217
kinetic energy 204, 246, 281, 285
kingdom 98

L

lactic acid 62
lamp 261
landfill 117, 123
land use 118
lattice 158, 160
law of conservation of mass 166
laws, Newton's 319, 332
LDR 261, 267
leachate 241
leaves 16
Le Chatelier's principle 210
LED 261
lens 5
Leslie cube 340
levels of organisation 109
life cycle assessment (LCA) 242
light 343
 ultraviolet 343
 visible 343
Light Dependent Resistor 267
light microscope 5, 9

limitations, of particle model 157
limiting reactant 173
limit of proportionality 248, 304
Linnaean system 98
lipases 26
lipid reactions 63
lipids 24
liquid 156, 280, 277
lock and key theory 24
longitudinal waves 330
LPG
lumen 30
lung cancer 36
lungs 16, 29, 31
luteinising hormone 74, 77
lymphocyte 32, 50

M

magnetic
 core 347
 fields 347
 poles 346
magnification 9, 376
mains electricity 271
malaria 44, 48
malignant tumour 37
malleable 161
mangrove trees 108
manufacturing aluminium 189
mass 166, 170
mass number 138, 288
mean 111
 rate (of reaction) 200
measles 45
measurements 169
measuring cylinder 279
measuring wave speed 334
median 111
meiosis 82, 83
melanoma 36
melting 280
 point 157, 222
memory cells 50
Mendeleev 142
menstrual cycle 74
meristem 8, 13, 14, 38
metabolism 24, 63, 78
metal carbonates 181
metallic bonds 150, 155
metal oxides 178
metals 143

EXAMINATION TIPS

When you practise examination questions, work out your approximate grade using the following table. This table has been produced using a rounded average of past examination series for this GCSE. Be aware that actual boundaries will vary by a few percentage points either side of those shown.

Combined Science: Trilogy

Grade	5–5	5–4	4–4	4–3	3–3	3–2	2–2	2–1	1–1	U
F Tier (%)	59	54	50	44	37	31	25	19	13	0

Grade	9–9	9–8	8–8	8–7	7–7	7–6	6–6	6–5	5–5	5–4	4–4	4–3	3–3
H Tier (%)	66	62	58	53	49	44	40	35	31	26	22	19	14

1. Read questions carefully. This includes any information such as tables, diagrams and graphs.

2. Remember to cross out any work that you do not want to be marked.

3. Answer the question that is there, rather than the one you think should be there. In particular, make sure that your answer matches the command word in the question. For example, you need to recall something accurately in a describe question but not say why it happens. However, you do need to say why something happens in an explain question and should include a connecting word such as 'so', 'but', therefore', or 'because'.

4. Physics involves a lot of equations. Some are on the equations sheet but most need to be recalled and then used. Forgetting or failing to learn these will cost you a lot of marks. Learn the formulae well and be able to use them confidently. Also, make sure you learn a method for rearranging equations and know the correct SI units.

5. Show all the relevant working out in calculations. If you go wrong somewhere, you may still be awarded some marks if the working out is there. It is also much easier to check your answers if you can see your working out. Remember to give units when asked to do so and follow instructions about standard form or significant figures.

6. Plot the points on graphs to within half a small square. Lines of best fit can be curved or straight, but must ignore anomalous points. If the command word is sketch rather than plot, you only need to draw an approximate graph, not an accurate one.

7. You could be asked to draw and label a diagram from an image. Draw only what you see, making it as big as space allows. Draw lines that are clear using a sharp pencil, don't use shading. Draw label lines with a ruler; the end of the line should just touch the item to be labelled.

8. Make sure you do not mix words and symbols in chemical equations. You will be given full marks if you are asked to write a word equation, but you give the correct balanced equation instead. This does not work the other way round! Check that all the numbers of atoms, ions and charges balance in symbol equations. Remember to include state symbols when asked.

9. Follow instructions carefully when writing or balancing nuclear or chemical equations. Check that all the numbers of particles and charges balance.

10. Remember that you may be asked to label a diagram or to complete a diagram. You may or may not be given the words to use.

11. Make sure you can recall experiments you have done or observed. About 15% of the exam is based on the required practicals.

Good luck!

 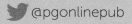